CLIMATE CHANGE, ENVIRONMENTAL REFUGEES, AND HUMAN RIGHTS IN THE MIDDLE EAST

CLIMATE CHANGE, ENVIRONMENTAL REFUGEES, AND HUMAN RIGHTS IN THE MIDDLE EAST

Mahmood Monshipouri

BLOOMSBURY ACADEMIC
NEW YORK • LONDON • OXFORD • NEW DELHI • SYDNEY

BLOOMSBURY ACADEMIC
Bloomsbury Publishing Inc
1359 Broadway, New York, NY 10018, USA
50 Bedford Square, London, WC1B 3DP, UK
29 Earlsfort Terrace, Dublin 2, Ireland

BLOOMSBURY, BLOOMSBURY ACADEMIC and the Diana logo are trademarks of Bloomsbury Publishing Plc

First published in the United States of America 2025

Cover design by Chloe Batch
Cover image © Agencja Fotograficzna Caro/Alamy Stock Photo
Background texture © iStock.com/desifoto

Library of Congress Cataloguing-in-Publication Data

ISBN: HB: 978-1-5381-6912-4
PB: 979-8-2163-7657-6
PDF: 979-8-2163-7659-0
eBook: 979-8-2163-7658-3

Typeset by Deanta Global Publishing Services, Chennai, India

CONTENTS

ACKNOWLEDGMENTS

For the past decade or so, my interest in studying the interconnected issues of refugees, climate change, and forced migration has significantly peaked. In undertaking this project, which is the culmination of several years of research and teaching and conducting research at San Francisco State and UC Berkeley, I have been fortunate to benefit from cooperation and assistance of many colleagues. At San Francisco State University, Professor Burcu Ellis has provided an unwavering support for my project in this area, rendering it possible for me to teach a course along these lines and contribute to the university's Certificate on Climate Change. Her tremendous encouragement and support for this project was a source of inspiration. Likewise, the feedback I received from my colleague, Professor Jonathon Whooley, was invariably most valuable and thoughtful. Teaching for the fall program for freshmen at UC-Berkeley has also sharpened my focus on the topic of climate change and forced migration.

I would also like to express my deep gratitude to Professor Ihsan Daği of Middle East Technical University (METU), Ankara, Turkey, who made it possible for me to stay on METU's campus and talk to students and experts about their views on the Syrian refugees. A sincere and special thanks must go to Dr. Muhittin Ataman, Dr. Bilgehan Öztürk, and Dr. Muhammed Hoseyin Mercan, who shared their valuable insights with me at the Siyaset, Ekonomi ve Toplum Araştırmaları Vakfı (Foundation for Political, Economic, and Social Research), known as SETA, in Ankara. I benefited immensely from their perspectives and views regarding the situation of Syrian refugees in Turkey. They also provided me with a plethora of sources (books and journal articles) on the issues of environment and climate change in Turkey. I also wish to thank Dr. Ali Zafer Sagiroglu, the director of Migration Policies Center at the Ankara Yildirim Beyazit University, whose views on the issue of refugees in Turkey proved to be incredibly helpful to me in writing the chapter on Turkey. My sincere thanks to staff and project managers of Bloomsbury, most particularly Mr. Arvind Harikumar, who carefully saw this manuscript through. I also wish to thank Alexandra Theodotou for her constructive and invaluable comments on this manuscript.

My visit to Masdar City during the late summer of 2024, an urban community in Abu Dhabi, the United Arab Emirates, that is widely known for setting the standard for sustainable urban communities, was highly educational and inspiring. The largest of its kind since 2009, Masdar City operates on several commercially viable types of solar panels to convert the sun's energy into electricity. I also wish to express my deep gratitude to San Francisco State University for granting me a sabbatical leave to finish this project. Without such a leave, I would not have been able to complete this project in a timely manner. Finally, I should note that any errors or omissions in this volume remain my sole responsibility.

PART I
CONCEPTS, FRAMEWORKS, AND THE RESEARCH AGENDA

CHAPTER 1
INTRODUCTION AND OVERVIEW
REFUGEES, FORCED MIGRATION, CLIMATE CHANGE, AND HUMAN RIGHTS IN THE MIDDLE EAST

In the past decade, millions of people have been displaced from their communities. In 2022, the United Nations High Commissioner for Refugees (UNHCR) announced that we had surpassed the 100 million mark for total displacement, indicating that over 1.2 percent of the world population had been forced to leave their homes. As of mid-2023, that figure exceeded 30 million refugees. Over half of those refugees came from just three countries: Syria, Afghanistan, and Ukraine.[1] Syria continues to be the world's largest refugee crisis as we enter 2024, constituting approximately a quarter of the total world refugee population. As of mid-2023, more than 6 million Syrians have sought refuge, primarily in Jordan, Lebanon, Iraq, Egypt, and Turkey.[2]

Just as persecution, war, and violence—rooted mainly in ethnic, tribal, and religious causes—are frequently cited as leading causes of refugees fleeing their countries, so too are human rights violations and both man-made and natural disasters, becoming more prevalent in the world due to climate change. The refugee crisis presents a multitude of problems, including a lack of access to adequate housing, inability to access public services and national resources, employment, environmental degradation, and ongoing trauma and violence.[3]

Climate change is one of the most serious environmental threats facing human society today. Most people, as one scholar reminds us, tend to treat climate change as a distant event, with little or no effect on their immediate personal well-being. "Largely intangible," Dan Rabinowitz writes, "its encroachment is insidiously slow and indirect, thus difficult to follow and easy to ignore."[4] Climate change is no longer an abstract idea, as more and more people around the world are living it now. Climate change refers to long-term shifts in temperatures and weather patterns, mainly caused by human activities, especially the burning of fossil fuels.[5] The effects of climate change appear across the globe in the form of various long-term changes in climate, including warming-driven sea level rise, heat waves, droughts, extreme weather, wildfires, deluges or severe precipitation, and ocean acidification.[6]

Human activity has been a powerful influence on climate and the environment. In the new Anthropocene epoch, species and ecosystems cannot avoid responding to climate change. Several questions come to mind: How can we reduce losses to the most vulnerable, including forced climate migrants, who have suffered from involuntary environmental displacement associated with climate change? How will climate change affect our most precious, life-sustaining resources—such as water? Can we reduce or even curb the risks to food security emanating from adjustments to crop yields, fruit output, herds, and fisheries?

Life on planet Earth will never be the same due to a changed and changing climate. Even if technological fixes materialize, from industry and private sector investment, governments—especially rich governments that are largely responsible for emitting far more carbon dioxide (CO_2)—must drive an effective response to guarantee a sustainable future.[7] The impacts of climate change will disproportionately affect populations in the most vulnerable positions due to poverty, gender, age, minority status, or disability. It is now widely believed that such effects will increasingly impinge upon human rights.[8] Environmental harm, largely caused by a dangerous anthropogenic interference with the climate, threatens a broad range of human rights, including rights to life, health, food, water, adequate housing, freedom of movement, and self-determination.

The average global temperature has risen by more than 1 degree Celsius since the mid-eighteenth century—that is, the Industrial Revolution. If it rises by 1.5 degrees Celsius, or 2.7 degrees Fahrenheit, it will markedly increase the likelihood of far more disastrous climate catastrophes that could aggravate the already difficult challenges facing the international community, such as hunger, disease, and conflict.[9] The need for collective action to grapple with this global threat at a time when rising populism, demagoguery, and nationalism run counter to confronting such crisis appears to be undeniably so urgent. Short of a global transformation, climate diplomacy is unlikely to succeed. While inaction is no longer an option, the question remains: Who will help the mostly developing countries that account for smallest share of the emissions when compared to countries, such as China, the United States, and some European countries that have produced the largest share of emissions?[10]

The purpose of this chapter is to set out a new framework for assessing climate change and its impacts on the region and people of the Middle East and North Africa (MENA), thus arguing for a broader and more complex pattern of security linkage. My approach proceeds from the assumption that if liabilities and exposure to climate change are not properly addressed, they are likely to intensify regional socioeconomic disparities, political instability, and sectarian and ethnic strife. "Environmental conflict," some experts remind us, "often travels under the guise of political turmoil or ethnic strife."[11] Human mobility is an important option for many people who seek to improve their living conditions. That option is indeed a human right, and its complex roots require thinking differently about fragile states and coming to grips with the exacerbating factors of climate stressors.

To effectively address the larger issues of climate change, forced migration, and displaced people, it is imperative to adopt a humanitarian ecosystem that depends on closer cooperation between and among different stakeholders, including states,

NGOs, the private sector, and humanitarian agencies. This approach also requires a shift of focus from Global North countries to the states of the Global South, where most internally displaced persons (IDPs), forced migrants, and refugees live.

In the sections that follow, I first define key concepts such as refugees, human migration, climate change, and climate refugees. This section is followed by exploring the main cause of climate change. My focus then shifts to outlining security implications of climate change, including water security and potential conflicts in the MENA region. In the ensuing section, I explore the issues of climate justice and human rights as inextricably intertwined notions. The necessity of intervention and adaptation is examined toward the end of this chapter with the aim of demonstrating the ways in which the risks of loss and damage caused by climate change can be prevented or significantly curbed.

Who Is a "Refugee"?

The 1951 refugee regime has faced sharp criticism because of its narrow and outdated definition of refugees—that is, those displaced persons who have crossed national borders while fleeing persecution and violence for reasons of race, nationality, religion, membership of a particular social group, or political opinion. Today, however, people are forced to flee their homes due to wide-ranging issues, including the threat of persecution, violence, political instability, climate change, poverty, and hunger. Regardless, the refugee regime has been influenced largely by "power relations among states more than the objective characteristics of displaced individuals."[12]

The 1951 Convention offers no protection for other types of forced migration such as displacement due to violence, human rights abuses, deluge, drought, famine, desertification, and earthquakes. There is a growing realization, and perhaps even consensus, in the international arena to recognize those displaced by environmental causes. Climate change is widely regarded as a real threat to the Earth and its inhabitants, and as such, many experts argue that it should be included in laws on asylum seeking.[13]

Wars and persecution are no longer considered the sole and key reasons behind the refugee crises across the world. It has become increasingly difficult to distinguish between terms such as migrants and refugees—though the conditions surrounding these terms are legally different—due in large part to climate change, economic deprivation, human rights violations, and the rampant violence. Political violence, too, often contributes to economic distress. Similarly, those who flee their country of origin for political reasons might also depart from their host country due to economic motives.[14]

The International Committee of the Red Cross (ICRC) has adopted a much broader description of migrants: "migrants are people who leave or flee their usual place of residence in search of safety or better opportunities abroad, and who may be in distress and in need of protection or humanitarian assistance. Refugees and asylum seekers, who enjoy specific protection under international law, fall into this category."[15]

Whether one adopts a narrow or a broad definition of what constitutes a refugee, Gillian Brock notes, it is generally assumed that a person becomes a refugee when her/his basic human rights are being trampled upon. Violations of human rights are widely regarded as the core causes of forced migration and displacement. Safeguarding human rights is thus crucial to effectively addressing the refugee issues. Being an asylum seeker or refugee renders individuals far more vulnerable to human rights abuses. Asylum seekers and refugees are likely to experience numerous threats to their lives, liberty, and security. As a group, refugees are among the most vulnerable people in the world, with their rights always endangered.[16] Refugees face many obstacles along the way to seeking asylum. These include, but are not limited to, racism, xenophobia, national or ethnic tensions, and intolerance. Refugees are frequently exposed to abuse and assault by both state and non-state actors. Young men may even be coerced into fighting in civil wars, and women may be subjected to physical and sexual abuse.[17]

Human Migration

According to the First Intergovernmental Panel on Climate Change (IPCC) report in 1990, the greatest single impact of climate change may be on human migration.[18] It is estimated that by 2050, approximately 150–200 million people could be displaced by climate change-related events, such as desertification, water scarcity, floods, and storms.[19] Yet human rights law provides no clear remedy to the status of such displaced populations.[20] On a broader level, and despite the inherent link between changes in climate and human rights, human rights treaties have yet to recognize a specific right to a safe and healthy environment.

The establishment of an entirely new global regime for the protection of "climate refugees" faces conceptual, political, and legal obstacles. There is little agreement on what constitutes "climate refugees," given the multiple causes of displacement and the slow process of unfolding climate-induced displacement.[21] Politically, states are reluctant to open their borders to climate refugees or climate migrants. Thus, states are unlikely to adopt a treaty on climate refugees any time soon. Despite a growing literature generated by scholarly and academic circles on the subject of "climate refugees" and its forms—temporary or permanent, internal or cross-border, or a combination—"climate refugees lack legal protection under contemporary international law."[22] From a legal standpoint, it is unclear what human rights obligations states have in relation to mitigation of the effects of climate change, although these obligations are much more evident with regard to adaptation.[23]

In some parts of the world, climate change poses a real threat to states, and its effects are palpable in terms of food insecurity and increasing vulnerabilities. A good portion of the Middle East is largely arid or semiarid, making fresh water a scarce and precious resource there. The combination of scarce freshwater resources and rapid population growth significantly increases the vulnerability of the region to future climate change.[24] In poor African countries such as Somalia, Sudan, and South Sudan, climate change is one of the major causes of generalized human displacement. Yet this should not be confused with the economic causes of migration. The fact remains that most migrants from Africa are economic migrants.[25] The impact of changing climate and environmental conditions on mobility will often be region specific or space specific. However, there is a broad consensus that climate change is influencing, in varying proportions, migration patterns globally.[26] This underlines the significance of enforcing existing human rights protections, while at the same time moving toward building a consensus around the norms and principles of protecting the "environmentally displaced."[27]

Defining Climate Change

Climate, as one expert reminds us, has long been a significant influence on human settlement and migration patterns. "For the two hundred thousand or so years that *Homo sapiens* has been on the planet," Robert A. McLeman writes, "climate and its interactions with other environmental processes have helped shape human population numbers, distribution densities, longevity, and well-being."[28] In fact, migration has been an ongoing process of adaptation. It is within this context that one can better understand the role that climate has played in influencing migration. Among other things, McLeman argues, climate has clearly shaped human population movements by mapping out the human habitat—that is, the parts of the world to which it is biologically safe to migrate and settle.[29] Interactions between climate and other facets of the natural environment, McLeman goes on to note, "shape the availability and distribution of resources and ecosystem services critical for human survival and well-being."[30]

Climate-related migration occurs when weather and environmental conditions and events force people to migrate from one place to another, but they need not be the primary or sole cause of the migration event. Climate changes can be prompted by a sudden onset (earthquakes, volcanic eruptions, and floods) that cause local displacement of great magnitude or long-term shifts in climatic regimes, including drastic changes in temperatures and weather patterns (drought, soil erosion, and forest loss) that extend over the course of decades or even centuries.[31] Since the Industrial

Revolution (*c.* 1750), human activities have been the main driver of climate change, releasing large amounts of CO_2 by burning fossil fuels such as oil, coal, and gas, as well as other heat-trapping gases into the atmosphere. The effects of human activities (also known as anthropogenic effects and processes) have significantly contributed to altering the Earth's climate. In addition to human factors, natural processes, such as the sun's energy and volcanic eruptions, have also precipitated climate change.[32]

Atmospheric temperature estimations provide robust documentation of climatic changes since 1970. Climatological observations from the Earth's surface remain the main source of information for global temperatures and precipitation back to 1970.[33] Between 1970 and 2017, global warming increased by ~0.9 degree Celsius. Global temperatures spiked more sharply during the 1990s and since 2001 than during the period in between.[34] From 1970 to 2015, precipitation increased over midlatitude land areas and the tropical zone, while it decreased in subtropical regions. Furthermore, since the 1980s, many glaciers across the world have been retreating. As a result of both warming and meltwater inflow, global sea level has risen by ~10 centimeters since 1970.[35] Global climate models indicate that most of the global surface temperature increase since 1950 can be attributed to greenhouse gasses (GHGs). The period since 1970 has been witness to many significant climatic events, including the Sahel droughts of the 1970s and 1980s, change of European Winters around 1990, the 1991 Pinatubo Eruption, the El Niño of 1982–3 and 1997, and the subtropical droughts and midlatitude heat waves in the new millennium.[36]

Today, sea level rise, extreme weather patterns, desertification, drought, and water scarcity have rendered large areas of the world less habitable than in the past. Climate change is widely regarded as a contributing and exacerbating factor in migration and conflict. Climate disruption is likely to uproot large segments of the global population, resulting in long-term mass migration over international borders. Consider, for example, the case of drought (2006–10) and the ongoing Syrian civil war. Consecutive droughts in the Horn of Africa have led to crop failures, water shortages, famine, and the death of livestock. The World Bank has estimated that almost 3 billion people will live in slums by the 2030s.[37] Yet today, neither of the two main international laws on refugees—the Geneva Convention and Refugee Protocol—cover climate refugees. The living conditions of 800 million people could seriously diminish, putting added strain on countries of origin, as well as on destination countries and the routes that migrants travel.[38]

Furthermore, there is clear and very telling evidence that demonstrates an increase in sexual and gender-based violence against women and girls both during and after disasters. Gender-sensitive policies must be identified and prioritized to help build climate-resilient, sustainable development. The Global Compact for Safe, Orderly, and Regular Migration (adopted by 164 countries—not including the United States—in December 2018) called on countries to make plans to prevent the need for climate-caused relocation and support for those forced to relocate. These agreements, however, are neither legally binding nor sufficiently developed to support climate migrants—particularly migrants from South Asia, the MENA, Central America, and the Horn of Africa.

Climate Refugees

"Climate refugees," or sometimes called "climate migrants," are a subclass of "environmental migrants" who have been coerced to flee owing to the gradual and/or sudden onset of changing climatic regimes related to the impacts of environmental factors.[39] "Climate refugees" have been defined as people who have to leave their homes, instantly or in the immediate future, "because of sudden or gradual alterations in their natural environment related to at least one of three impacts of climate change: sea level rise, extreme weather events, and drought and water scarcity."[40]

Climate refugees have added a new layer of complexity to the increasing refugee crisis emanating from civil unrest in the MENA region following the 2011 Arab Spring uprisings, and yet there exists no broad agreement on the definition of climate refugees—both legally and from a political standpoint. Also known as "environmental refugees," climate refugees are persons who have left their traditional homeland because of drought, desertification, deforestation, sea level rise, soil erosion, water shortage, and natural disasters including cyclones, floods, and storms.[41] The growing and aggregate impact of long-term effects of climate change is likely to lead to political violence, uprisings, poverty, and acute socioeconomic disparities, all of which might lead to forced migration.

Some experts point out that under the 1951 Geneva Refugee Convention, a refugee is a person who has fled her/his country of nationality (and is currently outside that country) owing to a well-founded fear of persecution for reasons of race, religion, nationality, membership of a particular social group or political opinion. Critics argue that women who escape domestic violence are often in a vulnerable position. The Refugee Convention makes no explicit reference to sex or gender as a protected ground.[42] Apparently, the construction of the refugee regime and its application were shaped largely by power relations among states more than the actual situation of displaced individuals. This has fueled a debate over how far to stretch the boundaries of who should be legally recognized as a refugee.[43]

Furthermore, as one expert reminds us, a person fleeing her or his country because of climate change is not facing persecution for one of the five grounds for protection given in this definition. In fact, none of those five specific types of persecution applies directly or could be extended to include those who cross the border due to climate change.[44] Furthermore, there is extensive opposition to revisiting discussions on the Refugee Convention because of a fear that the idea might be outright rejected. Countries of potential destination are likely to vehemently oppose the expansion of the definition of refugees—not to mention that there are various conceptual difficulties and legal uncertainties in determining the role of climate change in forcing people to leave their home countries.[45]

Thus, efforts to expand the current refugee convention are unlikely to succeed in developing new global frameworks related to displacement stemming from the effects of climate change. Today, governance of international migration is fractured and unbalanced. There is presently little desire to develop binding international treaties in any field, and there are boundless drawbacks in defining "climate change-induced mobility" to affirmatively forecast much stride in this direction.[46]

Applying the principle of *non-refoulement* to international environmental law faces many obstacles, as it is still confined to the states' national jurisdiction. Under current international refugee law, climate change-displaced persons do not qualify for refugee status. Environmental or climate refugees lack such a legal status. Consider, for example, the European Court of Human Rights (ECHR), which has traditionally focused on individual rights, yet has to develop a system of laws that can forcefully protect and promote the collective rights of people—and particularly minority groups—who have disproportionately suffered from environmental injustice.[47]

Increasingly, however, environmental refugees have entered the debate surrounding climate refugees and forced migrants—regardless of whether they have left temporarily or permanently or due to a natural disaster and/or triggered by people.[48] Some experts caution that the designation "refugee" creates an entitlement to certain protections and assistance under International Humanitarian Law (IHL). That is why UN agencies have stopped using the expression "climate refugee" and instead use "environmental migrants" or "environmentally displaced persons." Although the UNHCR refers to "natural or human made disasters" as causes for forced displacement, it emphasizes state—not international—responsibility to deal with such crises. Such responsibility, however, is not legally binding.[49]

While some observers have argued for the extension of the definition of refugees to include climate refugees under the 1951 Geneva Convention Relating to the Status of Refugees, others have argued against such an extension and have instead called for a new legal instrument specifically tailored to the needs of climate refugees—a Protocol on Recognition, Protection, and Resettlement of Climate Refugees to the United Nations Framework Convention on Climate Change (UNFCCC)—as well as a separate funding mechanism, the Climate Refugee Protection and Resettlement Fund.[50]

There are certain actors, including the ICRC, along with the rest of the International Red Cross (Red Crescent), who pay great practical attention to "migrants." These humanitarian actors tend to avoid putting "people on the move" into legal boxes, even as they identify conventional refugees and war refugees as having certain rights. For the Red Cross network, however, the definition of "climate refugees" relates to a combination of two conditions: (1) trying to meet the practical needs of vulnerable "people on the move" regardless of legal label, and (2) referring to human rights vis-à-vis governments regarding the freedom of movement, freedom from arbitrary arrest, freedom from torture, and some such freedoms. At present, for the Red Cross and these kinds of actors, climate refugees are subsumed under the term "migrants."

Known for its expertise on the laws of war, the ICRC now has a doctrine and much practice regarding the use of human rights norms in situations short of armed conflict. Given that states are unlikely to adopt a treaty on climate refugees anytime soon, the international community must rely on using general human rights norms. These days, most governments are not very forthcoming regarding different types of migrants or refugees, except for some in the West that have shown great flexibility and willingness to pay money to have them housed elsewhere.

Turkey's president Recep Tayyip Erdoğan has welcomed many Syrian refugees. But he then politicized that subject—that is, using them as pressure on Greece and the European Union (EU). The EU pays him to keep refugees out of Europe. Under such circumstances, the prospects for new protective norms for persons on the move are slim to none, as is true for new rules for the laws of war. We have seen no diplomatic conference since 1977 on the laws of war. There is a void in progressive national leadership on protecting persons, which is why Responsibility to Protect (R2P) is, for lack of a better word, dormant, if not dead. For the practical purposes of this chapter, I refer to climate refugees as victims of climate change impacts and extreme weather patterns such as drought, deluge and storms, and water scarcity.

Security Implications of Climate Change

Increasingly, the MENA region and its population face a complex, tenuous, and nontraditional security system. "A changing climate," Mariëlle Snel, Nikolas Sorensen, and Reed Power rightly argue, "can harm human security by damaging livelihood systems or the infrastructure and ecosystems that support them."[51] Climate change, Joseph Romm writes, "will increase risks of violent conflicts in the form of civil war and inter-group violence."[52] Several droughts (2006–10) triggered public unrest, civil war, and mass migration in Syria, where many fled from civil war, poverty, chaos, and persecution. Climate change poses clear and present dangers to the national security of states and will most likely lead to "food and water shortages, pandemic disease, and disputes over refugees and resources."[53] In that sense, climate change appears poised to lead to dangerous challenges to governance and even failed states.[54] One study has concluded that the countries surrounding the Mediterranean are likely to run dry in the coming years and that human-caused climate change is now considered a major factor in more recurring Mediterranean droughts.[55]

While rich in oil and gas resources, the MENA region suffers from a dearth of renewable resources such as water and arable land. The ensuing water scarcity, desertification, and land degradation have already posed security challenges to the region. Increasing populations, industrialization, and urbanization have placed more and more demand on the food supply, giving urgency to the redefinition of security in the face of these pressing regional concerns.

As a nontraditional security challenge, global warming will lead to unprecedented levels of socioeconomic and political stress on authoritarian regimes in the region. The consequences of climate change could potentially ignite destabilizing and life-threatening conditions that will eventually erupt in widespread unrest and revolution. The arcane preoccupation with the notions of hard security has become increasingly extraneous, as human security, weather-related risks, and sustainable development have taken center stage. Rainless, hot summers, long droughts, and severe flooding have not only threatened food and water security but also the stability of livelihoods throughout the region.[56]

New studies have shown that climate change-related uncertainties and disasters that result in resource scarcity tend to enhance the risk of conflict breaking out in ethnically fractionalized countries, spurring intercommunal violence over stressed resources. Several developments have triggered large-scale migration from rural to urban areas in the developing world. These include, among other things, widespread famine, population displacement, and the growing decline of agricultural productivity across the developing world.[57] Of course it is equally important to remember that the historical, deeply rooted, and complicated causes of violence in Syria, Iraq, and Yemen cannot be solely or primarily blamed on the impacts of climate change and natural disasters.

Other countries in the region—including Iran and Turkey—have undergone a severe drought across the same timescale but have not descended into armed conflict or civil war. Yet the security implications of climate change should be taken seriously. Growing evidence indicates that it is likely that climate change will have disproportionate adverse effects on economically disadvantaged groups and communities, exacerbating existing poverty, heightening grievances, and elevating the risk of civil war.[58] Today, most experts in the region argue that the 2011 Syrian civil war was not entirely unrelated to the drought that hit the country hard during the 2006–11 period.

In Iran, environmental mismanagement has led to protests in southwestern parts of the country. These protests began on July 15, 2021, in various cities in southern Khuzestan province and spread to several other Iranian provinces. At the heart of these demonstrations was the issue of water shortages in a province that used to be water rich. Faced with repressive measures by the central government, the protesters' demands and slogans assumed a political tone directed at the Islamic Republic. Most of these protests, which initially took place in the cities of Ahvaz, Hoveyzeh, Shadegan, Hamidiyeh, Abadan, Khorramshahr, Bostan, Susangerd, Dezful, and Izeh, spread to other provinces, including Tehran, Kermanshah, Isfahan, Bushehr,

Lorestan, Kurdistan, East Azerbaijan, North Khorasan, and Alborz. During these street upheavals, many protesters were injured and some killed by live ammunition. The country's mainstream media attributed these cases of unrest largely to the water shortage precipitated by climate change.[59]

Critics, however, have put the onus on the government's mismanagement. For example, Esmaeil Kahrom, a well-known Iranian ecologist, said in an interview with the Khabaronline news agency: "Khuzestan is one of the water-rich provinces of Iran, but unfortunately some unqualified individuals have been at the helm, and the result is the current situation we face today."[60] Similarly, Somayeh Rafiei, the head of environmental affairs in the Iranian parliament, views the causes of the water crisis in Khuzestan along the same lines:

> The water problem in Khuzestan has much deeper and more hidden roots than a multi-year management period. We see a clear example of this in the management of the country's wetlands. . . . In fact, the main problem of water scarcity in Khuzestan and its related environmental issues, as well as the problems of many other areas of the country, are the officials for whose past failures and inadequacies the ordinary Iranians must pay a huge price today.[61]

While acknowledging the impact of climate change and several other factors—such as improper cultivation patterns, construction of a dam by nonprofessionals, irregular transfer of water to other provinces, and the existence of destructive environmental industries such as oil, petrochemicals, and sugarcane—Rafiei insists that officials' environmental mismanagement has significantly contributed to Khuzestan's water crisis.[62]

The oil-rich Arab countries of the Persian Gulf face multiple challenges. For one thing, their political stability is directly espoused by profits from selling petroleum and natural gas. For these countries, which are absolute monarchies that rely heavily on oil exports, climate action presents an existential threat.[63] Furthermore, the ongoing flow of migrant labor to these countries in search of work, which has made major contributions to the economic development of the host countries and brought highly invaluable remittances to their home countries, is likely to pose problems of cultural polarity and political ostracism for the host countries. These forced or voluntary migrant workers, along with refugees and IDPs, have added a sense of urgency for taking up these new challenges.[64]

These environmental and climate dynamics call for framing a new security architecture that involves broader strategic thinking. Critics have argued that these new and emerging security challenges must not be seen simply as vulnerabilities but as existential threats for the future that can cause broader conflicts in the region—conflicts that cannot be effectively dealt with through mere technocratic approaches. On the contrary, a wider range of options, including the securitization of new threats and developmental policies, must be utilized.[65] In the MENA region, environmental pressures particularly susceptible to displacement are the water scarcity, transboundary nature of this water, and sprawling urbanization.[66] As previously discussed, water has become increasingly securitized in the political discourse, even as it is still subordinated to putative high political priorities such as military conflicts and terrorism.[67]

Water Scarcity and Potential Conflicts

Climate change presents new challenges for transboundary waters, as it will change patterns of water availability. Existing transboundary water accords may be challenged because these agreements are often predicated on multiyear averages, as opposed to percentages of flow. At the same time, transboundary waters can offer cooperative opportunities to protect and benefit the environment, economies, and communities within the basin.[68] In 2013, for example, Palestinian, Jordanians, and Israelis signed a memorandum of understanding to support three major water-sharing initiatives, including the development and use of a new desalination plant in Aqaba, Jordan, as well as increased releases of water by Israel from Lake Tiberias for use by Jordanians. The areas of cooperation over the Jordan River, however, remained limited as Lebanon and Syria—two of the five riparian countries of the Jordan River Basin—were not included in that agreement.[69]

Water scarcity, exacerbated by global warming in the already dry MENA region, could likely lead to future conflict as well as significant agricultural decline. As urban areas face extreme heat waves, large cities will be at risk of becoming unlivable while experiencing power outages during long and rainless summer seasons. Irrigation systems will stumble, and the agricultural economy will degrade. Over the past four decades, approximately 38 million people in the MENA region have been affected by drought.[70]

One expert argues that the severe multiyear drought that began in the mid-2000s, combined with inefficient irrigation systems and water diversions by other parties in the eastern Mediterranean, including Syria, contributed to the displacement of a large number of people from rural to urban centers, food insecurity for more than a million people, and increased unemployment, culminating in political instability in the ensuing years.[71] While focusing on the interactions between climate change and security issues, other experts have shown that between 2006 and 2011 close to 60 percent of Syria's land experienced one of the worst long-term droughts in modern history. A drop in the rainfall during the 1971–2010 period led to significantly drier winters in Syria.[72] Additionally, extensive exploitation of groundwater, which caused considerable drops in water levels and, in some areas, contamination by salts and nitrates, rendered local wells (e.g., in the ar-Raqqah region) unsuitable for human use.[73]

The ongoing drought was only one factor among many, including food insecurity, unemployment, and other economic problems related to neoliberal economic policies, that contributed to the 2011 Arab Spring uprisings in Syria. Nonetheless, the drought, combined with the mismanagement of natural resources by the Syrian government—who subsidized water-intensive crops such as wheat and cotton while promoting poor irrigation techniques—led to devastating consequences. The Assad regime's problematic agricultural policies exacerbated the drought, which led to the collapse of farming in the northeastern region, propelling mass migration to cities and causing widespread discontent. The drought led to the displacement of 1.5 million people within the country.[74]

In Africa, as Jeffrey D. Sachs points out, all indications suggest that subhumid and arid areas will dry further, intensifying the food crisis for many of the world's poorest and most vulnerable people. The dramatic reduction in precipitation in the African Sahel over the past thirty years is seemingly related to both anthropogenic warming and air pollutants. The violence in Darfur and Somalia is related to food and water insecurity. Ivory Coast's civil war stems partly from ethnic clashes after people fled the northern dryland of Burkina Faso for the coast. Further chaos could certainly arise.[75] Most analysts overlooked the degree to which environmental and climatic factors leading to food insecurity played a key role in impoverishing Syria's rural areas. The Syrian government could have subsidized agricultural production that was less water intensive and could have accounted for climate and drought projections that were already available.[76]

Another major water-related issue in Syria relates to the Euphrates and Tigris rivers, which originate in Turkey and flow through Syria and Iraq. While mountain snowfall in Turkey contributes 88 percent of the Euphrates' water, neither Syria nor Iraq contributes substantially to the flow. Both countries depend on these rivers—Syria on the Euphrates and Iraq on both—but face water supply shortages as Turkey's demand for water increases upstream. Since the mid-1960s, Turkey has built several giant dams on the Euphrates. These dams have not only reduced downstream water flows but also led to increased water consumption in Turkey itself. Both Syria and Iraq are thus greatly concerned about the future impact of the dams on their populations and have even labeled Turkey's claims of absolute territorial sovereignty over the two rivers a violation of international law.

Nevertheless, a water war between Turkey, Syria, or Iraq is unlikely in the short- to medium-term. Each state has many other geopolitical concerns besides water, which limits the impact of water issues on relations between them. At the same time, however, it is important to heed other observers' suggestion that the Tigris-Euphrates River—significantly affected by border disputes, conflicts over the Kurdish issues, and the lingering crisis in Syria—could potentially lead to more significant transboundary confrontations.

Clearly, long-term solutions regarding the Tigris and Euphrates rivers are indispensable to any peace plan involving states within the two-river system, given their central location in the wider region and the current volatility of Iraq and Syria. Syria's civil war, both in terms of the influx of Syrian refugees to Turkey and the rise of terrorism in the region, has also adversely affected cooperation between the two countries, with Ankara openly supporting regime change in Damascus. As a result, Turkish-Syrian relations came to a halt. Yet the increasing water shortage due to the civil war has compelled Syrian leaders to maintain communication with Turkey to deal with water scarcity even as the war resumed unabated. The ongoing construction of the Orontes Friendship Dam, located between the Syrian village of al-Alani and the Turkish village of Ziyaret, for example, demonstrates the extent to which the water issue has been used as a leverage point for Damascus to maintain communication with Ankara, albeit at a modest and low political level.[77]

During the Syrian civil war, both the opposition groups and the Syrian government have attempted to capture the dams on the Euphrates River, most alarmingly the Tabqa and Tishrin dams. The frequent attacks on these dams have invariably intensified the water shortages, accelerating the growing demand for water. This situation has necessitated further Turkish-Syrian cooperation, keeping the line of communication between the two countries—at both the governmental and nongovernmental levels—effectively open and thus maintaining a context for easing regional tensions.[78]

On March 22, 2022, the United Nations Children's Fund (UNICEF) released some alarming information about water scarcity in the MENA region. Citing the effects of climate change, UNICEF's regional climate change adviser, Chris Cormency, noted that "Water scarcity is only going to get worse in this region."[79] With regard to water scarcity, UNICEF officials said "nearly half of the water is unaccounted for or lost in leakages" owing to weak water systems that fail to adequately preserve water being pumped.[80] The region's groundwater has been overexploited to meet agricultural needs since the 1970s when motorized pumps were introduced. The problem has compounded by "inadequate governance arrangements, including weak water resource management policies and a lack of regulation."[81] While agriculture on average takes up 70 percent of water use internationally, in the MENA region, this figure could be upwards of 80 percent or more, according to UNICEF.[82]

UNICEF added that eleven out of the seventeen most water-stressed countries in the world are in the MENA region: Qatar, Israel, Lebanon, Iran, Jordan, Libya, Kuwait, Saudi Arabia, the United Arab Emirates (UAE), Bahrain, and Oman. The relation of water scarcity to conflict zones in the MENA region deserves much attention. While many families have fled the violence raging in Yemen, it is important to note that they have left their hometown because of saltwater intrusion into their drinking water. Water scarcity has many consequences. The fall in groundwater in

Turkish-controlled northern Syria has caused prices to spike. Iran's water shortages have caused the government to seek help from the Taliban in Afghanistan.[83]

The disputes over the Jordan River have constituted one of the most severe international conflicts over water in the MENA region. The small size of the Jordan Valley and its riparian system notwithstanding, its water resources are immensely complex, partly because several countries—Israel, Jordan, Palestine, Syria, and Lebanon—share the waters of the Jordan River and its source tributaries. Attempts to use the water for different projects by different countries have resulted in constant friction. The Jordan River, with the lowest elevation of any river in the world, rises at the Syrian-Lebanese border where the Hasbani River of Lebanon and Banias River of Syria meet. From there it flows south through northern Israel into the Sea of Galilee and eventually empties into the Dead Sea. The river is considered the border between the State of Israel and the West Bank. From its source to its Dead Sea destination 124 miles away (200 kilometers), the Jordan covers 223 miles (360 kilometers) due to its winding course.[84]

As the only major river in the area, it is one of the most significant features of the Palestinian landscape. Aside from involving several hostile riparian states, the Arab-Israeli conflict has eclipsed efforts to reach an agreement on cooperative utilization of the water system. Efforts to balance the water needs of the local population versus those of Jewish immigrants into Palestine complicated the matters. The question of economic assimilation of a large-scale Jewish settlement in Palestine permeated and indeed politicized the debate over the water management. The water issue became even more acute in the 1930s when Jewish immigration to Palestine increased. Worse yet, the UN partition proposal of 1947, and its ensuing resolution (181), which recommended to divide Palestine into Jewish and Arab states, overlooked the gravity of water problems.

The 1948 Arab-Israeli war exacerbated the difficulties of managing the water issue.[85] For many of these new Jewish immigrants, agriculture constituted the most efficient way to retain their livelihoods. Due to the shortage of rainfall in the area, the only alternative to increasing the land under cultivation was conversion to irrigation agriculture. The only substantial enduring surface water source was the Jordan River and its tributaries.[86] The war of 1967 militarized the water dispute between Israel and its Arab neighbors. Israel enhanced its hydro-strategic position by occupying the Golan Heights and the West Bank, making it impossible for the Arab states to divert the Jordan Headwaters. The 1967 ceasefire lines afforded Israel control of half the length of the Yarmuk River compared to 10 kilometers prior to the war.[87]

By the 1980s, Jordanians charged that Israel exploited the diversion of the Yarmuk's flow downstream to pump unauthorized amounts of water into Lake Tiberias. The canal input dispute, revolving around the issue of excessive Israeli extraction, as well as the failure to reach an agreement allowing the construction of the Maqarin Dam, has kept the Yarmuk a key boiling point for the tensions between Israel and Jordan for many years.[88]

The dispute between Israel and Jordan was settled in the peace agreements (1995), which provide for the supply of water by Israel to Jordan and joint development of water resources. Israel pumps water from the Sea of Galilee through its Movil Artzi water carrier to be used for irrigation of the Negev and other areas. Israel uses underground water sources in the West Bank, which are the main sources of water for the Tel Aviv metropolitan area. Without tapping into ground aquifers, Jordan's main sources for fresh water have been annual rainfall and water from the Jordan and Yarmuk rivers, both of which have been reduced significantly in recent years. Although plenty of seawater is available, Jordan does not have the financial resources that wealthy countries like those in the UAE have to build adequate desalination facilities.[89]

Another controversial issue is the disparity in water consumption between Israel and the Occupied Territories. The Mountain Aquifer, which provides one quarter of Israel's water needs, is fed by rains that fall on the mountains of the West Bank. Israel has used its political clout and military power to secure this water resource by limiting Palestinian access to it, including placing restrictions on the drilling of new wells. As a result, the Palestinians living in the Occupied Territories face a serious water shortage. While the total per capita water use in the West Bank is 22 cubic meters a year, the equivalent of 60 liters per person a day, which falls far short of the World Health Organization's (WHO) recommended minimum consumption of 100 liters per person on daily basis, Israelis' per capita consumption is five times greater, at 120 cubic meters a year.[90]

Israelis have in fact become the region's hydro-hegemon, arguing that "water is a question of existence for the Jewish state" both as a vital part of the country's development of its agricultural sector and as a key part of the strategic security of the state, driven by the rising water scarcity crisis and increasing water shortages. Since the 1967 war, in which Israel seized the territories of Gaza, the West Bank, Sinai Desert, East Jerusalem, and the Golan Heights, half of the Israeli water supply comes from water resources located outside of internationally recognized borders. Thus, Israelis argue, maintaining access to a maximal allocation of water from the Jordan River is a vital strategic issue for Israel.[91] Israeli-Palestinian conflicts have since intensified this water crisis.

Under the Oslo process, solving water issues has become subservient to resolving other disagreements including borders, security, settlements, refugees, and East Jerusalem. The two-state solution has failed, and there is no other solution in sight. The humanitarian crisis in the Occupied Territories is inevitable, while both Israelis and Palestinians are witnessing the pollution of increasing amounts of shared waters. Given that Israel has excess water at its disposal, sharing water more equitably is no longer deemed a disadvantage for Israelis. A final water accord will be a win-win situation for both sides. For Palestinians, it would enhance freshwater availability in every household; for Israelis, it would clean up pollutants from rivers and creeks that flow through its main cities.[92]

The October 7, 2023, Hamas attacks on Israel have only exacerbated the problem of water shortage in the Occupied Territories, leading to the Israeli military retaliation on Gaza with enormous loss of lives, mostly unarmed civilians, and an unprecedented humanitarian crisis stemming from forced displacement of the Palestinians. This humanitarian disaster has grown even worst by famine and starvation of the population at the hands of the Israeli soldiers. As of mid-September 2024, according to Gaza's Health Ministry, more than 41,000 Palestinians—mostly women and children—have been killed and 95,125 injured in Israel's air and ground offensive on Gaza since October 7, 2023, the day that Hamas-led militants attacked Israel, killing 1,200 people and taking 250 hostages.[93] Further, the outbreaks of polio may have already made its way into Gaza, where it could find the propitious conditions—many unvaccinated children packed together in unsanitary situations—that allow it to grow. At a time when outbreaks can be effectively stopped by using the oral vaccine, such vaccination efforts have become complicated due to the displacement of Palestinians in the ongoing war.[94]

Another contentious regional issue regards the potential security tensions between Iran and Afghanistan over the Helmand River. With the Taliban back in power in Kabul since August 2021, the old tension over the Helmand River between Iran and Afghanistan has amplified. The 621-mile-long river, which flows across Afghanistan and empties into lakes and wetlands in southeastern Iran, is an essential source of water for both countries. It has been critical to agriculture, fishing, and human consumption for millions of people in Afghanistan and Iran's southeastern Sistan and Baluchistan province. Since 2021, the Taliban regime has created several hydroelectric dams that have increased artificial reservoirs on some of the Afghanistan's rivers, including the Kajaki Dam on the Helmand River.[95]

The Iranian side has faced negative consequences, as dry soil has also left Sistan and Baluchistan vulnerable to flash floods. The Indian monsoon, which largely affects the Indian subcontinent and its surroundings, has also adversely affected this region. In January 2020, three days of heavy rain led to the displacement of thousand people in Sistan and Baluchistan region, killing at least three and destroying crops, homes, and infrastructure. The floods caused at least $53 million in damage, while also revealing the Iranian government's incapacity to provide relief to this region's people.[96]

Recent tensions between the two countries point to further conflicts over water rights. Critics argue that the Islamic Republic has failed to invest in improving the water management system by missing numerous opportunities. Over the past four decades, Iranian authorities have not only ignored experts' warnings but also have arrested and imprisoned environmental activists who have raised legitimate concerns about the water crisis.[97] Absent a lasting agreement between Iran and Afghanistan, we are likely to see mass migration and further conflicts between the two countries in the future due to water and climate change.[98]

Climate Justice

Also known as environmental justice, climate justice refers to the ways in which the benefits and burdens of climate change and responsibilities to cope with climate change can be equitably distributed. Without climate justice, marginalized communities, which have contributed the least to global warming and GHGs, unjustly suffer from the worst impacts of climate change. Underscoring the necessity to uphold human rights in the face of myriad threats that climate change may bring on, Anja Mihr writes that "Climate justice aims to hold stakeholders accountable and recognizes duty-bearers at various levels of society, including private, corporate, public and individual actors."[99] Notwithstanding the fact that obligations have yet to be defined precisely, Mihr goes on to argue, "the first step toward achieving climate justice is to identify the duty-bearers who carry the responsibility for the causes and impacts of climate change."[100]

Others have appealed to the moral obligations that humanity has toward future generations. It is within this context that intergenerational justice is invoked to protect people from present and future climate change-induced disasters with the goal of saving the planet for posterity[101] (more on that in the next section). Still others argue that the relation between climate change and migration is profoundly political and that like all forms of knowledge that link climate change to migration are never neutral and universal but rather always partial and situated.[102] It follows that achieving climate justice requires more than simply engaging human rights in abstract. It requires paying attention to the specific historical and geographic context within which and about which such claims to climate justice are made. In short, "the knowledge of climate change, migration, and human rights is not universal but situated."[103]

Answers to these challenges must be found in the lived experiences of different communities. Climate justice requires swift action to cut emissions and a serious attempt to explore pathways to a cooler future. It is widely believed that eliminating emissions by about 2050 is a difficult but achievable task. But that alone, as one expert repeatedly and ruefully reminds us, is unlikely to solve the climate crisis. Carbon removal, according to one expert, is surely the safest path to permanent cooling, but solar geoengineering may well be able to cool the planet this century with the least environmental risk and less socioeconomic disruption. Ultimately, if we are to reduce risks from the emissions already in the atmosphere, we must look to a combined approach that properly implements carbon removal, solar geoengineering, and local adaptation.[104]

The idea that "the causes and consequences of global environmental degradation cannot be addressed without tackling inequality and injustice" has steadily gained considerable momentum and force at the global level.[105] Increasingly, the concept of "environmental justice" has been locally expanded to include unequal exposures by class, race, and ethnicity. Since the 1970s and 1980s, environmental justice has progressively secured its position as the ascendant rallying cry for the poor- and middle-income countries. The leaders of these countries raised such concerns at the 1992 Rio Earth Summit. The issue of global climate change, however, has remained a key divide between rich and poor nations.[106]

At the very heart of this division over the debate over climate justice lies the question of who is most responsible for climate change. The United States has released more than 509 GtCO2 since 1850 and is responsible for the largest share of historical emissions, according to one study, with some 20 percent of the global total. China is second, with 11 percent, followed by Russia (7 percent), Brazil (5 percent), and Indonesia (4 percent). A few European countries have risen the rankings due to their coal-fueled industrialization. These included France, Germany, and the UK. Although these countries have markedly reduced their emissions in recent decades, they remain among the largest contributors to historical warming today.[107]

One of the most controversial issues relates to the question of "Is it environmentally possible for poor nations to pursue economic development as rich nations did?"[108] This question highlights a troubling anomaly facing developing countries—namely, the justice-sustainability paradox. Many countries in the Global South seek economic growth and development via the industrialization route known to them on the one hand, while also cognizant of the fact that climate change is likely to further exacerbate the effects of population growth, urbanization, and poverty on the other. Absent a long-term adaptation plan, climate change is likely to plunge these countries further into deeper poverty, dramatically undermining the opportunities for sustainable development to combat poverty.[109]

Closely related to the issue of climate justice is a growing awareness that the climate crisis is indeed a humanitarian crisis. Climate change is now considered the main driver of humanitarian need.[110] More attention must be given to the collaboration between national and local authorities on the one hand, and between and among private sector, new humanitarians, and local agents. The need for a holistic and urgent approach toward addressing the issues of climate change and climate justice could not be clearer.

Climate Change and Human Rights

There is general consensus that climate change directly affects human rights conditions in that "the enjoyment of virtually all protected rights can be undermined as a result of climate change."[111] While "principles shaping environmental (climate) justice, human security, and sustainable development overlap to some extent with the human rights framework,"[112] and while such broadly based norms provide additional lenses to address the human impact of climate change, taking a human rights approach toward climate change should be based on specific entitlements whose absence would jeopardize human freedom and would likely limit choice.[113]

Similarly, it is important to protect the collective rights of communities that have been exposed to climate impacts and have been forced to permanently relocate. In situations where climate-induced ecological change renders entire communities uninhabitable and threatens the lives of community residents, the right to relocate must be fully realized. "The right to relocate," one expert insists, "is an entitlement when relocation is the only feasible solution to protect the human right to life, as well as the right to basic necessities inherent in living a life of dignity."[114] This right also includes the right to movement when affected by natural disasters or threatened by other environmental events.[115] Given the ongoing climate vulnerability, "human rights protections need to infuse the relocation process."[116]

If not properly managed, one human rights expert cautions, climate change is likely to endanger various rights.[117] Dealing with climate change is not just a matter of rights but it also relates to the duty of the present governments to protect the environment for the human rights of future generations.[118] Richard Hiskes urges attention to the fact that environmental human rights are real and belong first and foremost to future generations. It is important to recognize that many of the actual harms of climate change—that is, their impacts on our planet and the lives of people—will take place in the future. We are living with the risks, but future generations will live with the devastation. All rights exist to safeguard people from harm. Environmental human rights and the rights of children are arguably inextricably intertwined.[119]

Thus, we have an obligation to future generations not to destroy the environment via climate change because those are harms against which they have rights. And, in return, we enjoy environmental rights too and the benefits of beginning to clean up the environment even though we are often living with risks only, not resultant harms that are still to manifest. It is worth noting that the reciprocity that benefits us really comes from our own actions, not the actions of future people, for obvious reasons. This so-called reflexive reciprocity is real and benefits us and those who come after us. That is why this should be seen as "intergenerational justice," because like all forms of justice, this one too involves reciprocity.[120]

Besides underscoring the importance of a justice framework, we need to turn to the issue of equity. One of the most disturbing issues of climate change is inequity. States and communities that have contributed least to global warming, writes Sumudu Atapattu, "are bearing a disproportionate impact without ever reaping the benefits of the economic development that resulted in these greenhouse gas emissions."[121] Africa faces desertification; South Asia and the MENA region face water scarcity; and Latin

America faces massive displacements caused by hurricanes, floods, drought, and loss of crops and food security. It is the poor and vulnerable communities that will suffer the most. Climate change acts as a threat multiplier and will amplify the intensity of current inequalities and poverty.[122] That may in part explain why many human rights scholars have shifted their focus away from conventional and customary international law to matters relating to social justice and human security on the grounds that climate change threatens the entire human family.[123]

From a different perspective, some experts have called our attention to the underlying political economies of local communities, noting that the interplay between environmental change and the changing socioeconomic and political context of affected communities is of profound importance. The "threat" to human rights comes not from specific weather events or climate change, but rather from the dynamics in political economy that influence the extent to which people can cope with and survive such threats. While certain rights are imperiled by the expected impacts of climate change, an ongoing lack of democratic and economic rights will amplify those threats. In short, the human rights implications of climate change are better understood if the context within which such rights are realized. Political inequalities are likely to migrate with affected communities or affect the availability of the range of options for community members.[124]

Not all human rights abuses, these experts add, are explicitly related to climate change as they seem to be complicated by issues of development and justice. Climate change policies that disregard the "justness of the development plans" will exacerbate—rather than ameliorate—the climate change-related disasters.[125]

Intervention and Adaptation

To address the potential refugee crises and forced migrations that could result from climate change effects, intervention must occur at three levels: prevention, protection, and solution. Insofar as prevention is concerned, international and humanitarian organizations should continue to offer technical advice and support to states for the domestic implementation measures needed at the national legal and policy levels. To provide adequate humanitarian assistance to IDPs, states should develop capabilities to forecast, gauge, and respond effectively to the various needs of these populations. States should also ensure that all relevant public authorities, including armed and security forces, are aware of their responsibilities toward facilitating prompt access for humanitarian organizations to IDPs.[126]

To prevent future droughts and water crises in the MENA region with limited political cooperation on water issues, it is important that local, regional, and international organizations and their policymakers zero in on improving water-use efficiency in agriculture while focusing on reaching effective international agreements for managing and sharing the rivers and other water sources that cross political borders. Because water is a transnational good, regional and collective actions are imperative.[127]

In terms of adaptation to the impacts of climate change, it should be noted that local and regional solutions must be sought by considering drastic changes to agriculture. The MENA countries have for too long grown crops that are ill-suited to the region's agricultural conditions: cotton, rice, and wheat. A shift to crops that use less water will be imperative. Saving water through wastewater recycling will be important, although its environmental hazards have yet to be fully determined. Water conservation methods practiced in Jordan and the subsidizing of efficient irrigation equipment put in place in Morocco are two preferred mechanisms to sustainably manage scarce water resources.[128]

In most parts of the MENA region, by contrast, subsidy of agriculture continues to be the bedrock of agricultural policy. The region, as Ashok Swain and Anders Jägerskog rightly point out, has indeed the highest level of agricultural subsidies for water in the world measured as a portion of gross domestic product (GDP).[129] In both Saudi Arabia and Yemen, Swain and Jägerskog go on to note, the heavy subsidies on fossil water and diesel, respectively, have had negative effects on the water situation in those countries as they have allowed cheap pumping of scarce water to irrigate the water-intensive crops. These water subsidies have led to untenable agricultural practices as well as a gargantuan misuse of this scarce and valuable resource. Such agricultural policies have minimized the impetus to find innovative and more water-efficient techniques for irrigation practices, while also attenuating the efforts to provide effective economic incentives to move to less water-intensive crops.[130]

Perhaps the biggest challenge facing the MENA region is phasing out fossil fuel usage while simultaneously investing in renewable energy sources, such as wind, solar, and nuclear power. "By the 2030s," experts remind us, "solar power will likely be the largest source of electricity on the planet."[131] Within the region, ruling elites and opinion-makers frequently talk of the need to be prepared for carbon-free and clean energy solutions in the inevitable post-oil era. Yet little of such talk has thus far been translated into actual reality, with the notable exception in Dubai. How would the rest of the region adopt the Dubai model, and whether by doing so this will hamper or reinforce the need for other political changes, remains to be seen.[132]

Conclusion

Escaping persecution, war, violence, poverty, hunger, natural and human-made disasters, millions of people across the world cross borders in search of a new place to survive. The contemporary forced migration and refugee crisis have engaged both the academic and policy circles with some of the most challenging political and practical questions of our times. It is not unreasonable to assume that violence, poverty, and climate change are among the leading causes of common threats to the entire humanity. The reaction—or the absence of reaction—to these causes involves elements of justice, equity, and human security. These threats impact on the enjoyment of human rights, especially socioeconomic and cultural human rights.

Climate change is widely regarded as a threat multiplier due in large part to the fact that it exacerbates existing inequalities and poverty. The nagging question persists: Are migration and mobility a threat to the security of states? While irregular migration is likely to provide new security challenges, such as the entry of potential terrorists into host countries, the rights of forced migrants and state sovereignty need to be balanced against each other. If and when properly managed, as some experts remind us, migration can contribute to greater openness, wealth, and human development by reducing global inequalities over the longer term. As developed countries open their borders to immigrants or migrant workers, gains from remittances, repatriation, and the return of skilled labor will have positive impacts on the development of the countries of origin.[133] The international liberal order, they argue, faces a serious pushback from the resurgence of a right-wing populism, as democracies are retreating from their global agendas, and open, liberal societies are at risk of disappearing from the Western world. At the same time, however, rising levels of human displacement—due to conflict, poverty, deprivation, and climate change—demand a response from the international community, especially the countries of the Global North. To deny any responsibility to bring under control refugee and humanitarian crisis in the Global South is a recipe for disaster.[134] It is unclear, however, how states' obligations toward global warming can be established, nor human rights treaties recognize a specific right to a safe and healthy environment.

Furthermore, human rights law provides no clear answer to IDPs as a result. Many states in the Global North appear reluctant to adopt a treaty on climate refugees any time soon. Customary international law—predicated on state practice and opinio juris—provides no legal protection to climate refugees. The so-called environmental refugees is not seen as a legal category and its mandate remains limited. The answer to the question of whether international human rights law, along with state moral obligations, can provide a useful framework to deal with climate refugees remains vague at best.

Yet the fact remains that the impacts of climate change have wide-ranging implications for the effective enjoyment of human rights. Increasingly, climate-induced displacement and migration affect the basic rights of inhabitants. Climate change effects in the MENA region have presented existential threats to several countries in the region. The number of refugees and IDPs has increased in the Middle East, especially in Iraq and Syria, where lingering civil wars have pushed many people from their homes.[135] Forced migration from Iraq in 2007 and Syria in 2011 has presented serious regional challenges. Amid these new challenges, it has become increasingly difficult to contain human mobility as an important option for many people who seek to improve their living conditions or escape chaos and political uncertainty. Human mobility is indeed a human right, and its complex roots require new thinking about fragile states and coming to grips with exacerbating factors such as climate stressors.

Organization of the Book

This book is divided into three parts. Part I sets up both analytical and normative frameworks for guiding research and informing policy. It focuses on explaining refugees' basic needs and dignity, such as employment, food security, and human rights. This part engages with the problem of climate change by elaborating on how the world's climate is changing at an unprecedented rate in recent human history. The four chapters in this part pay special attention to a new category of refugees generally known as "climate refugees." While such a category still lacks legal clarity and protection, because under current international refugee law, climate-displaced persons fall far short of qualifying for refugee status, there is a growing recognition that such refugees have become a common feature of global politics.

There can be no doubt that climate change has increased environmental degradation while also impacting political stability across the MENA region. Variables such as water scarcity and food security have already caused an influx of refugees and migrants crossing borders. Chapters 1–4 deal with introduction, conceptual and legal challenges, food security, and the refugees' right to work.

Part II, which covers Chapters 5–8, examines the case studies of Turkey, Egypt, Syria, and Yemen, where the impact of the refugee crisis and climate change, in some cases due to ongoing outside interventions, has resulted, to varying degrees, in political tensions and refugee uncertainties. All these countries have experienced crucial influxes of refugees, primarily due to ongoing regional conflicts, often resulting in significant numbers of asylum seekers, with Syria becoming a major source of refugees for the other three said countries.

Part III addresses mitigation, intervention, and adaptation measures adopted by both large and small states of the Middle East, underlining environmental, agricultural, public health, and political solutions to deal with climate change challenges, including low rainfall, high temperatures, and dry soil. Absent solutions, environmental difficulties will most likely pose serious challenges to the political stability and security of the region. These challenges must not be underestimated. Chapters 9, 10, and 11 direct our attention to the cases of Iran and some member states of the Gulf Cooperation Council (GCC)—namely, Saudi Arabia, Qatar, and the UAE—and Morocco. Chapter 12 concludes by looking into future and providing a balanced conceptual framework to demonstrate not only the current and future regional dynamics (given the interplay of climate change and water scarcity and food insecurity in the MENA region and climate-related causes of refugee crises, the Syrian civil war, the Yemen humanitarian crisis, etc.) but also the growing need for addressing climate change over the longer term.

On balance, the book seeks to achieve these goals by addressing the challenges of climate change in the MENA region. Open to further research are the questions of whether international human rights law, along with the moral obligations of good governance, can provide a useful framework to deal with climate refugees. How soon will Middle Eastern countries move away from fossil fuels and adopt environmentally friendly measures to meet the challenges of climate change? What roles do civil society, social movements, and the private sector play in addressing climate change? I hope to answer some of these questions in this volume.

CHAPTER 2
A POLITICAL FRAMEWORK
ENVIRONMENTAL SECURITY AND ACTIVISM

The urgency of dealing with climate-related vulnerabilities and potential security threats will most likely affect the governments' policies in the Middle East and North Africa (MENA) region. Regarding providing water and food security, it is worth noting that this region is already experiencing violent conflicts and faces potential conflicts in the future. Drought-related water and food scarcity requires long-term plans for managing such uncertainties. Increasingly, water has been securitized and weaponized in the region. Water scarcity and food security have been closely related to civil wars in Yemen and Syria—two countries whose severe water scarcity has prompted competing actors to deploy this precious resource as a tool in violent conflict to the detriment of millions of civilians.

At the same time, the MENA region is currently facing one of its most acute migration challenges, as the region has become the contemporaneous producer of and host to the world's largest population of internally displaced people (IDPs). Even though much of the debate over migrants surrounds the impact of refugees on European states, most of these IDPs live within the borders of the MENA region. The burden of care for these IDPs falls on regional actors and neighbors, some of whom lack the stable and robust socioeconomic and political conditions necessary for hosting and integrating large numbers of refugees.[1]

The wars in Afghanistan, Iraq, and Syria have produced the greatest share of the Middle East's refugees in recent years, but many more have fled wars and failed states in Libya, Somalia, Sudan, and Yemen. Neighboring states have faced severe challenges in absorbing millions of refugees, while North African states and Turkey have emerged as key transit hubs for refugee flows into Europe. Additionally, inevitable climatic shifts have exposed vulnerable populations to increasing risks in the region, necessitating the imperative of collective action to tackle climatic impacts.

This chapter is an attempt to review and assess the impact of environmental activism and security in the MENA region. It proceeds from the assumption that climate change often exacerbates—if not initially causes—preexisting problems such as social inequality, political corruption, an underdeveloped economy, and physical insecurity. The cascading risks of climate change are likely to spread tensions and conflicts among neighboring countries, and create rivalries between competing communities, groups, and even states.[2] Given this reality, the 2011 Arab Spring uprisings should not have come as surprise. Climate change would most likely prolong the existing conflicts, leading to the internal displacement of many people and refugee flows.

In the sections that follow, I try to unpack the connection between climate change, environmental security, and environmental activism in the MENA region by examining wide-ranging effects of climate change, including the new waves of refugees, floods, rising sea levels, severe heat waves, droughts, water scarcity, virtual water trade, food security, and land acquisition. I revisit the nexus between climate vulnerability and security throughout the chapter. My basic conviction is to demonstrate how environmental and climate-induced insecurities impact many societies and communities in the MENA region by disrupting urban and rural socioeconomic infrastructures, food supplies, economic development, while also damaging mental and physical health of their residents. The chapter concludes by demonstrating how these ongoing risks have triggered environmental activism on a scale not seen before, especially when such risks tend to aggravate existing socioeconomic, political, and gender inequalities and tensions.

The New Waves of Refugees

The current waves of refugees from the Levant can largely be traced to the 2003 US invasion of Iraq and the subsequent civil war, which produced an estimated 6 million refugees and a similar number of IDPs. The ethnic and sectarian transformation of Iraq triggered region-wide sectarianism and galvanized jihadist movements. The population movements put enormous strain on the infrastructure of Jordan, Syria, and other neighboring states, while introducing a semipermanent presence of international organizations into the everyday governance of these noncitizen populations.

The UN estimates that more than 6 million Syrians have fled the country, while another 10 million have been IDPs. The size of the displacement of the Syrian population, both internally and beyond the country's borders, can distract from other parallel catastrophes. The collapse of the Libyan state has produced a wave of Libyan refugees, while migrants from sub-Saharan Africa, whose passage Qaddafi once served to interdict, continue to transit through Libya and depart from there and other parts of the North African littoral on the way to Europe. The war in Yemen has also produced massive internal displacement, even if a naval and land blockade has prevented most from fleeing the country. The situation of these and other refugee populations in the region will continue to have an impact, not only on regional security but also on the regional political economy and political development into the foreseeable future.[3]

The presence of refugees in the MENA region will exert further pressure on countries already under great stress from previous waves of refugees, such as Lebanon and Jordan, where refugees constitute more than 25 percent of their populations. As in the past, refugees have fled to other countries, such as Turkey, Greece, Italy, and other European countries.[4] The Lebanese government, despite a policy crisis in the initial stage of the Syrian civil war, attempted to develop more concrete policies to confine the number of Syrian displaced persons on its territory and to contain the negative repercussions of the Syrian political turbulence.

The policy changes made subsequently failed to lead to further protection and integration of Syrian refugees due to a lack of capacity, ideological concerns emanating from the delicate sectarian balance, and the recent history of Palestinian refugees in Lebanon. International politics also affected Lebanese authorities' decision as they engaged in negotiations to receive more humanitarian development aid. The country's low absorption capacity, as well as factors having to do with national security and domestic politics, however, had a much greater impact on refugee governance than global politics.[5]

In addition to education and health sectors, the Lebanese public services infrastructure has been under stress due to the population increase with the Syrian refugees' presence. The most affected services have been electricity, irrigation canals, water networks, and solid waste management. Although the UN agencies and international NGOs have also supported the Lebanese public institutions to provide necessary services, their support has remained insufficient to meet ongoing needs.[6]

Housing presents one of the most difficult challenges for Syrians in Lebanon. The Lebanese government has avoided proposing alternative sustainable options for shelters to camps and has remained opposed to the building of nonpermanent Refugee Housing Units. Due to the lack of camps, millions of refugees have struggled to meet the basic needs of their communities. In agricultural regions in Bekaa and the plains of Akkar, many refugees lived in camp-like settlements that were set up by refugees themselves but lacked drinking water and a sewage system. This situation has had negative health and environmental impacts. The rented houses have remained below the minimum secure standards, as the demand has been soaring, making the rent expensive despite shabby conditions. On balance, the socioeconomic situation of many displaced Syrians was reported to be under serious threat, and nearly the entire population of Palestinian refugees from Syria were severely economically vulnerable.[7]

Rising Heat Waves

In addition to war and violence in the region, food production has been severely affected by the climate change, with about one-third of arable land in the MENA region hit by extreme heat and weather patterns, rising temperatures, diminishing rainfall, unbridled food insecurity, and socioeconomic and political unrest in the MENA region, suggesting that climate change can further intensify insecurity, precipitate an agricultural crisis, and cause mass migration. The region is currently the hottest and driest region on the planet Earth. Even worse, climate change would likely render some areas uninhabitable in the coming decades with temperatures possibly reaching 60 degrees Celsius or higher. The ramifications of climate change throughout the MENA region would be grave. "Future decades," Dan Rabinowitz writes, "will see millions across the Middle East lose their livelihoods, their ways of life, and indeed their lives to global warming."[8]

Climate change-related chronic water shortages, heat waves, drought, and extreme weather patterns are certain to adversely affect food production and agricultural output. According to one study, by the end of the twenty-first century, nearly half of the MENA population (approximately 600 million) could be exposed to annually recurring smashing and ultra-extreme heat waves. Most of the exposed population, who will live in urban centers, would find it imperative to cope with these socially and economically disruptive weather conditions.[9] At the current level of GHG emissions, the MENA region will suffer sweltering heat waves and unprecedented living conditions. Major urban centers around the Persian Gulf region, Arabian Sea, and Red Sea, such as Dubai, Abu Dhabi, Doha, Dhahran, and Bandar Abbas, are expected to experience severe temperature spikes on a more regular basis.[10]

Perhaps the most immediate consequence of spiking heat wave hazards will be water scarcity, as it poses one of the greatest challenges to the MENA region. While higher population growth and economic development will place great demand on water consumption, climate change will be the primary driver for the most notable changes in surface water shortage across the region.[11] Some of the biggest challenges caused by climate change—reduction in rainfall and increasing temperatures—will unfold in countries already facing politically and environmentally frail situations. Low-lying deltas, such as the Nile and the Shatt al-Arab, are noticeably vulnerable to the impacts of climate change, as have low-lying coastal area, such as Morocco's Mediterranean coastal zone. Sea level rise is likely to cause saltwater to penetrate freshwater aquifers. Coastal areas where groundwater is overutilized are distinctly at risk to saltwater intrusion into aquifers, because unrestrained groundwater extraction renders space for saltwater to flow into freshwater aquifers.[12]

According to one study conducted in 2016, insofar as macroeconomic factors, such as trends in the gross domestic product (GDP), national output, inflation, employment, spending, and monetary and fiscal policy are concerned, climate change will have long-term implications for economic growth. The rising heat waves and other adverse impacts of climate change could lead to declines in growth rates equivalent to 6–14 percent of the MENA region's GDP by 2050.[13]

Floods

Floods are the most prevalent natural disaster in the MENA, a region which, according to the World Bank, is among the most vulnerable regions on earth to rising sea levels. Between 1981 and 2011, some 300 floods hit the MENA region, killing 19,000 people and adversely affecting more than 8.6 million people.[14] The 2008 floods in Yemen caused $1.6 billion in damages—amounting to 6 percent of the country's GDP. Some of the driest parts of the region have also suffered from flooding. Jeddah in Saudi Arabia has suffered almost annual floods since the late 1960s, caused by sudden violent storms.[15] The 2009 floods in Jeddah led to $1.36 billion in losses. The 2004 floods in Djibouti resulted in 230 deaths and $11.1 million in losses, negatively affecting 100,000 people.[16] Further economic growth and rapid urbanization close to surface water bodies have exposed more households and assets to flooding. Forecasting a 0.5-meter rise by 2099, a World Bank report warned that low-lying coastal areas in Tunisia, Qatar, Libya, UAE, Kuwait, and Egypt are particularly vulnerable.[17]

Record rainfall in late March 2019, according to Amin Dezfuli, caused unprecedented flooding across several Middle Eastern countries and led to vast damages and casualties, particularly in Iran. These potent rains made the 2018–19 rainy season (October–March) the wettest in the past four decades, a sharp contrast to the previous year, which was the driest over the same period. This event, which was viewed as an example of rapid dry-to-wet transitions and boosting of extreme weather, can be attributed to climate-induced flooding.

An atmospheric river (AR), originating from the tropical Atlantic Ocean, accounted for the heavy precipitation that initiated the floods. Moisture transport by AR was equivalent to more than 150 times the aggregated flow of the four major rivers in the region, namely: Tigris, Euphrates, Karun, and Karkheh. Specific brief weather conditions, including tropical-extratropical interactions of the atmospheric jets, and exceptionally warm sea surface temperatures in all surrounding basins provided the necessary elements for the formation of such a massive AR.[18]

As in the case of droughts, the urban poor are the most vulnerable to floods. They lack the assets necessary to protect and retain their property. Their livelihoods are often at great risk from extreme rainfall. It is imperative that governments in the MENA region invest in flood control measures. Flood relief and insurance programs must be installed to effectively respond to soaring flood risk in the region.[19]

Rising Sea Levels

The MENA region is also one of the most vulnerable regions to rising sea levels triggered by climate change. The mean global sea level, according to some studies, is set to rise 30–122 centimeters (1 to 4 feet) by the end of the century. As a result of sea levels rising, seawater can intrude into coastal aquifers and wells, thus salinizing the water there while also having devastating consequences for littoral agricultural communities. Coastal flooding could also make some waterfront cities and towns in the MENA region unlivable, contributing to economic problems and migration-related stresses on already besieged governments.[20] Some cities in North Africa are particularly vulnerable. Algiers, Benghazi, and Alexandria, for example, will be exposed to even modest increases in sea level by 2050.[21]

Temperatures, the study goes on to demonstrate, are expected to rise in the region by at least 4 degrees Celsius by 2050—that is, if GHGs continue to increase at the current rate—and heat waves are set to be experienced ten times more frequently by the same year.[22] The region will see a significant rise in sandstorms, and by extension desertification in the coming years. Ultimately, as some experts suggest, governments in the MENA region must adopt a more inclusive process of political and socioeconomic reform from below that will better position their societies to combat the adverse effects of climate change. Adopting a holistic approach, which involves civil society, government regulatory agencies, private sector, and media, must depart from current top-down paradigms toward a climate adaptation move from below.[23]

Droughts

Water scarcity is caused by inefficient and wasteful use of water resources where demand consistently exceeds supply. While proper water management policies can address long-term water shortage challenges, such policies cannot prevent droughts from occurring. Droughts can have wide-ranging ramifications, including the destruction of agricultural production and forced displacement of communities. Droughts tend to lead to reduced crop production, increased fire hazards, increased livestock mortality, and decreased water levels.[24]

The MENA region has been experiencing a high frequency of some of its most severe droughts in 900 years. The drought peaked between 2006 and 2010 by an ongoing set of new droughts and, ironically, floods. The result has led to the erosion of topsoil, making the farming of crops extremely difficult. As a result, several countries have experienced political instability caused by climate change-induced conflict. In both Syria and Iraq, drought coupled with water mismanagement has led to widespread mass protests. In Iraq, in the southern city of Basra, water shortage has become a key driver for protests. In northern Iraq in the Iraqi Kurdistan region, growing tensions have emerged over the water issue between the Kurds and the Iraqis, as well as between the Kurds, Iranians, and Turks, who share either borders or rivers, or both.[25]

In Syria, when the drought hit the country hard during the 2006–10 period, farmers drained any groundwater available to them to tend to their crops. Some experts show the extent to which the ongoing droughts have been more damaging to wheat production than the armed conflict.[26] Having depleted the groundwater, which was not replenished by rain, many of the farmers migrated to larger cities like Hama, Homs, and Damascus in search of work to support their families. They found no jobs, while struggling to make ends meet. When in 2011 unrests broke out into full-scale civil war, they joined street protests and battles against the Assad regime.[27]

Jordan witnessed its sweeping anti-government protests in recent years in the 2018 summer, with water shortage as its core contributing factor. The protests led to the resignation of Prime Minister Hani al-Mulki, but the country is likely to see an ongoing wave of unrest in the coming years. Lebanon is on the verge of a water crisis owing to a combination of government failures and climate change. Yemen faces one of the worst humanitarian crises of the world. The ongoing war and power struggle have been linked to drought and resource mismanagement.[28]

The 2011 Arab Spring uprisings in Egypt were, to some extent, tied to climate-induced water scarcity. The looming instability in Sudan was due in part to resource mismanagement and a lack of water availability. The latter has been directly linked to climate. A 2005 report showed that the UAE had the highest per capita consumption of water in the world. This, coupled with the fact that the water table has dropped about 1 meter per year for the last thirty years, is cause for concern.[29] The long-term climate forecast points to more heat and more water scarcity—developments that could portend more conflict, instability, and refugee crises in the future.[30]

Diminishing Water Supplies

The Middle East is the most arid region in the world. The water shortage, magnified by climate change, is likely to make some parts of the MENA region inhabitable and negatively affect crop yields. The resultant food scarcity and dependence on imports of basic food have made the region's food systems vulnerable to price fluctuation in global markets and could lead to a rise in the cost of food. Because 70 percent of MENA's agricultural production is rain-fed, this makes the region largely exposed to changes in temperature and rainfall resulting from climate change.[31]

The population growth has led to overgrazing, disorderly land use, and soil exhaustion. These have led to desertification, which in 2010 threatened 20 percent of the MENA region's territory.[32] Water scarcity entails a significant financial burden, as studies suggest MENA will suffer the most of any region around the world, costing governments 7–14 percent of their GDP by 2050.[33]

The ensuing food and water insecurity will heighten the region's dependency on food imports. Rapid population growth coupled with a rising trend toward urbanization will further enhance this dependence. In 2019, according to one study, more than 50 million people in the region were chronically undernourished;

they were living in the conflict-affected states, notably among the region's refugee populations. Covid-19 also intensified these trends. High reliance on imports also meant that these countries were especially vulnerable to surging global food prices, as many crop-producing countries grew susceptible to temperature and precipitation changes in their own environments.[34]

Food dependency has made the Middle East immensely vulnerable to external sources and vagaries of the market. For example, 90 percent of Egyptian wheat comes from abroad, mostly from Russia. In 2010, Russia had to take its grain off the international market due to wildfires and a heat wave. As a result, food prices in Egypt rose 30 percent.[35] It is hard to deny that this situation was entirely unrelated to the following year's uprising in Egypt, especially in light of the fact that protesters' key slogans were "Bread, freedom, and social justice."[36] Furthermore, scientists predict that a temperature increase of 2 degrees Fahrenheit to 5 degrees Fahrenheit would expose 6–25 million North Africans to coastal flooding.[37] Rising sea levels would also lead to increased salinization of fresh water sources. Those that depend on coastal aquifers, such as the Gaza residents who depend solely on a single aquifer that they share with Israel, will be particularly vulnerable.[38]

Future Water Security: Virtual Water Trade

The recent attention given to demands made on the water environment has altered the discourse. Virtual water, also called "embedded water" or "indirect water," is the water entailed in the production of a particular commodity, service, and process that people buy and use every day. Although virtual water is hidden from the end user of a product or service, that water has been consumed at every step in the value chain of a given product, service, or process, making it possible to create that commodity or service. By contrast, direct water use is the water that is seen, felt, and used in a given time and location to produce a commodity or service (e.g., "tap water"). When taken together, all the steps in which direct water is used amount to the total water necessary to get a finished product to consumers. Those finished goods, services, and processes are often referred to as virtual water.[39]

Today, many commodities and services are exchanged through global trade. Importantly, water-scarce countries have come to depend largely on the water resources of other countries to supply their consumer commodity needs. Thus, a country with limited water resources often imports water-intensive goods like cotton textiles rather than having local farmers cultivate cotton crops at excessive cost to their local water conditions.[40]

Underlining the importance of virtual water, John Anthony Allan, who coined this new concept, reveals the hidden facts of our real water consumption. All the goods we buy, from food to clothing to computers, Allan notes, entail a water cost in the form of virtual water. He raises a question: How much water do you have for breakfast? A cup of coffee or a tiny espresso takes nearly 140 liters of water. That is in fact the exact amount of water used in growing, producing, packaging, and shipping the beans one uses to make their morning coffee. Forty liters of water are expended to bring a slice of toast to the breakfast table. An apple is 70 liters. The eggs are 120 liters, and a glass of milk has 240 liters. A one portion of bacon contains 480 liters. Altogether, a full English breakfast has cost about 1,100 liters. That amounts to a little over a cubic meter of water.[41]

On the positive side, Allan notes that, except for minor skirmishes during the 1962–4 period, between Syria and Israel, no one has waged a war over water. "Local water scarcity," Allan goes on to argue, "has been ameliorated by invisible and politically silent trade in water-intensive commodities."[42] Several countries in the MENA region, according to one study, are among the top thirty importers of virtual water. These include Egypt, Saudi Arabia, Algeria, Iran, Israel, Jordan, and Tunisia. No country in the MENA region is among the top exporters of virtual water. In those countries where regional and individual incomes are meager, virtual water trading (except in the form of food assistance) is not practical as local resources are likely to become swiftly exhausted.[43] Virtual water imports in the MENA region increased by more than 150 percent between 1986 and 2010.[44] The imports came from every region of the world. The United States, Argentina, Australia, and Brazil are among the largest exporters of virtual water to the MENA region.[45]

Given that water and food dependency on imports is rising, both commodities have come to be viewed as issues of national security. With the ongoing and pervasive risks associated with food price fluctuations and transportation, virtual water trade is likely to face mounting obstacles. These risks have the potential to affect food security and conceivably political stability of some countries in the MENA region.[46] How best to tackle the threat to our planet's most precious resource remains a key question facing the global community. The responsibility rests with both individuals and governments to effectively manage water use in a more sustainable and planet-friendly fashion.

Food Security

Water scarcity and limited arable land have combined to pose one of the most fundamental challenges to the region: food insecurity.[47] At the same time, overreliance on imported food has made the region painfully susceptible to oscillations in international commodity markets, affecting both the food accessibility and its cost. In a region adversely impacted by climate change, water scarcity, poverty, and food import dependency, it is highly unlikely to easily tackle food insecurity. Increasingly, the region has been overwhelmed by the multiplicity of problems, directly or indirectly linked to the adverse effects of climate change.[48]

The issue of food security across the Middle East has intensified the debate on whether food dependency, unsustainable subsidy programs, and controversial investments in foreign agriculture (commonly known as "offshoring of food supplies") would undercut the food sovereignty of regional states. Food self-sufficiency, as one expert notes, continues to be an unreachable goal for Middle Eastern countries in large part because of their lack of water. Yet at the same time, an entirely trade-based approach to food security affords them no confidence in the coming years, as the vulnerability to food supply disruptions reminiscent of the global food crises of the recent past as well as the early twenty-first century is vividly recalled in the region.[49]

Food security and food systems in urban environments, as many observers insist, can be seemingly linked to political instability, unrest, and even revolt.[50] The urban poor are particularly exposed to rapid increases in food prices. Non-income determinants, including poor infrastructure, defective housing, failing basic services (especially water and sanitation), and exposure to environmental risks like pollution, also contribute to food insecurity and malnutrition in urban areas, both in the MENA region and globally.[51]

Between 2006 and 2011, food prices rose by an average of 10 percent annually in Egypt, Iran, and Yemen, and by almost 5 percent or more in Djibouti, Kuwait, Lebanon, Oman, Qatar, Saudi Arabia, and the UAE.[52] Many of these countries have relied on functioning world markets to secure domestic food needs. This process, characterized by the domination of national agro-policy of global marketplace, "has had a direct and negative impact on local food supply and domestic food production, and, as a result, has increased the risk of food security."[53]

Over the past half century, millions of migrant workers, mainly from Asia, have formed a major part of the labor force of the GCC countries. The vast majority of migrant workers engage in construction, retail, industry, and domestic labor. Migrant workers have staged strikes to protest unpaid wages, sexual harassment and abuse, and other forms of mistreatment, while facing the threat of detention and deportation. Some reforms notwithstanding, workers are guaranteed no means of reporting abuse, nor are they capable of changing their employers at will. Nowhere is the food security problem more acute than in the Persian Gulf region that is one of the most water-stressed parts of the world. The GCC countries' water consumption is among the highest in the world and can no longer be sustained at the current rate. While all GCC countries are water-scarce, large amounts of groundwater, ranging from 67 to 93 percent, are used for agriculture.[54]

Today, Saudi Arabia is host to one of the largest water desalination plants in the world. The plant consumes a great amount of energy and contributes to air pollution. It is worth noting that desalination poses a substantial threat to the local environment. Because of its extremely energy-intensive nature, as one study notes, it contributes greatly to air pollution. Moreover, chemical discharges into the marine environment from the desalination process, along with an ensuing steady increase in water temperatures and salinity, are among the palpable environmental threats associated with desalination in the Persian Gulf. Massive amounts of brine discharged into the Persian Gulf, in excess of 132 million cubic meters on a daily basis, are likely to undercut local biodiversity and seriously harm the ecosystem. A future rise in the highly saline Persian Gulf Sea level is likely to contaminate natural inland water resources and further salinate cultivable lands, also impacting local food security.[55]

The Arab countries of the MENA region as a whole import nearly 60 percent of their food needs. They are in fact among the largest grain importers in the world, and as a result are significantly affected by spikes in food prices.[56] During 2019–20, four of the top ten wheat-importing countries in the world were in the Middle East—Egypt, Turkey, Algeria, and Morocco (see Table 2.1).[57]

The GCC countries are highly dependent on essential food imports (wheat, rice, maize, soybeans, sugar, and barley are among the major agricultural import items) and thus seriously exposed to global food inflation and crisis. On average, according to one report, the GCC countries import 90 percent of food products from other countries. Qatar topped the GCC in terms of its dependence on imports at 97 percent, followed by Bahrain at 92 percent, Kuwait at 91 percent, the UAE and Oman at 89 percent each, and Saudi Arabia at 80 percent.[58] The GCC investors, the report finds, purchased more than 2 million hectares of land in the Sudan between 2006 and 2012. This was equivalent to three times the area of land they bought in Australia—the second largest recipient of the GCC investment in agricultural land.[59]

A quick look at employment in agriculture as a percentage of total employment (Table 2.2) illustrates that the GCC states are ranked among the lowest groups of countries when it comes to employment in agriculture.

With declining agriculture, a growing population, and scarce water resources, the vulnerability of GCC countries will surge markedly in the future. Likewise, declining oil prices will undercut policies aimed at subsidizing food and water costs in the

Table 2.1. Wheat-Importing Countries

Rank/Country	*Value of Import (US$)*	*Percentage of Total Imported Wheat*
1 Egypt	2.7 billion	5.6
3 Turkey	2.33 billion	4.8
7 Algeria	1.5 billion	4.2
10 Morocco	1.42 billion	2.9

Source: Daniel Workman, "Wheat Imports by Country," *World's Top Exports*, June 17, 2017, available at http://www.worldstopexports.com/wheat-imports-by-country/. Accessed on April 3, 2022.

Table 2.2. Employment in Agriculture as Percentage of Total Employment

Ranking	*Country*	*Percentage of Employment*
38	Sudan	38.37
52	Morocco	33.25
66	Yemen	20.62
81	Egypt	19.70
87	Iraq	18.27
88	Turkey	18.11
91	Iran	17.37
93	Libya	16.41
103	Tunisia	13.80
108	Lebanon	11.32
115	Algeria	9.60
129	Palestine	6.05
141	Oman	3.99
155	Jordan	2.47
157	Saudi Arabia	2.41
164	Kuwait	1.78
176	United Arab Emirates	1.39
170	Qatar	1.17
174	Bahrain	0.94
176	Israel	0.92

Source: World Bank, Employment in Agriculture—Country Rankings (2019), Global Economy, available at https://www.theglobaleconomy.com/rankings/employment_in_agriculture/. Accessed on April 3, 2022.

coming years. Many of these countries have implemented food subsidy programs for decades that are no longer sustainable. Under pressure from the World Bank and International Monetary Fund (IMF), many of these programs have been cut. The global economy has directly impacted local food supply and domestic food production, increasing the risk of food insecurity.[60]

Given that most of the population in these countries are expatriates, the increase in food prices or any future food crisis will most likely undermine the legitimacy of these governments, triggering social unrest and protests by migrant workers. To the extent that purchasing agricultural land and establishing intensive farming operations overseas have become part of a broader strategy that considers "food security" a top national security priority, it is important to pay attention to the way in which Arab countries of the Persian Gulf region have encouraged domestic corporations to invest in Canada, Argentina, and Brazil, as well as in economically poor countries around the world in the interest of contributing to agricultural sustainability for the MENA region.

Middle Eastern countries have taken different approaches to deal with the food security issues they face. Rising food prices in 2008 triggered riots among migrant laborers in the UAE and Bahrain, demonstrating the extent to which GCC members were vulnerable to the pricing and potential disruption in the supply of imported food stocks. Qatar has since forged closer trade and agricultural ties with India, Indonesia, Thailand, Vietnam, and the Philippines. The GCC countries have shown growing interest in trade with the highly populated but energy-poor food-exporting nations in Southeast Asia. As noted above, in Saudi Arabia, the desalination plants have proven energy intensive and very expensive. It is important to take account of the cost-effectiveness and consequences of different strategies for pursuing food security.

Many countries of the GCC, including Saudi Arabia and the UAE, have purchased land overseas. The Saudis have purchased farmlands and invested in agribusinesses mostly in Sudan, Ethiopia, Senegal, Pakistan, and Egypt, with the aim of directly supplying corn, soybeans, rice, palm oil, pineapple, and other produce for domestic use. In 2012, according to one source, the Saudis spent $11 billion on agricultural and livestock investments abroad. Likewise, Kuwait has invested in Asian countries such as Cambodia, Laos, and the Philippines to obtain its domestic rice and corn supplies.[61]

Both the Saudis and the Emiratis have bought land in Arizona and California to increase their food security plans. Saudi Arabia has, in the past, attempted to grow its own water-intensive crops for food rather than rely on farms abroad but reversed that policy to protect scarce water supplies. The Saudis turned their attention to the US Southwest to diversify supply. In 2014, Almarai Company paid $47.5 million for more than 9,800 acres in La Paz County, Arizona, a sparsely populated alfalfa-growing region that is exempt from formidable statewide restrictions on pumping water that have been imposed on Phoenix, Tucson, and other large Arizona cities under a 1980 state law designed to protect the state's aquifers.[62]

These overseas agricultural investments by some countries in the MENA region, as well as other parts of the world, including Africa and Asia, have invited much controversy. Egypt has grabbed land in Sudan and South Sudan. Saudi Arabia and the UAE have bought agricultural land in South Sudan and Sudan as well. Perhaps more contentious has been the growing phenomenon of overseas agricultural investments in Cambodia and Ethiopia by the small Persian Gulf sheikhdoms. Much of the debate over such policies centers on the extent to which these initiatives pose a threat and risk to the host countries.[63]

Land Acquisition

In the late 1970s and 1980s, the Arab countries of the Persian Gulf saw a larger flow of foreigners—worker migrants and investors—that significantly offset the lack of ability of domestic sources to meet the people's needs. Governments felt compelled to adapt to the evolving environmental and demographic pressures to curtail their vulnerability to disruptions in water and food supplies. Many of these countries turned to the option of leasing or even acquiring farmland abroad and growing food on such farms.[64] In 2008, then Pakistani prime minister Yousuf Raza Gilani offered hundreds of thousands of acres of agricultural land to the Saudis in exchange for oil.

Similarly, the Sudanese government set aside one-fifth of the country's cultivated land for agricultural use by Arab countries.[65] The Saudi private sector and the government have made extensive land investments for this purpose in many countries, including Brazil, Canada, Ukraine, and Sudan.[66] The UAE, Qatar, Kuwait, Bahrain, and Oman have also made similar land investments—albeit not to that extent—in other countries, including Algeria, Morocco, Egypt, Ghana, Indonesia, Namibia, Romania, Spain, and Tanzania.[67]

Investment strategies that have revolved around land acquisition and farming abroad alone are likely to create short-term cash gains for corrupt governments without offering much to citizens. The resulting resentment could endanger the investment climate and ultimately undermine political regimes. These global farmland transactions result in a "lose-lose" scenario, generating neither food sovereignty nor food security for the host or the investor country.[68] The risks involved in such deals far outweigh the benefits, providing little or no assurance that these food-insecure states' investments will pay dividends.

In the meantime, the oil-rich countries of the MENA region have become increasingly cognizant of the fact that the oil-based economy is no longer a sustainable strategy in the coming years. While it is still too early to predict with any certainty the future of the MENA region in the post-oil era, much remains unclear in the attempts and commitments of the region's ruling elites to diversify away from oil. Over-dependency on oil revenue, as marked by heavily subsidized economic models, appears to be a defining characteristic of the economies of these oil-rich countries. Except for the UAE and Bahrain, oil and gas production continue to account for more than 40 percent of GDP of the GCC states.[69]

Environmental Activism in the MENA Region

Some experts have argued that the Arab Spring uprisings combined concerns for the environment with more discernible concerns for socioeconomic justice. On balance, socioeconomic concerns often overshadowed the environmental concerns of communities in the Arab world. This may explain why individual membership in environmental nongovernmental organizations (ENGOs) was not a common engagement activity for the Arab youth.[70]

Others have offered a contrasting view, disputing the lack of concern for environmental issues emphatically. Jeannie L. Sowers, for example, argues that environmental activism has gained much traction across the MENA region over the past few decades, bringing to the fore key environmental issues that affect public health, livelihoods, and essential services. The increase in the spread of ENGOs, as well as the growing rise in popular resistance campaigns, has put the pressure on the government officials to respond. Although ruling elites and officials often tend to portray environmental mobilization as a threat to state and national security, they sometimes deploy their own discourses and interventions around climate change, albeit by focusing primarily on technocratic solutions rather than questions of political economy and environmental justice.[71]

Still others have described the 2011 Arab Spring uprisings as a "green spring" as it put the spotlight not only on political issues but also on environmental issues.[72] In 2013, for example, in Ain Salah, an oasis town in central Algeria, a collective movement flared up in order to protest against shale gas exploration by the French oil company Total. Algerian citizens confronted the environmental and health hazards of fracking as a form of neocolonialism. In 2015, the campaign led to the formation of a coalition representing local councils, the resignation of the energy minister Youcef Yousfi, and the suspension of the fracking operations there.[73]

A campaign opposing coal imports in Egypt in 2012 saw the formation of a loose coalition of environmental and health activists, government officials, and voluntary organizations that led to a sustained campaign to contest the government's decision to import coal to supply Egypt's cement plants. This campaign, which became known as "Egyptians against Coal," made effective use of new and old media, drawing upon the protest networks forged in the 2011 Arab Spring uprisings. The anti-coal coalition challenged the government and business community's claims that importing coal was necessary to meet Egypt's energy needs. Despite encountering many formidable hurdles, the campaign influenced media coverage and official discourse on energy issues, while also raising significant popular concerns about social justice and corruption.[74] Regardless, the Egyptian government proceeded to expand coal use to secure heavy industries' investment and related revenues.[75]

The 2011 Arab Spring uprisings clearly spawned much scholarly interest beyond the purely security considerations and understanding of the dynamics of social protest in that part of the world. Subsequently, from the protests in Istanbul's Gezi

Park, which spread to ninety other towns in Turkey in 2013, to the demonstrations against electricity shortages in the Gaza Strip in 2017, environmental activism has since become a vibrant cause of protest and activism. The increasing diffusion of activism indicates emerging opportunities for long-term structural changes in the areas of population growth, urbanization, education, and migration.[76]

The manifestation of environmental contestation has been most conspicuous in countries with semi-competitive political systems and long histories of legalized collective action, such as Lebanon, Tunisia, Egypt, Algeria, Israel, and Iran.[77] In March 2015, for example, protests against hydraulic fracturing, also known as fracking, for shale gas in southern Algeria turned violent when the police clashed with demonstrators outside a base run by the American company Halliburton near the town of Ain Salah. While ultimately quashed, the anti-fracking movement was emblematic of the extent to which local communities reacted to the negative consequences of energy extraction. Similar protests have also occurred in Israel during the years 2012–13 in Adullam.[78]

Since 2008, Egyptian environmental activists have organized frequent protests against fertilizer waste, gas drilling, coal burning, and other hazards. In August 2015, a Lebanese youth group led a protest movement, complaining about the political and institutional stagnation that allowed trash to pile up around Beirut. The so-called garbage crisis took a violent turn when protesters and the police clashed. In Iran, protests over water shortages erupted throughout 2018, as farmers near Isfahan and the countryside spoke out against malfunctioning pipelines and water pollution from oil and sugarcane refineries. These protests were preceded by a period of severe drought, in which protesters accused the Iranian government of water mismanagement.[79]

In 2012, the UN held its annual summit for its UNFCCC in Doha, Qatar. The event featured a coalition of Arab and non-Arab environmentalists who mobilized a more aggressive advocacy for policy change. In November 2016, the UNFCCC held its meeting in Marrakech, Morocco, which was widely regarded as a victory for many Moroccan environmentalists who sought to persuade greater political action and similarly used the event as a unique display for their advocacy.[80]

With respect to gender equality in the global climate government, it should be noted that climate impacts have had negative consequences for gender equality. Women and girls are disproportionately affected by the climate crisis, which accentuates existing inequalities and builds on archaic notions regarding gender roles. To be sure, the Paris Climate Agreement (2015) served as a historic and important milestone to become the first universal climate change agreement to call on Parties to promote gender equality and the empowerment of women when attempting to address climate change.

The COP22 (2016) in Marrakesh, however, took the lead in adopting a decision on an enhanced work program on gender, taking a massive step toward "integrating gender in UN climate policies by helping to ensure their implementation in an effective and cohesive manner."[81] This presented a unique opportunity to advance gender equality along the lines of participation, empowerment, and accountability as these three principles aimed at enhancing the mainstreaming of gender in both policies and practices. As a result, we have seen a significant improvement of women's participation in the UNFCCC processes.[82]

In the UAE, a young Emirati girl made history in 2023 as the youngest environmental activist to have attended COP28 held in Dubai Expo City. Thirteen-year-old Ghaya al-Ahbabi, known as "Green Girl," has emerged as a potent voice advocating for taking climate change and sustainability seriously. Emphasizing the climate crisis, al-Ahbabi has asserted:

> When people don't really care about climate change and climate crisis, and they neglect it and never focus on trying to find solutions for it, slowly and slowly we get natural disasters, which can cause our world to collapse, and when our Earth slowly starts to collapse, who will be the most effective people on this Earth? Us children. We have power but we [are not involved in] decision making . . . but we are the most affected.[83]

Al-Ahbabi has raised awareness about environmental issues while magnifying her broader message to "change the whole world for the better." She has encouraged young activists to use their voices and raise awareness about environmental issues worldwide. She has urged: "We shouldn't be scared of using our voice to speak up, to share our concerns and ideas with friends, family, children, local leaders and everyone."[84]

While environmental activism has also emerged to a lesser degree in authoritarian states such as Arab states of the Persian Gulf region, civic engagement and protests have become almost impossible in war-torn Syria, Yemen, and Libya.[85] In the future, environmental mobilization in the MENA region is likely to surge, as ecological challenges increase alongside restricted developmental plans, increasing population growth, resource scarcity, and stiffened constraints on migration.[86]

Standing in the way of environmental activism has been decades of state corruption that has led to the failure to modernize and introduce new ways of working and doing business with both private and public sectors. In addition to the lack of innovative policies to tackle poverty, the state's curbing of civil society activities and space has blocked all forms of mobilization and awareness-raising, especially those related to the environment.[87]

Conclusion

As this chapter has discussed, the urgency of dealing with climate-related vulnerabilities will most likely affect the governments' policies in the MENA region in the coming years, especially when it relates to providing water and food security. This region is already experiencing violent conflicts and will most likely face potential conflicts in the future. Drought-related water security and food scarcity require long-term plans for managing such uncertainties. As a result, water has become increasingly securitized and weaponized in the region. The water scarcity and food security have been closely linked to civil wars in Yemen and Syria—two countries whose severe water scarcity has prompted competing actors to deploy this precious resource as a tool in violent conflict to the detriment of millions of civilians.[88] Collective responses at the global level have become inevitable because water is a transnational good.[89]

The challenges for energy security in the region are no less formidable than the ones posed by water and food security. The subsidies and cheap oil in the region militate against any serious attempt to diversify energy security programs. Still, more than 97 percent of all electricity generation in the region stems from fossil fuels.[90] Investing in human development and renewable sources of energy—such as wind and solar energy—in the face of population growth and growing demands for employment and economic growth presents the most serious challenge to these governments' attempt to achieve sustainable development.

In a region that depends heavily on fossil fuels for its economic growth and energy security, such adjustments, including shifting away from oil and gas revenues, will be less attractive. The talk of the need to be prepared for the inescapable arrival of the post-oil era has failed to translate into actual policy—with the notable exception of Dubai, which has substantively invested in renewable energy policies and technologies.[91]

How to mitigate and adapt to climate change stresses is a question that warrants consideration. Some experts argue that adaptation in the MENA region is all about water and how climate change exacerbates water shortages, thus bearing serious implications for water availability. Crucially, water-dependent sectors, such as agriculture, ecosystems, cities, and human health, will be significantly impacted by such water scarcity.[92] Others argue that while food shortages have in the past been mitigated by increasing food imports, this option is less viable owing to increasing food insecurity, a less dependable food market, and perhaps more importantly, climate change-related unpredictability, which will mean that rainfall will become less reliable in some parts of the MENA region. The regional governments should apply technological innovation in the food system, use sustainable agricultural production methods, and rely on improved storage and transport infrastructure.[93]

In the face of global warming, many governments in the MENA region have come to recognize that adaptation measures have become inevitable. Lebanon is developing hill lakes to conserve and store water for irrigation. In Egypt, efforts are underway to build wave breakers to preserve wetlands and coastal installations from seawater intrusion. In Jordan, treated wastewater is now used to irrigate agricultural areas, one expert noted. According to Paola Mercogliano, the Euro-Mediterranean Center on Climate

Change Foundation's director of hydrogeological impacts, "A project in Morocco is empowering women to harvest water from fog, while in Jordan another aims to empower rural women to help tackle agriculture in the context of climate change."[94]

Several MENA governments have made pledges to mitigate global warming through a transition to renewable energy and green technologies. Some experts have underscored the importance of governance, arguing that policymakers in the MENA region need to bolster the ability of their societies to withstand and adapt to climatic stresses through greater political and economic inclusion and by prioritizing policies that protect the most vulnerable sectors of their citizenry.

Facing many formidable challenges, climate activists in the MENA region include multiple actors. Mostly, they are part of a broader environmental civil society that prioritizes the science of climate change over geopolitics while also highlighting government mismanagement. Across the region, academics, artists, and businesspersons are engaged in the enterprise of generating knowledge about the effects of environmental degradation. Modes of environmental activism, depending on whether it is led by the youth, entrepreneurs, NGOs, or the private sector, might vary from one case to another, but in general the aim remains similar: environmental and social mobilization requires broader support and proper institutions. Despite considerable progress made toward addressing climate change through the social media platform, as Aziza Moneer caution, the MENA region lacks proper environmental institutions and legal structures that are essential to reverse the unfolding environmental degradation.[95] Moneer reminds us of the hard times ahead: "Decarbonizing economies that are heavily dependent on fossil fuel revenues will not be an easy task."[96]

CHAPTER 3
A SOCIOECONOMIC FRAMEWORK

REFUGEES' RIGHT TO WORK

Refugee inclusion in the local economies of host communities matters greatly for creating substantive and sustainable change in the policy and practice of these countries and international donors. Historically, the politics and economics of refugee rights have been underpinned more by security and power considerations than by ethics, values, and norms. Political and economic pathways to refugee employment have been frequently blocked, even as concerns grow across the world that host countries have not done enough to protect refugees' access to work. Unless stronger norms emerge that guarantee the protection of socioeconomic rights of refugees, and unless such protection is broadly understood as an obligation, political interests and power calculations are likely to dominate refugee policies.[1]

It is noteworthy that refugees' right to work has been consistently recognized in regional and international agreements. The 1951 Refugee Convention, signed by 146 countries, protected refugees' right to work, freedom of movement, and access to services, among other things. Article 6-1 of the International Covenant on Economic, Social and Cultural Rights (1966) also recognizes "the right of everyone to the opportunity to gain his living by work which he freely chooses or accepts, and will take appropriate steps to safeguard this right."[2] The New York Declaration for Refugees and Migrants (2016) and the Global Compact for Refugees (2018) reiterated many of these commitments. Host governments, however, have often failed to meet these obligations both in law and in practice.[3]

Throughout the world, laws on the livelihoods of refugees have changed less and remain more restrictive on balance than laws on movement and territorial access.[4] Granting refugees' legal permission to work has been one of the most controversial policy decisions for host governments. At the same time, helping refugees to rebuild their lives through jobs and community engagement is essential to protect and promote their dignity. Refugees' access to work is a significant and crucial pathway to autonomy in exile, empowering displaced people to help themselves and contribute to their societies.[5]

The difference in work rights in law and in practice is obvious, as imposed barriers on refugees' right to work are widespread. According to one study, at least 62 percent of refugees live in countries where there is an adequate legal framework for work rights. Yet many of these laws are not fully enforced: at least 55 percent of refugees live in a country that restricts their work rights in practice.[6] Countries of the Global North often offer credible work rights—both de jure and de facto—to recognized refugees. They, however, prevent access to these work rights in practice by significantly restricting refugee status (i.e., asylum), limiting the work rights of asylum seekers, and restricting the territorial access of would-be asylum seekers in some situations.[7]

In this chapter, I argue that it is morally and practically important to allow refugees to use their skills, insights, knowledge, and experiences in their host communities. I contend that granting refugees the right to work bolsters their financial autonomy, self-esteem, and development potential to contribute to the economic health of the society in which they reside. The denial of the right to work has had disastrous and tragic consequences for many refugees, leading to a long-term attrition of their skills and aspirations. When these forcibly displaced people gain access to the labor market, they become more independent and less of a burden on the economy of their host communities.

Charity and foreign aid, however, cannot constitute a program to effectively deal with the refugee crisis in the Middle East and North Africa (MENA). Proper state policies, in coordination with other actors such as international agencies and donors, need to address the issue of refugee inclusion in their economies head-on. Although the right to work is legally preserved in international law, in most cases refugees suffer from a lack of accessibility to decent work in practice. This increases the risks for those refugees who turn to the informal economy in urban areas, earning lower incomes and facing poor living conditions. Not ironically, the result is a greater dependency on aid.[8]

Four overarching questions are raised in this chapter: (1) Why is the right to work vital for reducing dependency of refugees on aid and securing their dignity? (2) What are the constraints on refugees' right to work in the MENA region? (3) What are the perils of working in the informal sector? And (4) how can refugee law be improved to increase refugees' access to sustainable livelihoods? To address these issues, I discuss the spiraling burdens on refugee-hosting countries, followed by a discussion of climate-related displacement with a view toward understanding the new challenges it poses for hosting countries. For the refugees' rights and entitlements, I return to the 1951 Refugee Convention and its implications and drawbacks. I also explain the intersection of human rights laws and refugee laws. Finally, I contextualize the right to work in several case studies in the MENA region to illustrate opportunities and constraints to enjoying such a right.

Politics of Refugees' Right to Work

The majority of Syrian refugees displaced by the thirteen-year Syrian conflict remain in the MENA region. Some 3.7 million Syrian refugees live in Turkey, which also continues to be the world's largest refugee-hosting country. Syrian refugees also reside in Lebanon, Jordan, and Iraq. Moreover, states of the MENA region increasingly face the daunting challenge of addressing issues of climate-induced migration.[9]

Forced migration and IDPs have added a growing burden to the economies of refugee-hosting countries. A recent UNHCR report on progress in indicators under the Global Compact on Refugees (GCR) showed that many refugees still lack access to decent work. At the global level, only 38 percent of refugees live in countries with unrestricted access to formal employment, including wage-earning jobs and self-employment.[10]

Around the world, we have seen a few "success stories" of the right to work and freedom of movement, with Uganda serving as a model and example. In East Africa, writes Alexander Betts, Uganda is the only country that provides both the de jure (legal) and de facto (in practice) right to work. Kenya provides the de facto right to work in certain parts of the country, but not the de jure right to work. Ethiopia, by contrast, provides the de jure right to work but not the de facto right to work. In the meantime, Tanzania denies refugees both the de jure and the de facto right to work.[11]

Many sources in the West have described Uganda as "one of the best places to be a refugee."[12] While many refugee-hosting countries in Africa require refugees to live in camps and deny them access to labor markets, Uganda offers a self-reliance model that has three core parts: (1) its regulatory framework that allows refugees to work and choose their place of residence; (2) its assistance model allocates plots of land for refugees to cultivate within its villages and farms; (3) its model of refugees fosters integrated social service provision and market access.[13] These policies have made considerable difference to the socioeconomic conditions of refugees in four areas: "greater mobility, lower transaction cost for economic activity, higher incomes, and more sustainable sources of employment."[14]

On the whole, a restrictive approach to the right to work reigns, as most states appear reluctant to relax these constraints. "The majority of refugees," according to one study, "work in the informal sector, but under much less satisfactory and more exploitative conditions compared with nationals."[15] The study goes on to demonstrate that "informal labor markets are also constrained in countries with fragile economies which often host large numbers of refugees."[16] Yet, refugees' right to work has been regularly recognized in international agreements (the 1951 Geneva Refugee Convention and the 2018 UN GCR). The debate over the benefits of this right for refugees and their host countries alike has raged on, even as most refugees today encounter significant legal and practical barriers to full economic inclusion in the labor markets of their host countries. The countries of the MENA region are no exception, as the refugees' right to engage in legal wage-earning employment has faced similar obstacles.

Syrian refugees in Turkey, Jordan, and Lebanon, for example, continue to face such barriers, as most of them work in the informal sector. I argue that enforcing the proper policies is a necessary first step toward ensuring refugees' equitable access to fair and sustainable legal employment. Domestic laws should grant refugees the right to work and freedom of movement. The work permit should be viewed as an integral aspect of their status. Denying them work permits pushes them toward engaging in the informal sector, an eventuality fraught with exploitative workplace conditions, gender abuses, and many other labor-related violations.

Furthermore, this lack of clarity in practical terms will most likely create more bureaucratic delays and confusion among employers and refugees alike. Protecting refugees' right to work in practice and supporting legal aid for refugees who experience unemployment can be reconciled with the national interests of host countries. Arguably, external support, such as donations, can and should be linked to the refugees' right to work, mitigating opposition to labor market access for refugees. Hosting communities should stop treating refugees as a "burden" to their economy but rather view refugees as an opportunity. To argue that refugees have a right to work in a safe working environment is based not only on making a strong ethical and emotional appeal but also on a logical appeal. In fact, with a timely investment and prudent policies, offering legal employment to refugees can yield significant economic gains. Finally, this chapter proceeds from the assumption that migration is by far the most potent weapon in the struggle against global poverty.[17]

Climate-Related Displacement

Climate migration has left many people homeless due to extreme weather events, including droughts, floods, heat waves, and wildfires, as well as slower-moving climate challenges including sea level rise and growing water stress. Some experts have warned that "as climate change makes some parts of the earth uninhabitable, a climate migration crisis looms that international law is not prepared to address."[18] This form of migration is on the rise due to the fact that the world has been unable to reduce GHG emissions and cease global temperature rise, which leads to more climate disasters. Most climate migration is likely to occur within a country's borders, but cross-border migration will also rise.[19]

Climate change and its consequences, including extreme weather events, will likely force people out of their homes and communities, creating climate migrants and refugees who lack legal status and yet deserve to be protected from a human rights standpoint. Migration is a logical and legitimate means by which people adapt to climate-related, existential challenges.[20] There is growing consensus among experts that those displaced by climate change are eligible for refugee status. Given the increasing link between climate change, migration, and conflict, the notion of security needs to be redefined and reconfigured. Global governance and cooperation have become imperative to tackle climate-induced migration.

Additionally, environmental refugees or climate refugees encounter far greater political risks than traditional refugees who have fled their homes because of political persecution. Compared to traditional refugees, climate refugees are more likely to be sent back to their ravaged homeland or kept in a refugee camp. The most formidable challenge facing these refugees and forced migrants across the region is that they lack the basic right to work. Numerous legal, organizational, and practical barriers prevent their egalitarian access to job opportunities. Removing these barriers to equitable economic inclusion would afford displaced people greater agency and power, enabling them to more fully rebuild their lives and contribute to their host communities. Many observers have pointed out that it is vitally important to underline the need to implement globally agreed norms in this context—both in law and in practice.[21]

These climate or environmental refugees face greater political risks than traditional refugees who flee their homes due to conflict or political persecution. Unlike traditional refugees, climate refugees may be sent back to their ransacked homeland or forced into a refugee camp. As noted earlier, most climate refugees are internal migrants. Often, climate refugees are rural and coastal residents who have been forced to migrate to urban areas. These climate refugees must cope with multiple challenges. Their skills, primarily in herding and farming, are less in demand in urban areas. Rural farmers are often more self-sufficient than many urban dwellers; they are unfamiliar with depending on a corporation or other business for employment.[22]

Climate change impacts in the context of refugee movements and forced migration have raised a key question: Whose responsibility—states, humanitarians, or international organizations—is it to tackle climate-based displacement? Some studies have placed the onus on states, arguing that among the key remedies available to policymakers are improvements in water-use efficiency and productivity in agriculture, better management and monitoring of groundwater resources, and multilateral agreements on managing and sharing the rivers that cross political borders.

UN and Red Cross agencies face an equally huge humanitarian crisis under present circumstances. The United Nations High Commissioner for Refugees (UNHCR), acting on the basis of current refugee law, appears helpless in the wake of these new crises. While the UNHCR promises to do its part to help those affected by natural disasters, such as droughts and floods, it is not within its legal mandate to do so. Thus, the responses of relief agencies, NGOs, and donors, as well as other forms of foreign aid, could prove vital.

Large-scale migration from the MENA region to Europe and the rest of the world following the 2011 Arab Spring uprisings led to major disruptions in those areas. While the primary drivers of this forced migration in the MENA region were economic upheaval, violence, and civil wars—especially in the cases of Syria, Yemen, Libya, Afghanistan, and the Sahel countries—it is estimated that over the coming decades, climate change will increasingly contribute to transborder refugee flows.[23]

In the MENA region, during 2021 alone, approximately 3 million people were forced to leave their homes due to natural disasters and climate-induced events. These displaced people represent the emerging reality of climate refugees.[24]

In this chapter, I focus on refugees' right to work, exacerbated by climate-induced displacement, with politically destabilizing and disruptive consequences. The challenge for states in the MENA region is to design policies aimed at offering the refugees lawful employment while also improving their adaptive capacity to deal with climate change. If managed properly, these policies may very well mitigate the likelihood of civil unrest and forced migration in the coming years.

International Instruments and Access to Work

It is important to note that the right to just and favorable remuneration, supplemented by other means of social protection, including forming and joining trade unions, is internationally recognized by the Universal Declaration of Human Rights (UDHR). According to Article 23 of the UDHR: "Everyone has the right to work, to free choice of employment, to just and favorable conditions of work and to protection against unemployment."[25]

Similarly, Articles 17 and 18 of the 1951 Refugee Convention provide and protect refugees' right to work and access to labor markets. Article 17 reads: "The Contracting States shall accord to refugees lawfully staying in their territory the most favorable treatment accorded to nationals of a foreign country in the same circumstances, as regards the right to engage in wage-earning employment."[26] Article 18 stipulates a similar obligation on host countries: "The Contracting States shall accord to a refugee lawfully in their territory treatment as favorable as possible and, in any event, not less favorable than that accorded to aliens generally in the same circumstances, as regards the right to engage on his own account in agriculture, industry, handicrafts and commerce and to establish commercial and industrial companies."[27]

In the late 1940s, the response of the UNHCR to the refugee situation was to keep them in camps by providing food and shelter based on their needs. Today, however, the vast majority of refugees live outside these camps. In Jordan, approximately 85 percent of the refugees ignore the camps. Similarly, the 260,000 refugees living in Turkish refugee camps, according to one study, constitute approximately 9 percent of the overall Syrian refugee population. The remaining 91 percent are spread across Turkey, migrating from one place to another in search of economic opportunities. This reality renders the return of Syrians to their home country even less likely.[28]

What refugees have long sought is the autonomy that comes from holding a job. Almost half of the Refugee Convention focuses on socioeconomic rights such as the right to work and freedom of movement. One of the most important components of autonomy is the right to earn a living. Some experts have argued that "Refugees—as refugees—need and should be entitled to expect three things: rescue, autonomy, and an eventual route out of limbo."[29] The refugee system has become a *humanitarian* system, and it lacks expertise in development or the tools to guide market-oriented solutions that can promote autonomy.[30]

Economic Inclusion: Agency and Power

A general consensus exists on the notion that refugees and forced migrants are "entitled to enjoy the full range of civil, political, economic, social and cultural rights set out in international and regional human rights treaties and customary international law."[31] Consider, for example, Articles 6 and 7 of the International Covenant on Economic, Social, and Cultural Rights (ICESCR) that calls for "the right to work, as well as the right to just and favorable conditions of work."[32]

In reality, however, refugees and forced migrants live in tough situations and prolonged periods of encampment with no more permanent solution in sight.[33] It is worth noting that human rights treaties are *living* instruments that evolve over time. This suggests that in the future, new groups of forced migrants may be able to safeguard further protection as the contemporary conception of human rights is likely to expand.[34] It is possible, and indeed probable, that significant constitutional and legislative developments may transpire in the future, furthering the human rights of refugees and forced migrants.

One study has found significant diversity in legal provisions and constraints on refugees' right to work. The majority of refugees work in the informal sector, albeit under much less decent and more exploitative conditions compared with nationals. Informal labor markets are also restricted in countries with fragile economies that often host large numbers of refugees. This study underscores the importance of further national and international coordination, sharing in the responsibility to deliver access to adequate work, to formulate labor market policies, and to provide training and education aimed at supporting sustainable livelihoods and effectively engaging refugees' social capital.[35]

Throughout the world, refugees and other forced migrants lack the basic rights to work, move, and thrive. A range of legal, administrative, and practical barriers prevent their equitable economic inclusion. Removing these barriers would give displaced people greater agency and power, enabling them to more fully rebuild their lives and contribute to their host communities. Historically, on average, international wars have lasted only six months. In contrast, the average civil war has been much longer, with estimates ranging from seven to fifteen years. The priority of refugees is no longer simply emergency food and shelter; it is earning a living. The camps run by UNHCR meet the basic needs of refugees, but they provide few opportunities to earn a living. Consequently, they leave families bereft of autonomy. Paying for 4 million refugees to live without work for ten years is manifestly unsustainable.[36] In the case studies that follow, I examine those Middle Eastern countries that have received the largest numbers of refugees.

Laws on Livelihoods in the Middle East

Throughout the MENA region, the rights of refugees, migrants, and IDPs to gain access to legal employment are compromised at best and blocked at worst. In Libya, for example, state officials, militias, and armed groups have regularly subjected refugees and migrants to pervasive violations, including unlawful killings, indefinite arbitrary detention, forced labor, torture, rape, and other ill-treatment. While thousands of people intercepted at sea and arriving in Libya were forcibly vanished, thousands of others were evicted from the country's southern borders without a chance to seek asylum.[37]

At the border between northern Morocco and the Spanish strip of Melilla, security forces on both sides used brutal means, killing thirty-seven people from sub-Saharan Africa and injuring many more. In Algeria, authorities arrested or expeditiously deported dozens of refugees and asylum seekers. In Iran, security forces opened fire at Afghan nationals crossing the Iran-Afghanistan border, and arbitrarily arrested and tortured others before ousting them illegally. Saudi Arabia violently returned many thousands of Ethiopian migrants after arbitrarily detaining them in inhumane conditions because they lacked valid residency documents and subjected them to torture and other ill-treatment.[38]

In Iraq, Foreign Residency Law No. 2 of 2017 and Foreign Residency Regulations No. 18 of 1987 clearly explain the law and the regulations that stipulate the conditions under which foreign nationals, including refugees, can work in Iraq. All foreign workers, including refugees, require permits to work legally in Iraq. Historically, compliance with the work permit and residency rules has been very low; however, enforcement has increased in recent years. It is, however, difficult for refugees in Iraq to obtain work permits. Foreign workers must obtain work permits. The same rules apply to refugees and other migrants. Foreign workers must be sponsored by employers. A visiting visa or a religious pilgrimage visa, which is frequently used by Syrian refugees who overstay, does not meet the requirement for a work permit application. There are no reciprocal treaties between Iraq and Syria relevant to Syrians working in Iraq.[39]

In the Kurdish Region of Iraq (KRI), the legal guide covers the laws, regulations, and procedures governing employment rights in the KRI for Iraqi and foreign nationals, including refugees. Regulation No. 7 of 2017 on Visas and Residency for Foreigners was issued by the Kurdistan Regional Government Ministry of Interior in accordance with Iraqi Law No. 118 of 1978 on Residency of Foreigners in Iraq. This regulation specifies the conditions under which foreign nationals, including refugees, can reside and work in the KRI. Anyone who wants to work in the KRI must have a valid residency card. There are no special rules for refugees, foreign workers, seasonal workers, skilled or unskilled workers in the KRI. Syrian refugees in KRI with valid residency permits are currently permitted to work in private employment and as self-employed persons on a de facto basis, although this right is not set out in any law or formal legal framework.[40]

The Arab states of the Persian Gulf, including Bahrain, Kuwait, Oman, Qatar, Saudi Arabia, and the UAE, by contrast, have been reluctant to label Syrian refugees as "refugees." These countries have donated tens of millions of dollars to help Syrian refugees in places like Jordan and Lebanon. Saudi Arabia claims it has admitted half a million Syrians since 2011. Syrians are welcome to come, Saudi officials maintain, even if they are not legally registered as refugees.[41] The rationale behind Gulf refugee policies is that in the smaller Gulf states like Qatar and the UAE, migrant workers already far outnumber nationals, a demographic reality that engenders feelings of anxiety mixed with xenophobia.[42]

Working Outside of the Camps: Jordan

Jordan is not a party to the 1951 Refugee Convention, nor is it a party to the 1967 Protocol Relating to the Status of Refugees, but it is a party to the 1966 ICESCR. In Jordan, according to one study, the vast majority of refugees live dispersed in host communities in cities and towns, while approximately 18 percent of the registered refugee population reside in camps.[43] The 1952 Constitution limited the right to work to Jordan's citizens. The 1973 Law instructed Jordanian nationals and companies not to employ foreigners without valid residence permits, even as in principle it allowed some exemptions for persons connected with humanitarian concerns or those seeking political asylum. The 1998 Memo of Understanding with the UNHCR also underlined that a legal resident refugee may "work for his own account whenever the laws and regulations permit," while "refugees holding credible degrees could practice liberal professions if the laws and regulations permit."[44]

In the ensuing Labor Code of 1996, "refugees" or "asylum seekers" were not mentioned anywhere. The law made clear that all bearers of temporary passports, including Palestinians displaced from Gaza since 1967, had to obtain permits to work legally. Jordan's Labor Code required employers to apply for a worker's permit, which was valid for one year and was renewable.[45] For Syrians, however, work permits have been free since 2016.[46]

According to the International Rescue Committee, under the Jordanian Constitution the full right to work was reserved only for Jordanian citizens. The Ministry of Labor imposed severe restrictions on access to work for all refugees and foreigners and can severely limit which professions are open to foreign workers. While national labor laws have been designed to safeguard all workers from exploitative practices, no laws existed in the past to categorically guarantee the protection of the rights of refugees, migrant workers, or other vulnerable groups. Up until recently, Syrian refugees were seldom granted the work permits required to gain access to formal and legal occupations; instead, many refugees found employment in informal sectors unchecked by labor laws and without sustainable protection against unfair treatment or unsafe work practices.[47]

In recent years, however, Jordan has issued record numbers of work permits to Syrian refugees. That country leads global efforts to give both refugees and host communities access to decent work. Jordan hosts 760,000 refugees and asylum seekers registered with UNHCR.[48] Of those, some 670,000 are from Syria, making Jordan the second largest host of Syrian refugees per capita globally behind Lebanon. The 62,000 work permits issued to Syrian refugees include 31,000 flexible ones that allow refugees to move between similar jobs in the same sector, as well as among employers and governorates. UNHCR Jordan works closely with the General Federation of Jordanian Trade Unions to keep Syrians informed about the available benefits.[49]

UNHCR notes that the current initiative puts Jordan at the vanguard of global efforts to give both refugees and host communities access to decent work, as promoted by the GCR. It is worth noting that previously, Syrian refugees in Jordan were largely permitted to work only in agriculture, construction, and manufacturing. In 2020, some were given exemptions to work in other sectors, including as healthcare professionals as part of the campaign against the Covid-19 pandemic. However, since July 2021, Syrian refugees have gained work permits in all sectors open to non-Jordanians. This means they can now work in services and sales; crafts; as skilled agricultural, forestry, and fishery workers; plant and machine workers; and in basic industries.[50] Given that over than 80 percent of refugees went to live in cities and given that some Syrian refugees were not officially permitted to work, they became part of the informal economy. The same pattern emerged in Turkey, as some refugees went to the camp, but most entered the informal urban economy.[51]

Refugee Dilemma: Turkey

Turkey is a party to the 1951 Refugee Convention and the 1967 Protocol but applies geographical limitation by granting "refugee status" only to individuals who fall within the definition of the 1951 Convention, meaning that they flee persecution occurring in Europe.[52] It is also a party to the 1966 ICESCR. Broadly speaking, refugees fall under three status types. First, Turkey provides "refugee status" only to those who are subjected to death or torture in their country of origin if returned and come from a European country of origin."[53]

Second, persons who fall within the refugee definition of the 1951 Convention but come from a non-European country of origin are classified as "conditional refugees." These refugees are allowed to temporarily reside in Turkey until they are resettled to third countries in coordination with the UNHCR.[54] Third are the vast majority of refugees—or more accurately called "foreigners"—who are at risk of violence, persecution, or torture due to conditions of war or internal armed conflict. These refugees, who come largely from Syria, qualify for "temporary protection."[55]

While the vast majority of refugees in Turkey are from Syria, the country also hosts significant numbers of refugees from Iraq, Iran, and Afghanistan. Turkey does not classify Syrians as "conditional refugees," but instead offers them temporary protection.[56] It is worth noting that all three durable solutions—that is, local integration, resettlement, and voluntary repatriation—are available to temporary protection recipients. For "conditional refugees," however, resettlement and voluntary repatriation are the only available durable solutions.[57]

Turkey shares the longest land border with Syria, which is around 800 kilometers (497 miles), so one of the main reasons for the large Syrian refugee population is the simple fact of geographical proximity. Thus, Syrians come to Turkey because it is easier to reach. Additionally, cultural, familial, religious, and historical ties play an important role. As of 2017, fewer than 8 percent of refugees lived in refugee camps. More than 90 percent lived in urban areas.[58]

Refugees can apply for a work permit six months after applying for international protection status. Those refugees who are employed in seasonal agriculture or animal husbandry are exempt from work permit requirements, but the Ministry of Labor and Social Security can place geographical restrictions and quotas on such permits.[59] Refugees and temporary protection recipients who have been residing in Turkey for three years, or are married to Turkish citizens, or have children with Turkish citizenship are exempted from such restrictions. Once employed, refugees cannot be paid less than the minimum wage.[60]

Inexpensive labor provided by refugees has induced some employers to replace natives with Syrian refugees. It appears that the rise in illegal labor in Turkey is linked to the length of the conflict in Syria, the longer wait to be admitted to the refugee camps, and the increasing desperation of Syrians, who are unable to meet their families' basic needs and may not be receiving any assistance from NGOs or the Turkish government. This has brought on humanitarian issues in relation to human trafficking and exploitation of workers who tend to be willing to work below the minimum wage.[61] This has also significantly enhanced competition within the unskilled labor pool, minimizing the chances of Turkish workers who were working for minimum wage to hold on to their jobs, replacing them with Syrian refugees who were willing to work for less than minimum wage.[62]

This situation has resulted in growing social tension. For refugees, the right to work and access to the labor market are essential for allowing them to secure sustainable livelihoods and maintain some degree of self-sufficiency. It also reduces vulnerability, enhances resiliency, and enables a dignified life. As noted above, restrictions on the right to work force the majority of refugees to work in the informal economy. The informal sector entails precarious income, the absence of contractual rights, and greater risks of exploitative working conditions.

Since 2016, refugees in Turkey can obtain a work permit through their employer. Syrian nationals under temporary protection in Turkey have the right to work in that country. There are various types of work permits that a refugee can apply for, but the most common ones are for wage employment and self-employment. Yet, Turkey faces many problems. Government agencies provide no regular periodic reports or accurate information about the reality of Syrian workers in that country. According to the Confederation of Revolutionary Trade Unions of Turkey (known as DISK), Turkish law allows the employment of one Syrian worker for every ten Turkish workers.

Despite having ratified the 1951 Convention and the 1967 Protocol relating to the status of refugees, Turkey does not grant Syrians the "right of asylum," preventing them from legally entering the labor market. Instead, it has limited the geographical locations from which refugees may enter European countries only. Turkey ratified the 1967 Convention but placed a reservation on it: that it would remove only the temporal restriction and not the geographical one, so the only recognized convention refugees are from Europe. There are a handful of countries with such reservations to the 1967 Protocol. It is, therefore, legal that Turkey only provides temporary protection to "guests." Syrians do not get asylum, but Turkey did offer work permits to a fraction of the Syrian refugees, thus it has the right to remove those temporary protections any time it wishes to do so.

As a result, Turkish authorities were content to grant Syrian refugees a temporary residence card known as "Kimlik" in Turkish, without providing them with travel documents that would allow them to enter and exit Turkey or enter the labor market legally.[63] As a consequence, refugees become trapped in a cycle of informal work where the risk of exploitation and abuse is high and wages are low. The lack of decent wages for adult refugees pushes many refugee children into the job market as well, instead of attending school. According to the report, 41 percent of all Syrian children in Turkey do not go to school. Because employers must pay a work permit fee of 537 Turkish Liras (118€) in order to employ a refugee, they are often reluctant to do so.[64]

Social acceptance for Syrians may not be sustainable in its current form. The huge cultural gap between Turkish society and the Syrians also presents problems. Turkish citizens are not sympathetic at all toward conferring citizenship to Syrian nationals. This makes local integration complicated, if not virtually impossible. Furthermore, the Syrian refugee crisis has also had significant repercussions for Turkey's relationship with the EU. Turkey-EU relations have been diverted from fundamental issues such as democracy, human rights, freedom of expression, and have largely focused on keeping refugees in Turkey. Security has become a primary prism through which Turkey is seen, as a buffer zone for Europe. Turkey's relationship with the EU has come down largely to the refugee crisis.

The Residency Permit: Lebanon

Lebanon is not a party to the 1951 Convention relating to the Status of Refugees and the 1967 Protocol, but it is a party to the 1966 ICESCR. The majority of refugees there are Syrian refugees, along with a large presence of Palestinian and Iraqi refugees. Palestinian refugees are settled in and around camps. Due to an official "non-camp policy," most Syrian and Iraqi refugees live scattered in host communities in cities and towns, some in rented housing, others in unofficial tent settlements.[65]

Palestinian refugees have limited socioeconomic and civil-political rights, restricted access to the government's public health and educational facilities, and no access to public social services. As refugees without a formal state, Palestinians in Lebanon are subjected to the constraints of the reciprocity principle, which is a requisite for the vast majority of occupations and professions that foreigners are allowed to practice in Lebanon, particularly where the law requires proof of a bilateral agreement between Lebanon and the foreigner's state of origin. As such, Palestinians, like all non-Lebanese, should be allowed a work permit, but they are entirely dependent upon a system of national preference and reciprocity agreements.[66]

In 2010, however, Palestinian refugees in Lebanon were granted the right to benefit from the retirement provisions under the National Social Security Fund, yet they still lack the right to illness, maternity, and family indemnities, despite having to pay the same contributions as Lebanese nationals. All refugees and asylum seekers operate under certain restrictions. In principle, non-Lebanese are regularly excluded from entering professions practiced through association, such as medicine, law, and accounting.[67]

The Lebanese government has also applied stringent conditions to Syrian refugees' access to work, linking it to their residency permits. They have two options for residency permits. First, Syrians can renew their residency permits contingent on the UNHCR registration certificate, in which case they must sign a notarized "pledge not to work," making it impossible for them to work. Second, the residency permit can be based on a Lebanese sponsor—that is, a Lebanese employer declaring a "pledge of responsibility"—allowing them to work as migrant workers and not as displaced persons.[68]

It should be noted, however, that the 2003 Memorandum of Understanding with UNHCR identified Lebanon as a country of transit. As such, refugees recognized by UNHCR are required to be resettled in a third country within a six-month period—a timeline that can be extended once. The Lebanese government, however, has framed the current Syrian crisis outside the scope of the Memorandum of Understanding and granted minimal legal authority to the UNHCR recognition of refugee status. As noted above, the Lebanese government has labeled Syrian refugees as "displaced" rather than as "refugees."[69] Since 2015, however, many restrictions, including a $200 residency fee and a binding agreement to refrain from working, have been dropped for Syrians with UNHCR cards.[70]

These changes, however, according to Human Rights Watch, are often erratically implemented.[71] Furthermore, there are many other constraints on refugees' legal status, including workplace violations, such as physical and sexual harassment, poor working conditions, child labor, and wage inequality. Additionally, when applying for permits, refugees are likely to face excessive fees and resistance by the Lebanese government out of fears of competition with nationals. Consequently, 92 percent of working refugees are compelled to find work in the informal sector where they lack social protections and typically earn less than minimum wage.[72] In general, the Lebanese government makes no effort to assist refugees in job placement, nor does it furnish vocational training services such as language and skills training. The responsibility of job placement often falls on the shoulders of NGOs that open the door for refugees to gain access to the labor market.[73]

Temporary Work Permit: Iran

Iran is a party to the 1951 Convention relating to the Status of Refugees (but with reservations) and the 1967 Protocol. It is also a party to the ICESCR. The majority of the refugee population arrived from Afghanistan, especially after the 1979 Soviet invasion of that country and before 2001 when they were given refugee status on a prima facie basis. A smaller number of Iraqis were also registered in 2002 as self-evident refugees. Only 3 percent of Afghan refugees live in camps located in rural areas.[74]

According to the latest figures released by the Iranian government, 762,000 refugees live in Iran, of which 750,000 are Afghans and 12,000 are Iraqis. It is also estimated that some 586,000 Afghan passport holders with Iranian visas also live in Iran; some of whom are former refugee Amayesh cardholders. In 2022, the Iranian government undertook a headcount of all undocumented Afghan nationals in Iran, including those who arrived in Iran following the abrupt US withdrawal from Afghanistan in August 2021, through which some 2.6 million Afghans were registered and received an Amayesh card. According to Iran, 500,000 undocumented Afghans did not participate in the headcount. According to the same report, a total of 1 million Afghans have fled to Iran, resulting in a dramatic surge in global protection needs.[75]

Registered refugees are entitled to apply for temporary work permits. Under Iran's Labor Code, work permits issued by the Ministry of Cooperatives, Labor and Social Welfare are valid for twelve months. The permits can be extended and renewed, depending on a written agreement of the Ministries of Interior and Foreign Affairs. The 1963 regulations recognized refugees' right to employment only in the fields authorized for foreign nationals and those fields considered relevant. Although a party to the 1951 Refugee Convention, Iran has placed reservations on Article 17 (right to engage in wage-earning employment) and on Article 24 (right to labor protections and social security on the same footing as nationals).[76]

In 2000, Iran passed a law under Article 48 of the government's five-year development plan that established the parameters for the repatriation of Afghans working without work permits unless they could demonstrate that the principle of *non-refoulement* (prohibition against expelling a legitimate refugee to his or her country of origin) would apply to them. To obtain a specific permit to engage in business, refugees must abandon their status and return to their country of origin to obtain a passport (if they did not have one already) and an Iranian visa. Afghan Amayesh holders can, in principle, open bank accounts and rent properties, including agricultural lands. Refugees can own movable property, but not immovable property (such as real estate) unless they obtain specific approvals.[77]

The Need for a Paradigm Shift

Recent years have seen massive demands for the potential employment of refugees, displaced persons, and forced migrants in host countries. There can be no doubt that for refugees and forced migrants, the right to work and access to labor markets, as well as the freedom of movement, are key to becoming self-reliant, building dignified lives, and allowing them to contribute to their host communities. Yet the orthodox view still persists: host countries have severely constrained refugees' access to work on the assumption that they would pose an existential threat to their already struggling economy.

As a result, lawmakers and policymakers in host countries have given little serious consideration to refugees' contributions to the economy, which may explain the presence of existing gaps in current domestic frameworks and legal as well as practical barriers to refugee work rights. Examples include, among others, inadequate local legal frameworks, many cases of ongoing workplace violations, restricted freedom of movement, bureaucratic barriers, excessive fees for residency permits and work permits, sponsorship requirements, and large asylum backlogs.[78] The prevalent de facto barriers to work rights in practice include, among others, discrimination and xenophobia, low pay, poor work conditions and exploitation, lack of the knowledge of work rights laws in the host country, and scant access to language and skills training.[79]

The question is: Will host countries be willing to open their labor markets to refugees while accelerating access to decent work for these individuals? Some experts have underscored the importance of a dramatic shift in thinking about the issue of forced migration. "The world needs," they argue, "a flexible, inclusive, and rights-centered paradigm to protect people who have been forcibly displaced—and the United States can help build it."[80] The world must reinvent the notion of forced migration as a human rights and humanitarian challenge, the solution to which is predicated on protecting human dignity and ensuring comprehensive rights.[81]

Others have directed our attention to the need for a profound shift from "humanitarian aid" to "livelihoods support." The refugee supply shock has a negative impact on the wages and employment of natives in the informal sector. Working in the informal sector results in low wages, leaving many Syrian families dependent on their children's income. Nearly 10 percent of all Syrian refugee children in Jordan, Lebanon, Iraq, Egypt, and Turkey are working.

Still others have noted that the right to work is a necessary but insufficient condition for formal employment of refugees. According to one study, the vast majority of refugees—irrespective of status and right to work—are employed in the informal sector. This is not only due to difficulties in accessing formal employment but also indicative of the fact that the majority of refugees are hosted by low- and middle-income countries, where the informal sector plays a significant part in the economy. Too often, the conditions under which they work are, however, unsuitable and more exploitative compared with nationals. It is thus important to not only look at the right to work but also the rights at work.[82]

On a different but equally important issue, refugee laws need to be refashioned to better cope with the refugee crisis. Current laws and processes are heavily weighted toward countering and denying the rights of refugees to work in some countries of the MENA region. More specifically, new policies and legislative interventions aimed at safeguarding refugees' access to work must be implemented. Preventing asylum seekers and refugees from earning a living will render them dependent on state welfare and benefits, leading to further huge public expenditures.

Conclusion

Do governments in host countries have the right to keep certain kinds of restrictions on refugees' right to work? There are no easy answers to this question, nor any hard and fast rules in this regard. There are many economic and security considerations involved in such decisions. At the same time, the issue of refugees' right to work is hard to ignore. The literature and case studies reviewed in this chapter tell us that refugees' right to work has enjoyed a surge of attention in academia and policymaking circles, in part because access to work is now widely regarded as integral to refugees' ability to rebuild their lives, achieve self-sufficiency, and engage with their host community.

Needless to say, all work has tremendous value, and all workers—refugees or not—have a justifiable claim to be treated with utmost respect and sensitivity on the job. More importantly, this right is a fundamental part of safeguarding human dignity, while also leading to the fulfillment of other sociocultural and civil rights.[83] Similarly, the right to work, together with the freedom of movement, as recent research has shown, tends to effectively contribute to the socioeconomic development of their host countries.[84]

Refugees' right to work has been clearly articulated in international legal instruments, and these rights have also been upheld in national laws and constitutions. This chapter has outlined the way in which countries of the MENA region recognize such a right. As a solution, some countries in the region have ensured that work permits—either temporary or long term—are lawfully issued to refugees (Jordan, Turkey, Lebanon, and Iran). Others, such as the oil-rich Arab countries of the Persian Gulf, have followed the long-standing policy of accepting "migrants" (as part of temporary labor admissions) not "refugees" (Bahrain, Kuwait, Oman, Qatar, Saudi Arabia, and the UAE).

The reality, however, is complex, less attractive, and probably not a noticeable occurrence in the face of the growing threat of illiberal nationalism and right-wing populism. Today, for example, refugees face a myriad of legal and political barriers to work, especially in light of the fact that populist leaders of a more authoritarian streak view the influx of refugees as socioeconomic, cultural, and security threats to their countries. The resurgent xenophobia and state fragility have undermined any sympathy toward accommodating refugees. Our attention is frequently drawn to the fallout against the influx of refugees and forcibly displaced people. "With rising structural unemployment exacerbated by recession, automation, and offshoring," Alexander Betts writes, "there may well be a significant global backlash against migration and mobility in general."[85]

Paradoxically, and in a decided contrast to this prevailing view, one can argue that the solutions may not be solely sought in political realms; rather, wise and prudent solutions for such complex and difficult problems must be sought in other normative and socioeconomic arenas. Making a compelling case for refugees' work rights, Betts notes that there can be no sustainable program to ensure self-reliance of refugees without a dramatic advance in the refugee system's ability to engage with the economics and politics of adjusting to new challenges.[86]

The future of a great number of refugees in the MENA region looks grim unless the politicians find a way to view refugees as an essential part of their developmental project—not just as a liability to their economy. Whether these countries will grant refugees lawful employment remains to be seen. One thing, however, is certain: refugees must be accommodated in labor markets if these countries' economic and political conditions are to survive the onslaught of the refugee crisis. Not surprisingly, dependency on charity and foreign assistance alone has provided no long-term solutions to the refugee crisis in the MENA region.

CHAPTER 4
A LEGAL FRAMEWORK
MAKING A CASE FOR CLIMATE REFUGEES

Extreme weather events, such as storms, floods, cyclones, and hurricanes, as well as slower onset processes, like droughts, heat waves, and sea level rise, have displaced and continue to displace millions around the world. A global rise in climate refugees has become a universal threat to humanity in the first quarter of the twenty-first century. Climatic disruption and breakdown continue to make millions of people homeless, and many thousands of others flee their countries of origin. As temperatures continue to soar, unexpected, uncommon, severe, or untimely weather will worsen, sea levels will rise, and the world's most defenseless segments of the population will face the dire consequences of years of climate inaction.

According to one study, between 2008 and 2016, 21.5 million people were displaced by climate-related disasters—far more than those fleeing war or persecution—as extreme weather events continued to cause death and destruction across the globe.[1] The climate change-related forced migrations and displacements, also known as IDPs, currently have no legal status or normative protection under existing laws and frameworks, such as the 1951 Geneva Convention on Refugees, to protect those fleeing environmental degradation—not to mention conflict and persecution.

UN secretary-general, António Guterres, aptly describes the current situation: "Climate change [is] now found to be the key factor accelerating all other drivers of forced displacement. These people are not truly migrants, in the sense that they did not move voluntarily. As forcibly displaced not covered by the refugee protection regime, they find themselves in a legal void."[2] The former UN secretary-general Ban Ki-moon even went further, noting that "Climate change not only exacerbates threats to international peace and security. It is a threat to international peace and security."[3]

The potential impacts of climate change, including the possibility of future violent conflicts, call for our immediate attention to environmental concerns and challenges that are characteristic of the Anthropocene epoch—a time when human actions have had an indelible and long-term impact on the environment. Climate change and the environmental risks it poses present a moral challenge by requiring states to balance national interests with shared global concerns, as well as existing economic needs with those of future generations.[4]

There were, according to the Internal Displacement Monitoring Center, 55 million IDPs across the world at the end of 2020, 48 million because of conflict and violence, and 7 million as a result of disasters. Additionally, the Covid-19 pandemic intensified IDPs' needs and generated new risks in 2020, while posing significant operational and financial challenges to governments and their humanitarian partners.[5] In the MENA, there were as many as 30.1 million IDPs as a result of conflict and disasters, of which 28.6 million were new displacements as a result of conflict and nearly 1.4 million new disaster displacements. The overall numbers of the latter category have been on the rise due to an increasing number of floods and an overall improvement in data collection and the monitoring of disaster displacement.[6]

Most disaster displacements during the last decade have been triggered by weather-related events, with floods causing more than half, and storms and cyclones accounting for more than 175,000. Floods "contribute to making displacement chronic, cyclical, and protracted in the region."[7] Sporadic earthquakes have also affected some countries in the region, prompting nearly 343,000 displacements. Desertification, salinization, sea level rise, and soil and coastal erosion are other examples of slow-onset hazards that will continue to strike the region, with climate change making their impact far worse. Data and knowledge gaps render it difficult to accurately assess the number of IDPs caused by these hazards.[8] Moreover, there are important differences between countries when it comes to desertification and water scarcity. Some, like oil-rich Saudi Arabia or the UAE, possess sufficient funds to access water sources or generate fresh ones through desalinization. Others, such as Yemen or Jordan, rely on groundwater but are financially strapped to effectively mobilize the necessary investments to obtain new water resources.[9]

This chapter seeks to address the magnitude and scope of the challenge of climate-induced migration and refugees by arguing that, even though there are currently little or no legal protections for climate refugees, a case can be made that there is a link between international refugee law and international environmental law. Several complexities have rendered it difficult to devise a coherent response and strategy to deal with this issue. These complexities stem from a wide variety of issues, including conceptual, legal, operational, and political problems in the context of disasters and climate change. The existing international legal frameworks neither expedite nor aid cross-border movement in the case of climate change-related impacts such as natural disasters.[10]

There are significant human rights gaps for displaced persons as international law fails to sufficiently address when disaster-displaced persons should be admitted into another country, their legal status during their stay, and the conditions under which such persons can be returned or find another durable solution.[11] This leads us to three persistent and implacable questions: (1) Who are climate refugees and what are their problems? (2) How best to protect the rights of "climate-displaced persons" given that such environmental IDPs have few legal protections? And (3) what ethical complexities and legal gaps stand in the way of protecting climate refugees?

In this chapter, I begin by examining the conceptual contestations surrounding the term "climate refugees," arguing whether there is a need to expand the conventional refugee definition. I then identify the international climate change regime while addressing the question of why many states in the Global North appear reluctant to adopt a treaty on climate refugees any time soon, and why customary international law provides no legal protection to the so-called climate refugees. After exploring the ethical complexities and legal protection gaps for climate refugees and migrants, I turn to the notion of "climate justice" to see if we can build a bridge to human rights in the context of climate changes and disasters. The chapter concludes by examining the relevance of domestic climate policies and state responsibility as part of a differentiated response depending on physical and social circumstances.

Climate Refugees

Climate-related events, including global warming and damaged ecosystems, have increasingly forced people to abandon their homes. In 1985, UN Environment Program (UNEP) expert, Essam El-Hinnawi, defined climate refugees—also known as climate or environmental migrants—as "people who have been forced to leave their traditional habitat, temporarily or permanently, because of marked environmental disruption."[12]

The term "climate refugee" implies that people who are described as such might be entitled to refugee status and broader human rights. Their status means that they should be allowed to enter and stay in a country because they have been forcibly evicted from their residence, and as a result, they are on the move. This term is neither universally shared nor legally recognized.[13] Some observers argue that the impacts of climate change have yet to be included in the Refugee Convention that specifies very clearly "a well-founded threat of persecution" as a key requirement for seeking asylum. Climate refugees, by their very definition, are broadly viewed as an international "security problem," as they cross an international border.[14]

Most climate-related displacement, however, is internal, prompting the notion among critics that the primary responsibility to protect these so-called IDPs lies with local and national governments. Given that climate change is likely to create internal movement, rather than refugees crossing international borders, the term "climate refugee" has fallen short of global approval. It is also important to note that the term "climate refugee" has not been endorsed by the UNHCR, which prefers referring to them as "persons displaced in the context of disasters and climate change."[15] Moreover, most forcibly displaced people as a result of climate change events are reluctant to be seen or labeled as "refugees" in part because they fear negative consequences of becoming known as refugees.[16]

Others, such as Jane McAdam, argue that the interconnectedness of environmental, socioeconomic, and political factors renders it extremely difficult, if not impossible, to determine whether climate change events are the overriding factor.[17] Living outside the legal framework for protection granted to those under the international refugee law (1951), these displaced people find themselves in a legally and perilously disadvantageous position. Yet regional refugee instruments—such as the 1969 Organization of African Unity Convention (OAU) and the 1984 Cartagena Declaration—offer a broader definition of providing protection to refugees fleeing conditions that "seriously disturb public order."[18]

These regional instruments long predate when climate change was on our public radar, but the intent was clear: to protect people's fundamental human rights when events beyond their control threaten their lives.[19] Some experts regard this as an issue of equality, equity, and justice. The risks associated with climate change pose a quintessential menace to basic human rights, in part because climate change disproportionately affects the most poverty-stricken, marginalized, persecuted, and defenseless people in the world who have played little or no role in contributing to the problem facing the planet, and yet will most likely pay a disproportionately high price.[20]

Some experts have argued that socioeconomic deprivation and punishment that resulted from environmental displacement can amount to "economic persecution." Flight from habitat destruction caused by natural disasters or the human-made risks of climate change should, therefore, qualify for the definition of refugee within the context of the Refugee Convention.[21] Critics argue that although adverse climate impacts such as environmental displacement and dislocation are harmful, they do not meet the legal threshold of "persecution" as currently understood in international and domestic law.

Furthermore, the difficulty stems not from the fact that the core rights affecting these IDPs are socioeconomic in content and scope, but rather in demonstrating that the violation has been predicated on the necessary discriminatory content of an IDP's complaint—for reasons of an individual's race, religion, nationality, political opinion, or membership of a particular social group—to amount to "persecution" for a Refugee Convention reason.[22]

The International Climate Change Regime

The international mechanism for addressing climate change is known as the UNFCCC. This framework is an international treaty agreed upon at the Earth Summit in Rio de Janeiro in 1992 as part of the United Nations Conference on Environment and Development (UNCED). The treaty constitutes the basis for several ensuing international climate agreements. The Kyoto Protocol (1997) and Paris Climate

Accord (2015) are such agreements. The UNFCCC was initially designed as a regime to begin and support a process for future and more comprehensive agreements about how to respond to climate change.

The UNFCCC was officially enforced on March 21, 1994, and has been ratified by 197 countries. Although the signing parties acknowledged the existence of a global problem and threat, and although they committed to cooperate toward resolving these problems, they fell short of undertaking their significant obligations to correct or alleviate these issues. The UNFCCC lacks binding policy commitments. It contains, as experts note, few specific requirements and, perhaps most notably, no enforceable mechanisms for signatories to reduce the emission of GHGs. The parties have pledged to work toward stabilizing GHG concentrations in the atmosphere at a level that would prevent dangerous anthropogenic interference with the climate system, and developed countries have agreed to adopt national policies to mitigate climate change with the aim of returning to their 1990 levels of anthropogenic emissions of GHGs.[23]

Following the Cancun Adaptation Framework in late 2010, which recognized human mobility as part of the adaptation challenges to climate change, an Advisory Group on Climate Change and Human Mobility providing technical assistance to the member states and advocacy was established. After Cancun, the issues of migration, displacement, and relocation have figured prominently in negotiations.[24] In the subsequent Conference of the Parties (COP), we have seen a growing recognition by states that "human mobility constitutes an adaptation as well as a loss and damage issue."[25] However, the ensuing COP21 in the Paris Agreement (2015) did not address questions of legal status, protection, and assistance for people moving in the context of climate change.[26]

Some analysts regard UNFCCC as an inadequate regime to deal with displacement, arguing that the framework lacks the necessary institutions and capacities to deal with climate change-related mobility given that it has already faced numerous difficulties in achieving progress on less contentious core functions.[27] Others have referred to the effectiveness of regional refugee instruments such as the 1969 OAU Convention, the 1984 Cartagena Declaration, and the 2014 Brazil Declaration, which have offered the possibility of opening the door to those displaced by natural disasters. These regional instruments could potentially provide mechanisms for the future protection of cross-border, weather-related displacements while also developing similar initiatives in other regions.[28]

One of the most important examples of how soft law initiatives on the environment-migration nexus can contribute to addressing the needs of affected populations that have fallen through current protection cracks is the Nansen Initiative. This drive was initially launched by the governments of Switzerland and Norway in October 2012, highlighting the fact that current international law provides no assurance that people forced by natural disasters to flee across international borders will be admitted and receive the necessary assistance, much less find durable solutions to their displacement. Absent any organization with a mandate to protect these displaced people, such dispersion creates not only legal protection gaps but also operational, institutional, and funding challenges.[29]

As an institution that aims to foster government action around cross-border mobility in the context of climate change, the consultation within the context of the Nansen Initiative has underscored the primary responsibility of states to prevent displacement when and where possible, and to find durable solutions for such a problem. The existing international and regional mechanisms, laws, and policies fall well short of sufficiently addressing the challenge of cross-border displacement caused by disasters. Hence, the need for improved preparedness by "bringing together stakeholders dealing with humanitarian action, human rights protection, migration management, disaster risk reduction, climate change adaptation, refugee protection, and development."[30]

The Nansen process has highlighted the important role that regional and subregional organizations can play to complement national efforts to identify solutions to this challenge by expanding and strengthening existing laws and mechanisms.[31] The efforts of the International Federation of the Red Cross to establish International Disaster Response Law is a giant step toward fortifying national capacity to respond to climate-induced disasters.[32]

Conceptual and Legal Challenges

Given that climate change would most likely create internal movement, rather than people fleeing the threat of violence or persecution by crossing international borders, climate refugees lack global legal protection and recognition under contemporary international law.[33] The term "climate refugee" is not endorsed by UNHCR, and it may be more accurate to refer to "persons displaced in the context of disasters and climate change."[34] Human rights treaties recognize no specific right to a safe and healthy environment. At present, state obligations regarding the prevention of "environmental flight" are not explicitly specified in treaty law. The legal and policy framework on migration and asylum insofar as the concept of the "climate migrant" is concerned, as well as access to food and water and security issues for "climate migrant," continues to be unsettled.[35]

Theoretically, Beatriz Felipe Pérez argues that the best option for the protection of all climate migrants is to create a new global ad hoc treaty. This option, however, seems unrealistic given that it is unlikely, at least in the short term, to effectively implement it. The question is: Will the states take on the commitment of attributing shared responsibilities with regard to safeguarding climate migrants? Clearly, allocating shared responsibilities and obligations to climate migrants is a difficult task on which to reach agreement. The key barrier to such initiatives is the "unwillingness of states to adopt new international obligations that limit their sovereignty and that bind them to provide international protection to additional persons."[36] The result is that forced climate refugees or migrants have failed to figure prominently in international refugee and immigration policy. Sumudu Atapattu puts it succinctly: "There is no 'home' for forced climate migrants, either literally or figuratively."[37]

Critics argue that international law cannot fix climate-induced displacement.[38] Absent broad and sweeping ratification and implementation, a treaty will not settle the humanitarian issues related to climate change. Even though there is a broad consensus on the reality of global warming, there appears to be much less agreement over its drivers, timing, and gravity. Still, economic growth and employment (for this generation) trump tackling climate change effects (for the future generation). Furthermore, the question persists: How can state obligations owed toward future generations be established?

Many governments show little interest in concluding treaties that expand personal rights or that create extraterritorial and transboundary obligations. The existing enforcement mechanisms offer limited opportunity to protect and enhance the interests of future generations. Some experts remind us that given numerous internal and external factors responsible for causing displacement, identifying climate change "as a sole and primary driver" for displacement is problematic. The traditional international responses, such as drafting new treaties and/or creating new institutions, may not be appropriate solutions in this context.[39] The current legal frameworks, which require identifiable violations and harms, provide no useful structure for establishing accountability or redress for situations arising from climate change, where responsibility and harm are largely transnational.[40] Another problem is that not all parties to previously agreed-upon climate change accords have ratified human rights treaties.

Some observers, such as Jenny Poon, have attempted to bridge international environmental law and international refugee law, especially the latter's principle of *non-refoulement*. The precautionary principle, that is "[w]here there are threats of serious or irreversible damage, lack of full scientific certainty shall not be used as a reason for postponing cost-effective measures to prevent environmental degradation,"[41] is applicable even prior to establishing a causal link by fully clear scientific evidence and even where in the absence of scientific evidence to demonstrate a causal link between emissions and effects.[42]

Scientific uncertainties notwithstanding, Poon argues that "there is a duty upon the State to take a precautionary approach to preventing climate change and thus a corresponding duty to assist those fleeing from its effects."[43] This principle not only provides an insurance policy to avert future harm, but can also be utilized to ascribe state responsibility to polluters, blending both a backward-looking (state responsibility) approach and a forward-looking (harm-prevention) appeal to bolster protection for climate change-displaced persons.[44]

Climate Change Governance

Although human-caused climate change is likely to destabilize the planet Earth in the long term, its responsibility falls solely on the shoulders of the current generation. Absent such a responsibility, it becomes immensely difficult and unbearably costly to initially curb and ultimately eliminate the risk posed by climate change. It is not unreasonable to ask: Who is most responsible for climate change? Developed countries, such as the United States, Canada, Japan, and many European countries, account for just 12 percent of the global population today and yet they are responsible for 50 percent of all the planet-warming GHGs released from fossil fuels and industry over the past 170 years.[45]

From the practical standpoint, one expert reminds us, "mitigating future impacts through reductions in GHG emissions now is considerably cheaper than simply trying to adapt to high temperatures and rapid climate change in the future."[46] Keeping total global warming stabilized below 2 degrees Celsius will require a considerable degree of climate intervention on the part of the human race. In line with this logic, the Paris Agreement proposed to keep the increase in global average temperature to well below 2 degrees Celsius above preindustrial levels and to pursue research programs to confine the temperature increase to 1.5 degrees Celsius above preindustrial levels. Its main aim was to endow the 2 degrees Celsius global temperature target with legal effect.

From the Kyoto Climate Change Conference (1997) to the Copenhagen Climate Change Conference (2009) to the Paris Climate Change Conference (2015), geoengineering has gained much traction to counteract the effects of human-caused global warming. These conferences have increasingly underscored the way in which climate change will "increase risks of violent conflicts in the form of civil war and inter-group violence."[47] One of the most serious challenges that climate change poses to human security is through intensifying "disputes over refugees and resources."[48]

Climate governance also factors into unprecedented environmental degradation that some countries in the Middle East have faced. In Syria, for example, mismanagement of water proved to be a key driver of human mobility. While climate change proved to be largely responsible for the prolonged drought that led to rural-urban migration and subsequent uprisings, failures of governance, caused by overuse of groundwater, and poor agricultural planning were among the culprits. Clearly, growing evidence links the 2011 Syrian uprisings to ongoing drought that destroyed the livelihood of many thousands of farmers. In fact, John Beddington, the UK government's chief scientist, has warned that by 2030, "A 'perfect storm' of food shortages, scarce water, and insufficient energy resources threaten to unleash public unrest, cross-border conflicts and mass migration as people flee from the worst-affected regions."[49]

Throughout the world, several countries stand as those with the greatest number of "climate refugees," or people displaced due to climate events and natural disasters.

These include Afghanistan, India, Pakistan, Ethiopia, Sudan, Bangladesh, Niger, Yemen, Kenya, and Vietnam.[50] The case of Yemen, as one analyst reminds us, highlights how food scarcity caused by the climate crisis can become magnified in conflict zones. The civil war, followed by the Saudis air strikes, in Yemen (2014) caused nearly 4.7 IDPs, many of whom were trapped inside the country. The lives of Yemen's displaced were further deteriorated by outbreaks such as cholera and malaria, and by the ongoing food scarcity and the threat of famine. And importantly, the reality of environmental degradation is interwoven with all of these tragedies. In addition, the prevalence of deluge in Yemen's coastal areas often renders roads impassable, limiting access to medication and other crucial resources.[51]

It is virtually impossible to precisely link Yemen's displacement crisis to the many climate adversities the country is facing. What is evident, however, is that Yemen suffers from fragile institutions and fractured governance, stemming from both the legacies of colonialism and of postcolonial intervention and extraction. As a result, the country is inadequately prepared to tackle the risks posed by extreme environmental changes and events. Yemen simply cannot afford to adopt mitigation and adaptation strategies to resist environmental degradation and its impacts.[52]

Similarly, the Israeli occupation of Palestine has entailed not only forced eviction but also seizing both their land and water resources, restricting Palestinians' ability to manage shrinking resources. In Gaza, where nearly 2.2 million residents are densely packed into a small swath of coastal land, Palestinians' mobility has been noticeably restricted. A lack of electricity due to a long-standing blockade has had a significant impact on public health, with at least one-fourth of diseases in Gaza being water related.[53] The Israeli intense bombing campaign, followed by a ground invasion of Gaza in retaliation for the October 7, 2023, Hamas's attacks on Israel, has made Gaza uninhabitable and graveyard.

The GCC states (Bahrain, Kuwait, Oman, Qatar, Saudi Arabia, and the UAE) could possibly be hosts to some of the 143 million climate migrants displaced by 2050 from Southeast Asia and sub-Saharan Africa, as well as Latin America, according to the World Bank.[54] Ironically, a wide swath of the Gulf region as a whole will become almost unfit to live in by 2050 due to rising average temperatures. According to the World Bank, by 2025, 80–100 million people in MENA, including the Gulf region, will be exposed to a significant level of water scarcity. In the Gulf region itself, humidity and heat will rise so high that some parts of the region will be seen as entirely uninhabitable by 2100.[55]

In the absence of any standard definitions of a climate refugee in international law, a distinction between man-made environmental change and "natural" changes in the environment (e.g., earthquakes, volcanic eruptions) that force people to leave their homes is instructive. The question persists: Can international human rights law provide a useful framework for dealing with climate refugees? Adverse climate impacts have all spawned starvation, disease, and widespread economic dislocation, with the capacity to threaten all of humanity.

The international community's reaction—or the absence of a reaction—to global destruction implicates not just national and regional security but also elements of justice, equity, and human security. Global warming's impacts on the enjoyment of human rights, especially socioeconomic and cultural human rights, can no longer be underestimated. Climate change is widely regarded as a threat multiplier, in large part because it exacerbates existing inequalities and poverty, which are key contributors to regional and global unrest.

As noted above, it is unclear how a state's obligations toward the victims of global warming can be established. Human rights treaties recognize no specific right to a safe and healthy environment. Many states in the Global North appear reluctant to adopt a treaty on climate refugees any time soon, and customary international law provides no legal protection. They are not recognized as a legal category, and anyway, most of them are IDPs, not international refugees, and so likely outside the scope of any future (however unlikely) international convention. Given this reality, I argue that the international community must rely on general human rights norms to safeguard the rights of the millions of people under threat from global warming. From a practical standpoint, governments should be encouraged to invest in clean energy technologies such as solar, wind, hydroelectric, tidal, geothermal, biomass, natural gas, and nuclear power.

Human Security and the Climate Change

Climate change, as Jon Barnett argues, is likely to increase human vulnerability, impede socioeconomic development, provoke reactions that may heighten the risks of conflict, such as forced migration and resource competition, cause statelessness, and undermine mechanisms of international cooperation. These risks are increasingly defined as perils to human security in contrast with the more traditional and abstract notion of national security.[56] Human security has become a particularly suitable mechanism for comprehending and acting on climate change.[57]

Global warming of more than 4 degrees Celsius above preindustrial levels by the end of this century, Barnett goes on to argue, will entail enormous risks to human security. At 2 degrees Celsius of warming above preindustrial levels, millions of people will suffer from increased water stress, declining supply of some ecosystem goods and services, falling productivity of crops in low latitudes, rising floods in coastal area, and significant increases in morbidity and mortality from diseases and extreme weather events. As a result, social systems will be forced to deal with consequential changes in environmental changes.[58]

The emergence of the human security concept has called into question the traditional ways of contemplating national security in terms of arms control and disarmament. With a people-centered approach to security, environmental security has become a widely accepted concept. Linking development with the human security agenda today has been expanded to include freedom from poverty, malnutrition, disease, natural disasters, and displacement. Most notably, "the freedom of future generations to inherit a healthy natural environment" has become inextricably intertwined with national security.[59]

Multiple factors account for the outbreak of violence in the MENA region. Ravaged by Islamic State in Iraq and Syria (ISIS) and violent conflicts, the region of northeast Syria—once known as the country's historic breadbasket—has been struck by prolonged drought and drying rivers, rendering its stability precarious at best. Climate shocks, such as droughts and rising temperatures, have resulted in intense conflicts in the MENA region. In some parts of the region, extreme poverty and lack of opportunity have contributed to young men joining the Islamic State.[60]

The US decision in late 2021 to withdraw its troops from Afghanistan and the return of the Taliban to power has seen the worsening of violence and governmental dysfunction, generating a new wave of "economic migrants" intent on border crossings. Intense poverty and desperation following the Taliban's takeover of the country will most likely kill more Afghans than the ongoing war of the past two decades, blurring a clear distinction between traditional refugees and economic migrants even further.[61]

The plight of Afghan refugees can be used as an opportunity to rework migration and asylum policies as the distinction between traditional refugees (those fleeing from a well-founded fear of persecution), economic migrants (those escaping poverty and economic misery), and climate migrants (those escaping environmental disasters and climate disruptions) becomes increasingly complex and conflated. The courts in the West have come to expand the scope of the definition of refugee to include persecution based on gender and sexuality, as well as those people forcibly displaced. Some 84 million people are displaced around the world, of whom nearly 26 million are refugees outside their own borders.[62]

The Undefined Legal Status of Climate Refugees

Defining a protection framework for those adversely affected by climate change-related hazards will require identifying who is on the move owing to climate change as opposed to other factors. Given numerous internal and external factors that ultimately cause a person to move, one expert notes that identifying climate change "as a sole or primary driver of displacement or migration is problematic."[63] The most dauting hurdle is ascribing the event (e.g., flood, drought, storm) to climate change as opposed to a natural disaster or human-made event. Science-based attributions as well as scientists' use of the term anthropogenic climate change—that is, referring to environmental change caused or influenced by people, either directly or indirectly—are still evolving and scant at best.

Moreover, in the case of weather patterns (compared to temperature rise), many other natural and human-made factors are at play. Human-made factors often contribute to the extent of displacement stemming from climate change-related effects. Consider, for example, the historic 2010 floods in Pakistan, which dislodged 9 million people and were apparently made worse by rapid deforestation in high mountain areas and the unfettered expansion of informal settlements into river basins.[64]

Those who cross borders as a result of climate change do not fit aptly into existing definitions in large part because no international agency or institution has been given the mandate to protect climate-induced migrants. Despite the fact that both the Office of the UNHCR and the International Organization for Migration (IOM) have done much by way of research and advocacy in this area, neither of these has established institutional responsibility or a UN mandate to promote strategies and mobilize political will and resources to effectively preserve and protect climate-induced migrants.[65] Nor have regional instruments, such as the 2006 Great Lakes Protocol on the Protection and Assistance to IDPs and the 2009 Kampala (African Union) Convention for the Protection and Assistance of Internally Displaced Persons in Africa, proposed a satisfactory method for national governments to determine whether a person has been affected by a climate-related hazards as opposed to a "disaster."[66]

Some legal scholars have attempted to devise legal ways to translate these arguments and complexities into practical and effective protection of rights by relying on scientific evidence. Defining legal thresholds that engage the responsibilities of the states will be the key to turning slogans into legal arguments actionable through domestic and international human rights institutions.[67] There is a great promise in the potential role of the law of state responsibility—as underscored by international human rights law—in bolstering the legal protection to those affected by climate change.

The law of state responsibility has automatically arisen in the past when a state fails to take measures to prevent future harm or to ensure the realization of human rights at home and abroad, "irrespective of whether any victim of the violations actively seeks a remedy for the damage or harm suffered."[68] Wherein a state's failure to take proper measures to prevent loss of life incurs an obligation to provide full compensation for the injury. The effectiveness of the law of state responsibility depends on the extent to which states cooperate to make operative the victims' right to a remedy and to reinforce the rule of law.[69] Absent administrative and legal schemes to protect climate refugees or climate migrants at the international level, reliance on the national plan of action appears to be inevitable. Developed countries can assist such plans with necessary financial and legal assistance.

At present, state obligations regarding the prevention of "environmental flight" are not explicitly specified in treaty law.[70] Regarding "internal environmental refugees," the Guiding Principles on Internal Displacement are relevant. Whether these principles constitute part of customary international law is not clear. It is, however, widely believed that they at the minimum reflect existing international obligations. Their value lies in demonstrating how human rights provisions oblige the state of origin to take alleviating measures during "internal environmental flight." They are widely regarded as valuable instruments in making national policies and laws. However, the inability and unwillingness of most states that are affected by internal displacement to deliver their responsibilities toward IDPs remains a problem.[71] Critics, however, are quick to point out that international law cannot fix climate change-related displacement. Absent broad and sweeping ratification and implementation, a treaty will not settle the humanitarian issue related to climate change.[72]

Ethical Complexities

Establishing a truly global ethic of climate justice in the face of the priorities of territorial states is a challenging task. Geopolitical factors typically trump moral issues as factors encouraging, if not justifying, the deployment of protection policies for people fleeing persecution and wars. Thus, framing the issue, Rawan Arar and David Scott FitzGerald note, to "include the maintenance of political stability in areas severely affected by climate change may carry more weight with policymakers and publics than moral pleas alone."[73]

Similarly, effectively tackling climate change-related displacement and migration, while also accommodating claims on behalf of future generations, presents hard ethical questions. The conflict between two simultaneously important but competing interests—that is, economic development and environmental sustainability—makes it difficult to choose between current human needs and obligations owed toward future generations. For many, regardless of where they were born, the choice is between limiting economic growth and preventing global warming, between job creation and protecting the environment.[74]

Furthermore, since the progressive harmful impact of CO_2 is not felt instantly, there is little incentive for countries to confine the use of such fuels. This is especially true even more acutely in the case of low-income countries that are bent on increasing their standard of living through economic development and modernization. Aside from the challenge of how best to reconcile the intergenerational trade-offs, crafting a morally apt policy may not be consistent with public policymaking that is often based on a multitude of capricious conditionalities on the ground. Thus, while scientists, educators, and policymakers have serious concerns about climate change, there appears to be less understanding and even much less solidarity over its causes, timing, and magnitude.[75]

Another ethical concern relates to how to designate the burden of pollution reduction among states. Most climate change accords have failed to establish binding obligations on member states to ensure that emission reduction targets are usefully harnessed and their responsibilities to the global community met. Different countries—developing, emerging economies of the South as well as the North's industrialized nations—continue to disagree on such burdens and obligations.[76] Still, the shift to nuclear power to curb fossil fuel use has raised serious ethical and safety issues. The increased reliance on nuclear energy poses equally controversial moral issues involving significant environmental dangers and risks.[77]

These ethical challenges raise further questions in terms of the linkage between climate change and human rights. Some of the most pressing questions in this area include: Will human rights provide a useful set of standards and legal principles for addressing all these difficult moral choices? Is the notion of collective responsibility invoked in global climate policy effective? What is the status of climate IDPs? Finally, how can state obligations owed toward future generations be established and protected?

Climate Change and Human Rights

Climate change and human rights intersect in several ways, whereby dramatic alterations in the environment will infringe significantly on specific human rights and challenge the realization of the potential of human rights more broadly.[78] That said, the scientific delay in establishing the harm caused by the climate change, as well as the fact that no single state's emissions can be identified as the cause of the problem, renders it difficult to establish notions of causation and proof. Climate change is caused by all states' emissions collectively and across generations. While human rights are generally enforceable territorially, environmental events and issues are transboundary or global in nature.[79]

The fact remains that under existing human rights laws and regulations, the structures and mechanisms for representing future generations are lacking. Yet, the scope of existing human rights must be expanded to correctly illustrate how human relations are closely tied to the environment and the way in which the interests of future generations are effectively protected. Current human rights law and jurisprudence, however, fail to establish state obligations toward future generations. More specifically, the existing enforcement mechanisms offer limited opportunity to protect and enhance the interests of future generations.[80]

Climate change caused by human activity has negative impacts on the full enjoyment of human rights, including those to life, adequate food, health and sanitation, water, adequate standards of living, adequate housing, development, and self-determination. Solidarity in the name of the universal recognition of equality and differences is one way to underline the significance of this link.[81] It helps to recall here that while civil and political rights are enjoyed individually, socioeconomic and cultural rights are collectively shared and depend on ethical, empathetic grounds. Protection against the risks of climate change is a collective responsibility, one that requires a collective response. It is worth remembering that the environment and its components constitute a "collective common" in which natural resources should not be exploited.[82] Not only are countries and governments responsible for saving the planet from heating up, but also everyone—regardless of where she/he lives, should be expected to bear such responsibility. To be sure, the key task ahead requires as many multiple inputs and contributions as possible.[83]

We are also increasingly witnessing how the human rights framework can be deployed in relation to mitigation, adaptation, and adjudication. Working on reducing emissions, stopping their effects, as well as lessening future consequences constitutes what is called mitigation. It is vitally important to work toward preventing catastrophes and protecting vulnerable communities and people. This practice is known as adaptation. Projects to mitigate emissions are more financially affordable than those aimed at adaptation.

In the preceding chapters, we have noted that frequent storms, droughts, heat waves, rising sea levels, warming oceans, and melting glaciers can directly harm animals and wreak havoc on people's livelihoods and communities. As the risks of climate change heighten, extreme and dangerous weather patterns are likely to become more frequent and severe. Some experts, such as Vicente Silva Didier, a program officer with the Global Initiative for Economic, Social, and Cultural Rights, UK, argue that the climate emergency is a human rights issue. Didier writes that the enjoyment of the right to health across the world is under serious threat because of the adverse effects of climate change. Growing evidence shows that climate change is one of the key drivers to a global health catastrophe, profoundly impacting those who already suffer from poverty, discrimination, and structural disadvantages, compounding enduring socioeconomic and gender inequalities.[84]

Many credible sources, such as the IPCC and WHO, have demonstrated that climate change is already affecting the right to health directly and indirectly—by affecting the underlying determinants of health, such as housing, water, sanitation and hygiene, food, and access to a clean and safe environment. According to the Report *Lancet Countdown* (2021), since 2000, heat-related mortality climbed by 53.7 percent in people above sixty-five years, resulting in a total of 296,000 excess deaths in 2018. The *Lancet* report goes on to warn that 67 percent of global cities expect their public health care systems to be damaged due to the climate harm on human health.[85]

Perhaps the most challenging part of establishing this link is the difficulty in identifying rights holders and duty bearers. While identifying a rights holder is easy (the affected citizen), locating a corresponding duty bearer for that "violation" is difficult. The inability to precisely determine duty bearers, as Sumudu Atapattu writes, poses problems for applying the human rights framework as it pertains to the damage caused by environmental displacement. "While the human rights framework is useful in relation to environmental damage within a particular state," Atapattu writes, "it becomes less useful either when the damage is transboundary in nature or when there are multiple states contributing to the damage."[86]

The intersection of climate justice—distributive, procedural, corrective, and social justice—and climate change, Atapattu notes, opens the possibility of linking human rights and climate change. It is in this context that Atapattu surveys the principles of international environmental law, such as Common but Differentiated Responsibilities (CBDR), that are relevant to a rights-based approach to climate change as well as climate justice. Elsewhere, Atapattu has underscored the importance of the social pillar of sustainable development for climate justice and human rights, embracing the issue of equity for current generations as well as for the well-being of future generations.[87]

Regarding institutional developments at the intersection between the environment, climate change, and human rights, it should be noted that there is a Special Representative on Human Rights and the Environment. The former Special Representative has adopted the Framework Principles on Human Rights and the Environment that outline precisely what obligations a Human Right to a Healthy Environment should entail. In October 2021, a new Human Right to a Healthy and Sustainable Environment was recognized in the UN Human Rights Council. This right will be discussed in the UN General Assembly with the aim of creating a new international treaty on a human right to a healthy environment soon. The UN Human Rights Council has clearly recognized the human right to a healthy environment, while also creating the mandate for a Special Representative on Human Rights and Climate Change. This implies that several institutional developments are under way to recognize that environmental challenges and climate change are particularly germane to human rights issues of our times.

While there is a consensus on developing a human right to a "sustainable environment," such a right is not adequately defined or identified in its substantive content and methods for adjudication and enforcement.[88] The question of the extraterritorial application of state obligations in relation to climate change remains unresolved. There exists, according to a World Bank study, major differences of view between and among states on the importance of human rights link to the climate change context as well as the comparative value and basis to be granted to human rights obligation within and beyond national borders.[89]

Some observers argue that prioritizing moral issues over economic issues and public policymaking might risk overwhelming an already loose climate change agenda.[90] Human rights have sometimes been characterized as a source of mistrust between developing and developed countries with certain developing countries expressing concern at human rights being deployed as a way of either fending off their development (should mandatory emissions reduction targets be applied to them) or operating as conditionalities on climate change adaptation funds. Similarly, certain industrial countries have equally expressed concern at the possibility of an official recognition of human rights and climate change linkages to strengthen the case for extraterritorial human rights obligations or a collective and absolute right to a safe and secure environment or otherwise be used as a political and legal tool against them.[91]

Regarding the status of the climate displaced, there are no "refugee like protection" mechanisms for the climate displaced. Within the context of the EU, there exists no applicable legal instruments regulating the protection of climate-displaced individuals at the moment.[92] Slow-onset climate displacement caused, for instance, by rising temperatures and related drought conditions is not recognized as a basis for applying a right to refuge.[93] Today, substantial divisions prevail within the European community as to the most proper method of responding to climate-induced changes in patterns of migration. While some states, such as Sweden and Finland, have taken the initiative to grant temporary protection to those displaced by environmental calamity, others, including Ireland, have offered no such temporary protection to climate migrants, other than the pledge that "humanitarian considerations" will be taken into consideration when deciding on deportations.[94]

Lacking a fully legal identity, the climate displaced are not entitled to adequate valid protection where their "irregular" status leads to "routine and publicly legitimated legal violence against them."[95] Similarly, at the international level, there exists a lack of clarity as to what types of legal protections should be extended to those displaced by inimical climate conditions. The international community has yet to address significant "protection gaps" in existing legal arrangements.[96]

Implications of Climate Justice

The implications of climate-induced displacements for the full enjoyment of human rights of those who are vulnerable owing to geography, age, gender, indigenous or minority status, or disability can no longer be ignored. The debate about the intersections of human rights and climate justice continues, although views vary on its validity and remain unsettled and challenging. Justice is both a basic human right and a means to protect other human rights. Access to justice is inseparable from having access to freedom. Several questions in this context arise: How best to protect human rights in the face of climate change? How might environmental damage violate human rights? How could we establish institutional accountability and responsibility for environmental malpractice?

Climate justice is a process of dealing with climate change by holding the major polluters—especially fossil fuel companies and oil-producing states—accountable. The principles of rights, equality, and justice continue to be linked to those most vulnerable to climate impacts. Climate justice demands that those who have contributed the least to climate change should not suffer unjustly and disproportionately from its harm. Addressing the burden sharing and equitable distribution of the costs and benefits of climate change, as well as how to determine responsibilities to curb negative consequences of climate change, deserves considerable attention.

The ECHR on occasion has found such rights as a right to life or development, as well as a right to participation, violated by polluting parties. In the case of *Lopez-Ostra v. Spain*, in 1994, it ruled that the pollution and fumes from a particular waste treatment plant forced the plaintiffs to move from their homes. In the case of *Oneryildiz v. Turkey*, in 2005, the European Court ruled that Turkey had failed to take measures to preserve a healthy environment and prevent an explosion at a waste dump and as a result had violated the right to life.[97]

Likewise, the African Commission on Human and People's Rights has recognized polluting practices as a violation of human rights to life, health, and property, all of which are safeguarded under the African Charter on Human and People's Rights. In 2001, the African Commission ruled that Nigeria's participation and failure to stop substantial oil spills and water contamination violated the Ogoni people's right to life, health, and property.[98] In June 2015, the District Court of the Hague, relying on the obligations embedded in the European Convention on Human Rights and the Dutch Constitution, ordered the Netherlands to reduce its GHG emissions by 25 percent by the end of 2020.[99]

Critics argue that the human rights framework, which requires identifiable violations and harms, provides no useful structure for establishing accountability or redress for situations arising from climate change, where responsibility and harm are largely transnational.[100] One suggestion is to mobilize solidarity and collective responsibility—not just to protect human rights, but to right climate wrongs.[101] The proponents argue that the Paris Agreement's preamble points to Parties' human rights obligations, even as they lack legal enforcement. The reference to human rights in the agreement is widely regarded as groundbreaking. Arguably, the frequent references made to sustainable development in this agreement can be read in a way that enables human rights organizations, scholars, and activists to introduce human rights into the Paris Agreement through the backdoor. There is an undeniable link between climate justice, the social pillar of sustainable development, and human rights as the Copenhagen Declaration of Social Development (1995) demonstrated.[102]

The fact remains, however, that not all Parties to previously passed climate change accords have ratified human rights treaties. It is also suggested that the United Nations' Universal Periodic Review could be tasked to explicitly highlight human rights concerns related to climate change, while becoming a means to see how Parties to climate change deal with human rights issues associated with climate change and seeking an effective way to disseminate best practices.[103]

Certain developments and pilot schemes have unfolded despite the fact that there has not yet been an appropriate solution to tackle climate refugees. In October 2017, New Zealand's then-new minister for climate change, James Shaw, announced that the country was considering issuing "an experimental humanitarian visa" category for Pacific Islanders evicted by the adverse effects of climate change. The new plan aimed at bringing approximately 100 people a year to New Zealand. This pilot scheme was dropped because Pacific Islanders refused the visas in the end. Instead, they called on the New Zealand government to institute a stepwise approach: reduce emissions, support adaptation efforts, provide legal migration pathways, and finally, if all this fails, grant refugee status.[104]

The Path Forward

Given that climate refugees are *not* protected by the 1951 Refugee Convention and that there is no legally binding global agreement in place to shore up climate refugees, it is more important than ever that governments and legal bodies reframe conditions attributable to climate change as a menace to human rights. A collective effort is required to protect climate-displaced persons. It is also critically important to underline the extent to which climate refugees face a real threat—even as that threat is not always as imminent as the dangers faced by refugees fleeing persecution and violence.[105]

Responses to climate change will have a profound impact on the full enjoyment of human rights for many people across the world. This will transpire through both direct impacts on human societies and indirect effects, such as the degradation of ecosystems and the diminishment of environmental resources upon which many lives and livelihoods depend. The question that needs attention is: Who is responsible for containing and even diminishing the adverse impacts of climate change and safeguarding the human rights of individual members of the society? States have obligations to respect, protect, and fulfill human rights, while also attenuating domestic GHG emissions, protecting citizens against the deleterious effects of climate change, and ensuring that responses to climate change have not materialized at the cost of abusing human rights.

Can international human rights law provide a useful framework for dealing with climate refugees? On balance, while there are growing numbers of agreements that recognize the need to protect climate migrants, soft-law resolutions fail to provide the necessary legal protections that migrants need. Yet, soft law can serve as a first step toward forging a consensus over time in the hope of paving the way for creating a treaty at some point in the future. Revising the 1951 Refugee Convention to account for climate refugees faces many difficulties; this is in part because most of those affected by climate change remain internally displaced, and the 1951 Convention provides no protections for internal displacement.

Some experts argue that there is an urgent call for an international legal agreement for the protection of climate refugees.[106] Others have even suggested that the international community should eventually aim to create a new convention.[107] Still others maintain that short of creating a new convention, there are several effective ways to protect climate migrants. These include signing the Global Compact for Migration, extending Temporary Protected Status, preventing deportations to disaster areas, reinstating migration and development aid, and making climate displacement a priority at the future COP meetings.[108] Adopted in 2018, the UN Global Compact on Migration has a specific chapter and articles on "Natural disasters, the adverse effects of climate change, and environmental degradation." This is the latest standard the international community could agree upon.[109]

In short, there are many new institutional developments recognizing the relationship between climate change and human rights, including the mandates of the UN Special Representatives on Human Rights and the Environment and on Human Rights and Climate Change. Framework Principles on Human Rights and the Environment were adopted in 2018, and they laid the basis for the newly recognized Human Right to a Healthy and Sustainable Environment that has been brought to the General Assembly for Adoption.

It is unclear, however, how states' obligations toward the victims of global warming can be established. Many states in the Global North appear reluctant to adopt a treaty on climate refugees any time soon, and customary international law lacks any legal protection in this regard. Climate refugees are not granted a legal status, as most of them are IDPs, not international refugees. They are seen outside the scope of any future international convention. Even in those handful of governments that have adopted binding laws on the treatment of IDPs, enforcement remains far from robust.[110] Taking into account all these realities, it is safe to conclude that the international community must rely on general human rights norms to protect the rights of the millions of people under threat from global warming.

PART II
POLITICAL TENSIONS, REFUGEE CRISIS, AND CLIMATE CHANGE

CHAPTER 5
TURKEY

REFUGEES, FORCED MIGRATION, AND CLIMATE CHANGE

Turkey is a vast country consisting of several key parts and multiple borders. In the west, known as Thrace, Turkey shares borders with Greece and Bulgaria. In the northeast, it has borders with Georgia, Armenia, Azerbaijan—the so-called South Caucasus region. In the east, Turkey shares borders with Iran, Iraq, and Syria. The rugged and often scenic landscape of Turkey, which is the result of complex and powerful land forces, has shaped Anatolia for many millions of years and still exhibits itself in frequent earthquakes and accompanying seismic events.[1]

The country's coastal borders are surrounded on three sides by the Black Sea in the north, the Aegean Sea in the west, and the Mediterranean in the south. The Mediterranean region and southern and western parts of the country, as well as snow-dominated eastern Turkey, are among the most vulnerable to climate change and global warming. Furthermore, political problems stemming from allocation of water in eastern part of the country between Turkey and the downstream countries of Iraq and Syria have further underscored the importance of gaining a better grasp of the climate change effects in Turkey.[2]

The vast majority of Turkey's nearly 85 million people live in urban areas, with only 12 percent living in rural areas. Like any other country in the Middle East, Turkey is vulnerable to environmental and climate changes, as it has become exposed to extreme weather events and rising temperatures over the past three decades. These climate-induced changes have negatively impacted the country's terrestrial, marine, and freshwater ecosystems.[3]

Turkey has a very heterogeneous landscape and topography, with some parts of it hot and others temperate. Those areas along the mountain ranges that run parallel to the coasts have a milder climate, whereas the inland Anatolian plateau experiences extremes of hot summers and cold winters with scant rainfall. Even the coastal areas experience substantial climate variability. The coastal areas of Turkey bordering the Aegean Sea and the Mediterranean Sea have a typically hot and dry summer climate with mild to cool temperatures, as well as wet winters. The coastal areas bordering the Black Sea have a temperate oceanic climate with warm, wet summers as well as wet winters.[4]

The 2021 UN IPCC report underlined the fact that Turkey would experience three upward trends: accelerating temperatures, dehydration, and rising sea levels.[5] Thus, the country is likely to experience more frequent and more severe weather conditions during the course of the year. By 2050, as one study suggests, temperatures are predicted to increase by 2.5 degrees Celsius in east and central Turkey, and by 1.5 degrees Celsius on the coasts. Temperatures surpassing 40 degrees Celsius are anticipated during summer times for extended periods.[6]

The country continues to rely on fossil fuels, as indicated by the fact that the import of Liquefied Natural Gas (LNG) reached a record high in 2019. The construction began on Turkey's first nuclear power plant despite international protest at the implicit risk from earthquakes. The steady decline in the costs of renewable energy technology and storage has meant that reliable power could have been obtained effectively without burning coal. Approximately 34 percent of Turkey's energy emanates from coal today, as the country falls behind only China and India in terms of global ranking of newly designed coal plants.[7]

In comparison with the EU average, Turkey's power, transport, and agriculture sectors are less carbon intensive—partly owing to the large use of renewable energy in the country's power system and low motorization rates. The country, however, continues to be greatly dependent on coal, and this reliance is likely to intensify further under current investment plans.[8] Interestingly enough, installation costs of solar power in Turkey are among the lowest in the world. Aside from the energy efficiency plans and renewable energy auctions, Turkey's progress toward implementing climate action has been very sluggish. Turkey's policies and actions, according to one source, are inconsistent with the Paris Agreement's 1.5 degrees Celsius temperature limit.[9] In a survey conducted by Konda in 2021, approximately 75 percent of Turkey's voters—including 81 percent of the ruling Justice and Development Party (AK Party)—considered global warming a real concern, a level of awareness that matched, and even exceeded, most European countries.[10] That said, fossil fuels still account for about one quarter of the country's energy portfolio. Further, Turkey is fiercely engaged in a dispute in the Eastern Mediterranean over natural gas there with Greece and Cyprus—a critical naval standoff with the potential for a real war that could draw in Egypt, Israel, and others in the region.[11]

The aim of this chapter is to evaluate the way in which climate change impacts Turkey, but more specifically how it affects the country's policies and practices as they relate to migration. The chapter is divided into four sections. In the first section, I examine the country's water issues and its prospects for food security. The second section turns to the impact of drought on farming and the expansion of urbanization. The third section critically assesses the pace with which the shift toward the climate change is occurring in Turkey. The fourth section examines the relationship between climate change and migration in Turkey, with specific focus on Syrian refugees in the country. Finally, the implications of the Russian invasion of Ukraine for Turkey are examined.

Turkey's Water Issue

Turkey, as one scholar observes, has the most potential for hydroelectric power among all European countries except Norway and Sweden. Devoid of oil and gas resources, it has turned to hydropower for energy. Several dams have been constructed in many separate river basins, including the Kizilirmak, Yeşil, upper Gediz, Sakarya, and Seyhan. The construction of the giant Euphrates dams after 1965, however, illustrated Turkey's great achievements in hydropower and extensive irrigation.[12]

Istanbul's freshwater sources have come under increasing threat from severe droughts, heat waves, urban sprawl, and a growing population. At the beginning of 2021, Istanbul reservoir levels dropped to 20 percent, leaving less than one and a half months' supply for around 16 million people.[13] In recent years, droughts and other extreme weather events have reduced the availability of clean water in many cities around Turkey, but especially in Istanbul. Istanbul's growing urbanization puts additional pressures on freshwater supplies and land occupation in water basin areas. The city contributes more than 30 percent of Turkey's GDP and accounts for 18.5 percent of the country's population.[14]

In the 1990s, local authorities constructed seven dammed reservoirs in Kırklareli, Tekirdağ, and along Istanbul's provincial border with Tekirdağ to end the water shortage in Istanbul in the coming years. The water problems in the ensuing years, however, continued. The 2007 drought compelled the construction of the Great Melen Project—a 189-kilometer water pipeline built to carry water directly from the Melen Stream to Istanbul.[15] The 2014 drought, however, proved that water issues linger, as the Melen River's water flow decreased by 50 percent and the reservoirs built outside of Istanbul in the 1990s had almost no water left in them. In fact, drought and growing urbanization have rendered the city's water infrastructure built after the 1990s defunct.[16]

Let us turn our attention to the Euphrates and Tigris. There exist 200 river basins in the world, and only one-third of these streams are managed by treaties, and only a few of them have the status of international streams, such as the Danube (it passes through nine countries: Germany, Austria, Slovakia, Hungary, Croatia, Serbia, Bulgaria, Romania, and Ukraine). A control example of stream not managed by treaties is that of Euphrates and Tigris. Indeed, Euphrates and Tigris spring up in Turkey and they flow through Syria and Iraq.[17] While mountain snows in Turkey contribute 88 percent of Euphrates water, neither Syria nor Iraq contributes substantially to the flow, yet both countries depend upon these rivers—Syria on the Euphrates and Iraq on the Tigris and Euphrates—while Turkey gets an increasing amount from both rivers before they reach the two downstream countries.[18]

Since the mid-1960s, Turkey has built several giant dams on the Euphrates. Turkey's impressive achievements in hydropower and large-scale irrigation came with the Southeast Anatolia Project (*Güneydoğu Anadolu Projesi*), which was implemented in the early 1980s. This project included twenty-two dams and nineteen hydroelectric installations—seven on the Euphrates and six on the Tigris. Turkey's blocking of the Euphrates flow, which has resulted in reduced water flows, as well as its increased consumption of water, has greatly concerned the downstream riparian states: Syria and Iraq. Both Syria and Iraq have accused Turkey of violating international law by claiming its absolute territorial sovereignty over water.[19]

Likewise, both Syria and Iraq have complained about their water quality. It is worth noting that the Euphrates and Tigris Rivers are the largest in the region and provide water to Turkey, Syria, and Iraq. Farmland has dried up across the region in Iraq, Syria, southeast Turkey, and Lebanon. Jordan consumes more water than is replenished by rain. Lebanon and Syria will be joining Jordan in that group by 2025. Turkey's water-related projects on the Tigris and Euphrates Rivers could reduce up to 40 percent of the water in Syria and up to 80 percent of the water in Iraq.[20] Syria and Iraq have yet to introduce modern technology and are still largely dependent on rainfall. But rainfall, or its absence, is not the only problem. The problems are political, not climate induced. Many Arabs believe that "Turkey is trying to assert itself as a regional superpower, and water is being used as a tool to advance that interest."[21]

Turkey will most likely experience more severe weather conditions in the coming years. In unison with a temperature increase and precipitation reduction, the frequency of extreme drought events is estimated to surge by the end of the twenty-first century. Thus, the appraisal of the impacts of climate change on water supplies and agriculture will become crucial for a sustainable future.[22] One of the core goals of Turkish policymakers, Ayşegül Kibaroğlue notes, is to reduce the country's dependence on external energy sources. Hydropower is thus seen as an essential domestic source in the water-energy security context. Increasingly, hydropower is viewed as "the most reliable and economically feasible energy source among the renewable energy sources available to Turkey."[23]

Most hydrological impact studies in Turkey have projected future streamflow decreases. One study has projected around a 40 percent decrease in streamflows and nearly 25 percent decrease in groundwater recharge by 2070.[24] Another study has reported surface runoff reductions by 20 percent by 2030, 35 percent by 2050, and 50 percent by the end of the century in the Gediz and Big Menderes River Basins in the western Anatolia.[25] These projected water supplies reduction will adversely affect all sectors of the economy. Most studies predict increases in future flood in the northern parts of the country.[26]

Farming, Drought, and Urban Food Security

Agricultural production, experts remind us, has consistently been the primary source of food security in the country. Although the share of agriculture in the Turkish economy has declined, it still accounts for a relatively larger share of total output and employment when compared to the EU and North American countries. The share of agricultural employment fell from 50 percent in 1980 to 20 percent in 2017. During the same period, the contribution of agricultural production to GDP decreased from 23 percent to 7 percent. However, thanks to significant export revenues generated from agricultural products, agriculture still constitutes a major place in the country's national economy.[27] Today, it should be noted that Turkish agriculture remains more vulnerable than ever to the impact of climate change.

Among the most serious food security problems that Turkey faces is the obsolete irrigation techniques of Turkey's farmers. Turkish officials, who have frequently pointed to a much larger neglect of farmers by the government, have noted that these systems suffer water losses of 35–60 percent through evaporation, seepage, and leakage. Turkey's farmers, according to one study, have long been victims of misinformed, outdated agricultural policy, as well as woefully inadequate subsidies. Coping with climate change in the face of the government's incompetence has added yet another twist to the business of farming in Turkey.[28]

A survey of 3,100 farmers carried out in 2019 by the Turkish agri-tech company, Doktar Agriculture, found that 70 percent of farmers were discontent with the government's agriculture policies, as nearly 79 percent of them said they continued to struggle on a daily basis with the effects of climate change. Unbridled use of groundwater and wasteful, inefficient, and nonviable irrigation techniques have exacerbated Turkey's water problems, with 60 percent of its land now facing aridity and erosion. Over the past half century, 2 million hectares of wetlands, including thirty-six lakes, have dried up. Poor irrigation policies and practices have increased soil calcification and salinity.[29]

The government-sponsored Southern Anatolia Project is emblematic of state mismanagement. The project that began in 1996 promised better harvests and higher income to farmers. By 2012, however, the region had lost 12 percent of its farmland to salinization. Moreover, stormy weather destroyed more than 1.9 million hectares of farmland in 2018.[30] According to a law passed in 2006, an amount equivalent to 1 percent of the country's GNP is to be handed out each year to the agricultural sector as subsidies. But, except for only one year, farmers have not received any such subsidies. Consequently, 3 million hectares of farmland have remained uncultivated. In 1980, agriculture employed half of Turkey's population. Ten years later, the number had dropped to 46 percent. By 2020, it had declined to approximately 18 percent.[31] The agriculture sector reached the lowest share of employment in 2023, amounting to 14.8 percent.[32]

Furthermore, climate experts have long warned that the entire Mediterranean basin, which includes western parts of Turkey, is at risk of dire drought and desertification. In recent decades, droughts have occurred more frequently in Turkey, due to the erratic nature and decrease in precipitation and high temperatures, both expedited by the effects of climatic fluctuations taking place over an extended period of time.[33] Although socioeconomic factors, such as population growth and economic development, will increase the demand for water, climate change will substantially affect the demand for domestic and industrial water, negatively impacting the amount of water allocated for irrigation in agriculture.[34]

Turkey's semiarid climate, extreme temperatures, and an annual precipitation of about 643 millimeters render it vulnerable to climate change. The country's water resources are not distributed equally among its different regions. Most of the water resources are used for irrigation. It is estimated, however, that the rate of agricultural water usage will decrease in 2030 from 75 percent to 65 percent, as modern irrigation methods become ubiquitous.[35] Water scarcity and depletion have reached pivotal points not just in agricultural and energy sectors but also in relation to water management, irrigation, drinking water, and other hydrological systems.[36]

In 2007, the governor's office of Izmir Province formed the *Water Council* to identify the problems and ideas for dealing with water resources and sustainable water use in that province.[37] Increasingly, however, the fears of droughts are substantiated by realities on the ground. Consider, the case of Lake Tuz. The 1,665 square-kilometer (643 square-mile) lake, Turkey's second-largest lake, has entirely sunk in recent times.[38] The lake (Salt Lake in Turkish) is a recent casualty of climate change-induced drought. One study shows that water levels at Lake Tuz started to decline beginning in 2000. The lake completely receded in 2022 due to rising temperatures, escalated evaporation, and inadequate rain. Several other lakes across Turkey, which have been impacted by decreasing precipitation and untenable irrigation practices, have similarly dried up or have significantly retreated.[39]

Climatic changes and water shortages will lead to greater expenses in the form of higher food costs in the region. It is important to remember that global food prices are estimated to surge significantly due to many factors, including population growth, a growing middle class, limited availability of fertile and productive soil, and dwindling crop yields owing to climate change, with wheat prices rising by 59 percent, maize by 106 percent, and rice by 78 percent from 2010 to 2050.[40] With the global food prices on the rise, Turkey, which once was one of the world's biggest exporters of cereals, is expected to become a net cereal importer. The ensuing higher energy and food costs will have inflationary impacts, particularly affecting the major agricultural areas of eastern Anatolia.

The negative impact of climate change on destabilizing rural farm incomes is likely to disproportionately affect the Kurdish economy and communities of southeastern Anatolian regions of Turkey, further inflaming existing Kurdish-Turkish tensions, especially in light of the fact that a main base of Kurdish Partiya Karkeren Kurdistan (Kurdistan Workers' Party—PKK) support comes from small-scale farmers who resent Ankara's policies of favoring large landholders.[41] This climate-induced migration to urban areas for economic opportunities will accelerate—both in terms of the scale and the pace of migration.[42]

Climate Change and Environmental Activism

As in many parts of the Middle East, Turkey's increasing GHG emissions have been driven by population growth and economic development. Despite hydroelectric, solar, wind, and geothermal energy sources accounting for 54 percent of the country's electricity generation capacity in 2021, its energy transformation beyond electricity remains lagging behind.[43]

In 2018, almost three quarters of Turkey's total energy consumption was derived from fossil fuels, a ratio that continues to soar because of spiking energy demand in the transportation and industrial sectors. While transportation relies on oil for 97 percent of its energy, the industrial sector receives 67 percent of its energy directly from fossil fuels. Turkey also spends nearly $7 billion annually to support petroleum and coal production and consumption, using tax breaks and state-owned enterprise investment for that purpose. During COP26 in Glasgow, Turkey refrained from joining voluntary commitments to retreat from using coal for electricity generation, or to gradually dispose of its reliance on oil and gas, or even to cut methane emissions.[44]

The country's shift toward climate adjustment has been slow but steady in the past decade. Ankara became a party to the Kyoto Protocol in 2009 and the Paris Agreement in October 2021 (entered into effect by November 10, 2021). Turkey then committed to achieving net-zero carbon emissions by 2053. Following the signing of the Paris Agreement, the Environment Ministry was renamed the Environment, Urbanization and Climate Change Ministry on October 29, 2021. The Climate Change Department was eventually established, which subsequently held a Climate Council in central Anatolia's Konya in February 2022 to design new climate actions and strategies.[45]

The country's new road map was determined in "seven different areas such as greenhouse gas reduction, green financing, carbon pricing, adaptation to climate change, local governments, migration, just transition and social policies, science and technology."[46] Council members have thus far taken several hundred new decisions in line with 2053 net-zero emissions (NZEs) and green development targets within the scope of combating climate change.[47]

Critical voices concerning public health and the environment, however, have emerged in recent decades with the government inconsistently translating them into policy enunciation. In 2009, protests against the proposed construction of the Ilisu Dam on the Tigris River in Turkey erupted in which local activists in conjunction with global networks of activists pressured foreign multinational companies and European export credit agencies to withdraw from the project.[48] Protests in Istanbul's Gezi Park and some ninety other towns in Turkey in 2013 pointed to new and further openings for environmental activism in the country.[49]

The most recent controversial project involves President Recep Tayyip Erdoğan's plan to build a shipping canal in Istanbul as an alternative to Turkey's internationally used straits: the Dardanelles and the Bosphorus. Known as the Kanal Istanbul, a shipping route running parallel to the strategically important Bosphorus Strait has become a key divisive issue in Turkey. The Istanbul Canal is said to be deployed as an artificial sea-level waterway on East Thrace, connecting the Black Sea to the Sea of Marmara, and ultimately to the Aegean and Mediterranean seas. Since 1936, passage through the strait has been governed by the Montreux Convention, a multilateral treaty that guarantees the free movement of ships across these straits during peacetime while also firmly restricting the movement of naval vessels.[50]

Turkish officials state that the new canal, which is created on the European side of the Bosphorus, will be safer and faster to navigate compared to the Bosphorus, rendering it a more expedient and attractive route for commercial ships, who will have to pay a transit fee to pass through. Some critics argue that Erdoğan would likely utilize the canal to circumvent the Montreux Convention, which has given Turkey full control over the straits and regulates the transit of all ships, including naval warships, making it legally possible for NATO allies to send their warships into the Black Sea to counter Russia. This canal, it is noted, will most likely attract Chinese investment.[51] Moscow considers the Montreux Treaty vital to the region's political stability insofar as it strictly controls the use of the Bosphorus by warships of non-Black Sea countries, while also confining the presence of the US Navy in the Black Sea.[52]

Similarly, some politicians, including Ekrem Imamoglu, mayor of Istanbul, from Turkey's main opposition Republican People's Party (CHP), as well as environmental experts, have raised serious concerns. The real cost of this so-called $12 billion project will be more than $75 billion.[53] Furthermore, they note, the canal threatens the city's water supplies and risks broader environmental repercussions given the intricate balance between the interconnected seas in the region, which already suffer from "sea snot." Additionally, the canal would bring polluted waters of the Black Sea into the Sea of Marmara, and eventually in the Mediterranean.[54]

Turkish officials, according to one source, have attempted to align their policies with those of international climate politics. Turkey has ratified the Paris Agreement and has been updating its second Nationally Determined Contribution (NDC) with a long-term, low-emissions development strategy, while also drafting the Carbon Border Adjustment Mechanism regulation. Other carbon pricing instruments have paved the way for changes in the country's national climate policy. Thus, the country has done a great deal to reach its goals of decarbonization and NZE by 2053.[55] It is worth noting that the core challenge Turkey faces, as experts remind us, is a lack of financial assistance in the wake of the increasing burden and damage of climate change.[56]

Turkey's Emigrant Identity

Many considerations render Turkey particularly exposed to human mobility and migration. These include its physical geography, history, social structure, and economic conditions. Moreover, Turkey's proximity to Europe, which is the preferred destination for immigrants and where nearly 87 million immigrants already live—that is almost 8 percent of the global immigrant population.[57] The Eastern Mediterranean route, where many irregular migrants use as a transit to Europe, connects Turkey to Europe via Greece. Also, the Syria-Turkey corridor, according to the World Migration Report 2020, has become one of the largest corridors with the arrival of more than 3.7 million Syrian refugees in Turkey in 2019.[58]

Given the impact of tropical storms, heavy rains, deluges, and droughts on migration movements to Turkey, it is reasonable to assume that migration trends to Turkey can be expected to occur largely from South Asia, the Balkans, the Black Sea, and the Middle East.[59] Increasingly, such considerations were included in Turkey's 11th Development Plan (2019–23).[60]

Turkey's migration identity has shifted from being principally a country of emigration and transit to becoming a destination for immigrants and people fleeing conflict. In response, Turkish policymakers have recently reevaluated migration systems to address this new status and create a more comprehensive migration and asylum policy. It is worth noting that Turkey is one of the original signatories to the 1951 Convention relating to the Status of Refugees, which defines refugees, their rights, and states' obligations to them. Unlike the majority of signatory countries, however, Turkey maintains a geographical limitation, affording refugee status only to individuals from European countries. These displaced persons may be legally defined as climate migrants and/or asylum seekers, but certainly not as refugees. Clearly, climate refugees are not covered under the 1951 Convention.

As a result, asylum seekers from non-European countries are expected to be resettled elsewhere or returned to their country of origin. Syrian refugees are therefore under temporary protection intended to relieve the current emergency humanitarian situation, but it is unclear whether Turkey will allow these refugees to remain indefinitely and integrate into Turkish society. With prospects for a near-term end to the Syrian civil war looking grim, Turkey will have to contemplate a plan to address the longer-term future for Syrians, who are unable to return home and unlikely to gain third-country resettlement in significant numbers.

To be sure, unrest and political instability in nearby countries have led to major humanitarian inflows from Syria, Afghanistan, Iran, Iraq, and Somalia. The Syrian crisis, which has pushed more than 6.6 million Syrians to flee their country, has placed mounting pressure on neighboring countries.[61] Between 2010 and 2014, Turkey spent more than $3.5 billion to establish and run refugee camps and provide services.[62]

Today, Turkey is host to the world's largest refugee population, with 4 million people, of whom nearly 3.7 million are Syrians. The Government of Turkey, the EU, and the World Bank have continued their partnership under the Facility for Refugees in Turkey (FRiT) program to tackle Syrian refugee problems in Turkey. FRiT has earmarked €6 billion in humanitarian assistance, education, health, municipal infrastructure, and socioeconomic support for Syrian refugees in Turkey.[63]

The Syrian Refugees

Turkey has become a safe haven for refugees, asylum seekers, and irregular migrants from across the world.[64] Turkey hosts the world's largest refugee population since 2014. There are nearly 3.7 million Syrians under temporary protection and approximately 330,000 refugees and asylum seekers under international legal protection. Afghan nationals have been the chief International Protection applicants in Turkey since 2019.[65] Since protests began in the 2011 Arab Spring uprisings, millions of Syrians have fled their homes in hopes of finding safety amid the civil war in that country. In total, Syrian refugees, asylum seekers, and irregular migrants constitute 6–7 percent of the total population in Turkey.[66]

The Turkish government, according to one source, plans to relocate approximately 3 million Syrians to a "safe zone" inside northeastern Syria, a previously Kurdish-controlled area, if it could control the 300-mile-wide strip border and would receive $26–$27 billion from the EU. Many EU officials appear skeptical about this plan, calling it unrealistic and pure fantasy.[67]

Within the context of curbing migratory waves emanating from Syria, Turkish authorities in the Ministry of Internal Affairs provide a contrasting view, arguing that safe zones serve three distinct purposes. First, they create a safe space for civilians, eliminating the need to head to the Turkish border directly. Second, the safe zones represent a normal situation whereby they can stem irregular migration. Finally, the safe zones enable Syrians to return to their country, and with foreign aid and outside financial support designed to promote economic development, serve to encourage Syrians to go home. In fact, the Turkish Interior Ministry has announced that some 462,000 Syrians have returned to their homes in the aftermath of creating these safe zones.[68] Turkish officials have noted that this policy has significantly reduced the number of terror attacks within and near the country's borders, while also sapping the motivation of Syrians living in these areas to migrate to Turkey.[69]

In 2014, the Turkish state issued a new "Temporary Protection Regulation," initiated based on the 2013 Law on Foreigners and International Protection, which provided asylum seekers from outside of Europe the first legal protection and a status—known as "conditional refugee"—in line with the standards of EU and international conventions. This act regulated the status, rights, and responsibilities of the people who arrived in Turkey in large numbers and whose applications could be processed individually. Under this regulation, Syrian refugees became the beneficiaries of "temporary protection."[70] This protection was further reinforced by *Ikamet*, a residence permit obtained by a foreign citizen legally residing in Turkey. This document is temporary, although it can be extended frequently. Additionally, foreigners can be issued a *kimlik* (identity card equivalent to domestic Turkish passport) for the purchase of real estate or other forms of investment in the Turkish economy.[71]

More specifically, Article 38 of this resolution provides the refugees under consideration with access to shelter, food, health care, social assistance, education, and other related services within interim accommodation centers. There is, however, no direct mention of the right to work, education, and social assistance for Syrians.[72] While Syrian refugees who held temporary protection beneficiary identity cards could apply for work permits under Article 29(2), these work permits were limited to a single place of employment, making it difficult to transfer one's workplace. The law also provided that the proportion of refugees and asylum seekers cannot exceed 10 percent of a company's workforce.[73]

It should be noted that Syrian refugees in Turkey face multiple challenges in gaining access to legal employment. Between 2016 and 2019, according to one study, a total of 132,497 work permits were issued to Syrians registered in Turkey. This figure also included renewals of already existing work permits. It was estimated that at the time approximately 1 million Syrians were working informally without legal protections, and 45 percent of Syrians under temporary protection were living below the poverty line.[74]

Syrian refugees who lack a work permit and are working in the informal market suffer not only from low-paying jobs but also from highly exploitative and physically demanding conditions. One study in 2020 demonstrated that Syrian refugees working in the informal sector worked, on average, five more hours per week than Syrians with a work permit and their Turkish counterparts without a work permit. Also, three in four Syrians earned less than the minimum wage per hour. The lack of educational opportunities for Syrian refugees' children, along with the absence of formal employment opportunities for Syrian adults in Turkey, raises alarming concerns for the long-term socioeconomic integration of these refugees.[75]

In January 2016, Syrians under temporary protection obtained the legal right to a work permit under stringently defined conditions.[76] It is worth noting that only 15,000 work permits has been issued for these refugees, a small number given that approximately 1 million of these Syrians would have been on the labor market.[77] Yet Syrian refugees, like refugees from other countries, under temporary protection in Turkey continue to face significant obstacles to accessing the formal labor market. This situation persists because the permit system requires employers to apply for work permits on behalf of the applicants. Syrians, however, have been able to open businesses in Turkey. In 2019, some 7,000 companies were formally registered by Syrians, many of whom were owned by displaced Syrians living in Turkey.[78]

Furthermore, the law gave the prospective employer the responsibility of submitting applications for work permits to the Ministry of Labor and Social Security. Yet many employers refused to do so, in large part because they could have easily exploited Syrian refugees for cheap labor. A combination of cheap labor, the barriers to accessing the labor market legally, and the large number of refugees has evoked a negative reaction among the local population, fearing that refugees have become a serious burden on national resources and that they would compete with them for jobs.[79]

The inclusion of Syrians in the informal labor market has ramifications for the host community, including reduced wages, displacement, and increased unemployment. Other issues, such as the high incidence of child labor among Syrian children as well as new patterns of internal migration among Turkish nationals, especially Kurds, have impacted the changing labor market.[80] As of 2017, nearly 40 percent of more than 1 million Syrian children at school age were not receiving any form of schooling.[81] Syrians' precarious legal status, as well as the restricted access to social rights, work permits, education, and health care services, continues.

Approximately 96 percent of Turkey's Syrian refugees reside outside of camps, in both urban and rural areas. More than half a million Syrian refugees live in Istanbul and thousands more are said to live there even though they have been registered in different provinces or have failed to be registered at all. The presence of vibrant textile, tourism, and service industries in Istanbul offers refugees and migrants attractive job opportunities.[82] The reality is, however, that access to work for Syrian refugees in Turkey is still unreliable. Low-paid jobs and high cost of living make dependency on the cash assistance under the EU-funded Emergency Social Safety Net (ESSN) necessary for covering basic costs, such as rent. Many are willing, however, to trade ESSN for a decent and long-lasting employment.[83]

Another key concern regarding the Syrian refugees in Turkey relates to the limited access to work for women. Nearly half of the Syrians in Turkey are women, and approximately 34 percent of Syrian refugee households there are female-controlled. Yet, as the UN sources indicate, merely 15 percent of Syrian women have gainful employment.[84] Syrian women who wish to work face a myriad of challenges, foremost among them are (1) a lack of childcare, (2) resistance by family members who disapprove of women working outside the home, and (3) the presumed moral obligation of women to care for the elderly people or people with a disability at home.[85]

Turkey-EU Deal

The onset of Syrian mass migration to Turkey in the aftermath of the 2011 civil war in Syria led to a critical juncture in EU-Turkey relations. These relations increasingly became framed within the context of migration and asylum seeking. In 2013, Ankara and Brussels signed the so-called EU-Turkey Readmission Agreement (EUTRA). The main goal of the EUTRA was to establish a road map for the rapid and orderly readmission of the persons who failed to fulfill the conditions for legal entry to, presence in, or residence on the territory of the other side. By 2015, a dramatic increase in irregular refugees, along with growing migrant deaths at EU external borders, had created a refugee crisis of the sorts in Europe. The lack of a timely and collective response to the crisis exposed the mounting difficulties of migration control and refugee burden-sharing systems within the Europe.[86]

The refugee crisis provided a unique opportunity for Turkey to use mass migration as an effective bargaining tool in its relations with Brussels. With the refugee crisis making its way to the top of the EU-Turkey negotiation agenda, Turkey gained considerable leverage vis-à-vis Brussels, a development that represented a game changer in the way that Ankara and Brussels treated each other. At the same time, the unprecedented increase in crossings from Turkey to Greece placed mounting pressure on European countries to introduce a new policy tool to address the refugee crisis. Hence, the signing of the EU-Turkey Statement (Deal) on March 18, 2016.[87] The eventual gain for Turkey appeared to be not full membership for Turkey in the EU, but rather differentiated and more practical cooperation in specific policy areas such as migration.[88]

The 2016 Turkey-EU deal—in which Ankara agreed to accept Europe's undesirable asylum seekers in return for gaining a long-sought goal of integration with Europe, as well as an additional €6 billion for refugee assistance—typified the Turkish strategy of establishing its political and normative influence regionally.[89] The statement of cooperation between the EU and Turkey was based on four key points: (1) Turkey would take any measures necessary to stop people traveling irregularly from Turkey to the Greek islands; (2) anyone who arrived on the islands irregularly from Turkey could be returned there; (3) for every Syrian returned from the islands, EU Member States would accept one Syrian refugee who had waited inside Turkey; and (4) citizens of Turkey will gain access to Schengen visas. In return, Turkey would receive €6 billion to improve the refugees' humanitarian situation there.[90]

Opinions vary in Turkey regarding the effectiveness of this deal. It should be noted that the EU channeled its aid of €6 billion—half from the EU budget and half from member states—into two installments to be given to Turkey in two equal tranches in 2016–17 and 2018–19. The second installment was given to selective NGOs accredited by the EU that dealt with refugee affairs—not the government of Turkey. Some Turkish experts are highly critical of the EU approach, as this clearly indicated a lack of trust in the Turkish government. EU had clearly real concerns about the way the Turkish government would allocate these financial resources.[91] Others argue that Turkish president Recept Tayyib Erdoğan's proposal regarding how the second installment can/should be spent fell out of favor with the EU leaders. Turkish proposal warranted further inspection as it failed to fulfill the EU conditions and specifications.[92]

Regardless, the deal was intended to curtail the pressure on Europe's borders and prevent the influx of asylum seekers and economic migrants from crossing international borders, slowing migration to Europe as effectively as possible. Turkey now hosts more refugees than all of the EU member states combined.[93] Between March 2016 and March 2019, the 18 EU countries participating in the resettlement scheme under the EU-Turkey deal had resettled 22,292 Syrians from Turkey.[94] The deal has been widely criticized by human rights organizations and International Crisis Group, arguing that this agreement has left Syrian refugees without legal protection and remedies in Turkey since Ankara labels them "guests" rather than "refugees." Absent such legal status and protection, refugees' physical and legal insecurity is likely to generate destabilization and violence not only in the refugee camps but also throughout the country. Such policies will have serious implications in large part because refugee protection and state stability are interconnected.[95]

Despite such criticisms, Turkish and European leaders have continued to show interest in maintaining core parts of this agreement. In the face of the burgeoning tensions in spring 2020, when Turkey threatened to let hundreds of thousands of migrants into Greece, this illustrated just how much the EU countries had relied on Ankara to tackle the region's refugee crisis. European Commission president Ursula von der Leyen underlined the key point, noting that the deal "remains valid and has brought positive results."[96] While the deal has occasionally shown signs of unraveling, its core mission has persisted. The number of migrants and asylum seekers has significantly dwindled, thanks to this deal, and the fact that the EU bloc's position toward the deal has remained intact, further rationalizing Europe's strategy of managing the refugee crisis for some time to come.[97]

The Impact of the Covid-19 Pandemic on Refugees

The Covid-19 pandemic pushed millions of Syrian refugees living in Turkey deeper into poverty, many of whom felt left behind and demoralized.[98] After an abnormally long period free of the virus and with multiple neighboring countries already battling Covid-19, Turkey documented its first case on March 11, 2020.[99] In late April 2020, Turkey was on the list of worst affected countries.[100] By May 2020, Turkey had 122,392 confirmed cases and 3,258 deaths from the coronavirus.[101] Although the infections spread rapidly, Turkey was able to provide over a million coronavirus tests to its citizens and significantly limit the spread of the virus.[102] Due to ample preparation and precautionary measures, within a day of the first confirmed case, all schools were closed and able to continue online classes within a week.[103]

Before the coronavirus reached Turkey, all eighty-one provinces were prepped and prepared for the outbreak.[104] Excessive stockpiles of medical supplies, new technological advances in Covid-19 testing, and preapproved national plans for viral containment helped make Turkey a leading actor in the fight against this pandemic. Although Turkey has exported coronavirus-related supplies to thirty countries and has received medical equipment requests from eighty-eight countries,[105] certain groups within its own borders are struggling to find support.

As stores began to close and the economy faltered, workers faced decreasing job prospects. Concurrently, the Turkish government worked to financially support its citizens. In March, Turkey started the Economic Stability Shield to provide $14.93 billion for taxpayers.[106] Citizens who were laid off from the formal sector, need unemployment relief, need to postpone loan payments, or need other types of financial flexibility and assistance can receive financial support through this fund.[107]

However, approximately one-third of Turkish citizens, along with a large majority of Syrian refugees, asylum seekers, and irregular migrants, work in the informal sector and do not qualify for the program.[108] People who work in informal sectors are exempt from paying income tax and hence are not registered as employed by the country. Unfortunately, a majority of jobs within this sector cannot be plausibly performed at home, and many people have been fired as a result.[109] With no jobs or government aid available, millions of people living in Turkey are left without an income.[110]

Predominantly employed in textile and construction sectors that increase their exposure to the volatility of economic recession as well as the novel coronavirus, urban refugees lack government-provided social security networks, sick leave, or unemployment assistance, increasing the chances that the workers will continue to attend to their jobs even if they are sick. Moreover, urban refugees in Turkey have restricted access to free health care assistance limited to their province of registration. Many refugees are registered in small towns but prefer to live and work in larger cities such as Istanbul and Izmir due to the availability of informal employment opportunities. Underreporting illnesses is very common due to fear of crackdown by government authorities on irregular workers or refugees outside of their province of registration.[111] Irregular migrants lack access to the public health care system according to domestic regulations, further compounding the problems of access and equity in the country's healthcare system. While the government has suspended some of these restrictions, urban refugees and irregular migrants are still reluctant to report to hospitals for fear of deportation.[112]

Although the Turkish government has attempted to stabilize the economy, many people residing within its borders have been overlooked. For example, protective masks were mailed through the national postal service to citizens that applied, but not to unregistered individuals.[113] All people in Turkey had to wear face masks in public places and markets,[114] but many unregistered refugees lacked the supplies and the money to adhere to the rules without help. There was—and still is—a large discrepancy between the type of people who were receiving support from the Turkish government and those having to make do without such support.[115]

While Turkey instigated limited lockdowns over the weekends and holidays for the working populations, both seniors and children had stricter shelter-in-place requirements. These regulations led to dense and uncomfortable living accommodations for extended periods of time, leading to higher rates of domestic violence and subpar care for the elderly and students. To make matters worse, schools in Turkey were closed for more than two months.[116] Education Minister Ziya Selçuk released televised and internet classes that were available to the general public.[117] Many refugees in Turkey, however, live in "crowded and often particularly squalid conditions."[118] Urban refugees in Turkey live in shantytowns or apartments at the outer edges of major cities and rely on public transportation to reach their place of employment during the weekdays, increasing their exposure to the virus. Many asylum seekers, refugees, irregular migrants, and Turkish residents living below the poverty line have been struggling to buy food, let alone have access to devices that can stream online lectures. With such a sharp economic decline, unstable employment, and poor living conditions, many families were unable to provide their children with educational opportunities for months to come.

The Russian Invasion of Ukraine

The war in Ukraine has further complicated Turkey's foreign policy approach. President Recep Tayyip Erdoğan initially adopted a seemingly neutral position toward this issue, carefully balancing the country's support for Ukraine and the Western alliance without antagonizing Russia. This position, however, failed to shield Turkey from the conflict's fallout. Indeed, rising fossil fuel prices have led to a higher energy import bill for a country where 99 percent of its gas is imported from Russia, Azerbaijan, and Iran. Similarly, about 93 percent of its oil is imported from Iran, Iraq, and Russia.[119]

Initially, Turkey condemned Russia's invasion and subsequently blocked Russian warships from passing through the Bosphorus and Dardanelles Straits and into the Black Sea. Turkey also restrained Russian military flights—but not its commercial flights—across its territory. In the meantime, Ankara provided advanced drones to Ukraine. Turkey, which is not a member of the EU, has honored EU and US economic sanctions on Russia. Turkey's central banks have suspended the country's use of the Russian Mir payment system—a card payment system for electronic fund transfers established by Russia's Central Bank adopted since 2017. Turkey has also facilitated the shipment of natural gas from the Republic of Azerbaijan to Bulgaria in order to compensate for the absence of Russian supply, which became Bulgaria's only source.[120]

The war in Ukraine has presented Ankara with both opportunities and risks. The latter involve its relations with the United States and the EU regarding supporting sanctions against Russia. At the same time, Turkey can seize an opportunity to achieve a long-sought objective of becoming a major gas transit hub for Europe. As Europeans appear more determined than ever to curtail their energy dependence on Russia, one analyst suggests, Turkey can take advantage of this opportunity to advance its long-perceived, new subsea pipeline between Israel and Turkey, allowing eastern Mediterranean gas to reach Europe through Turkey.[121]

Conclusion

Much of Turkey's vulnerability to climate change stems from its heavy reliance on fossil fuels. Turkey's carbon emissions have increased twenty-four-fold over the past sixty years, while its emissions per capita have increased eight-fold during the same period, currently far exceeding the global average.[122] From wildfires across southern coastal regions to deadly floods in the northeast, to the warming sea temperatures in the Mediterranean and other parts of the country that dried up dams, Turkey faces a myriad of challenges ranging from population growth, rural-to-urban migration, to water mismanagement that have had grave consequences for farmers' lives and farming business more broadly.

To prevent the negative effects of climate change on migration, Turkey must seek internal as well as regional cooperation and coordination now and in the future. The Turkish case of hydropower and irrigation development demonstrates that Turkey needs to forge a broader consensus among its major stakeholders, including farmers, the government, the private sector, civil society, and academia if it intends to bring about a positive impact on equity and efficiency in the management of its climate change policy. Turkey's recent water development policies have been directed toward maximizing the country's hydropower potential by developing more small-scale hydropower projects. These projects, however, have had negative sociopolitical consequences. Local people living near the construction sites have criticized these projects because they have not been involved meaningfully in their planning, construction, or implementation.[123]

In addition to these local and national challenges, concerted efforts are still needed to turn those efforts into durable, broader attempts to tackle problems related to the transboundary nature of management and protection, such as water shortage in the region. Water pollution control and safeguarding of water-based ecosystems, while increasingly viewed as significant, have yet to reach satisfactory levels.[124] There is a broad consensus among Turkish scholars that the key challenge facing the country is a lack of financial assistance given the increasing burden and damage of climate change.[125]

Regarding migration and refugees, it should be noted that in addition to adopting a human-oriented plan to deal with these asylum seekers and "guests," Turkey must protect its national sovereignty and the rule of law in a way that comports with human rights, especially children's rights and gender equality, and sustainable development. Undoubtedly, Turkey risks combating multiple ongoing natural hazards and disasters if it fails to effectively manage or curb the effects of climate change.

Turkey is also caught between the refugee problem stemming from conflict-prone crisis regions such as Syria and Afghanistan on the one hand and climate refuges—that is, IDPs—on the other. The Syrian-Turkey corridor is among the largest, with the entry of nearly 3.7 million Syrian refugees into Turkey in 2019.[126] The Syrian refugees, who make up the vast majority of over 4 million refugees and asylum seekers, currently face multiple problems, including gainful employment and fair and equal treatment before the law. Moreover, the sheer size of the refugee influx has led to mounting economic difficulties and surging anti-refugee sentiments throughout the country.

It should be noted, however, that there are many different reasons underlying these migratory trends. Immigrants coming to Turkey may come not only because of the regional conflict and violence or due to climate change, but also because of historical or cultural ties. Many immigrants from Central Asia, for example, come to Turkey to find jobs. Additionally, because Turkey is a transit country between sending and receiving countries, some immigrants remain in Turkey when unable to move toward their target destinations.[127]

In 2021, Afghanistan was the country with the highest number of irregular migrants living in Turkey, in large part because drought affected 80 percent of Afghanistan in that year. Drought was not the only reason for Afghan migration; the return of the Taliban to Afghanistan was also a major cause. Given the growing impacts of climate change—tropical storms, floods, torrential rains, and droughts in the region and beyond, Turkey should expect to see more immigrants from South Asia, the Balkans, the Black Sea, and the Middle East.[128] As explained in the preceding sections, and according to Turkish laws, persons coming from outside European countries can be legally defined as climate migrants, climate asylum seekers, or even displaced persons but not necessarily climate refugees.[129]

As noted in this chapter, many of the nearly 1 million Syrian refugees of working age are employed in the informal sector, a development that leaves Syrians in perilous socioeconomic conditions while fueling public resentment driven by dropping wages and spiraling unemployment. As a result, today's Turkish economy is much weaker than the initial phase of Syrian refugees' arrival. Turkish GDP per capita, according to the World Bank, dropped from its pinnacle of $12,614 in 2013 to $9,126 in 2019.[130] One suggestion to improve the Turkey-EU deal was for the EU to grant trade liberalization and concessions to Turkey, enabling Ankara to boost its agricultural exports to the EU. Such a deal would lead to the formal employment of Syrian refugees, which is consistent with ILO and EU labor standards.[131]

CHAPTER 6
EGYPT

THE GROWING RELEVANCE OF CLIMATE CHANGE POLITICS

Located in northeastern Africa, Egypt borders the Mediterranean Sea, with Sudan to the south; the Red Sea, Gaza Strip, and Israel to the east; and Libya to the west. The Mediterranean coastline is most exposed to sea level rise owing to its relative low elevation compared to the land around it.[1] The dominant feature of the northern coastal zone is the low-lying delta of the Nile River, with its large cities, industry, flourishing agriculture, and tourism. The delta and the narrow valley of the Nile comprise 5.5 percent of the area of Egypt, but over 95 percent of its agriculture is concentrated in the Nile region and many Egyptians live near the banks of the Nile River. Egypt's Mediterranean coast and the Nile Delta have been identified as vulnerable to seal level rise.[2] The Nile waters are highly sensitive to impacts of climate change, exacerbated by declining rainfall and rising temperatures.

The country's climate is semidesert, typified by dry and hot summers, moderate winters, and scarce annual precipitation. Most of the country receives little rainfall, as there is no significant rainy season. Rainfall occurs irregularly. The northernmost part of the country—that is, coastal zones—receives most rainfall, approximately 200 millimeters per year.[3] Since 1960, daily minimum temperatures have risen throughout the country, with a diminution in cool nights and an escalation in warm nights.[4] Over the past thirty years, Egypt has seen significant reduction of annual total rainfall. This has led to reduced water availability in some parts of the country and increased periods of drought and dryness. The rise in the frequency and intensity of extreme weather events, including heat waves, sand and dust storms, rockslides, and flash flooding, in recent years has come to characterize Egypt climate.[5]

Egypt has only one main source of water supply, the Nile River, which supplies over 95 percent of the water needs of the country.[6] As a result, agricultural development depends heavily on the Nile and its management. The Nile waters, however, originate outside Egypt, flowing through nine countries. Egypt has established its claim on the use of the Nile waters based on the Anglo-Egyptian treaties signed in 1929 and subsequently in 1959—agreements that have dramatically increased Egypt's and Sudan's shares to the detriment of that of Ethiopia. As the water needs of all the riparian countries along the Nile River grow rapidly, political tensions between and among these countries tend to increase. The increase in water consumption due to the high population growth rates has had profound impact on the country, rendering water management an imperative adaptation action.

Another key concern confronting Egypt has been land erosion both on the coastlines and along the Nile. This has begun to impinge on the limited amount of the area's agricultural land, enhancing Egypt's potential vulnerability and coinciding with significantly more pressure on the country's limited water resources. This has increased the need for further water purification plants and greater deployment of desalination technology, which is likely to result in more CO_2 emissions. The spike in temperature has contributed to the higher number of cases of malaria and some other diseases.[7]

Climate change researchers have long underlined the impact of climatic variation on food production and distribution systems as mediated through world trade. Climate change, they have argued, is likely to have direct effects on crop production and can also cause adverse changes in markets, food prices, and supply chain infrastructure. Given these realities, food-importing countries, such as Egypt and the Philippines, are most vulnerable. Similarly, a worldwide shift to biofuels would worsen the situation, crowding out at least some food production globally.[8] It is in this context climate change could compromise a sovereign state security. Climate change will likely render some "old" resources less available, possibly increasing the tensions between Egypt and Ethiopia over water security in the coming years.[9]

This chapter explores climate change within the context of the dispute over the Nile River. It begins with the issue of how scarce supplies of water and agricultural land affect food security in Egypt, and next examines climate governance and the kinds of challenges that it presents to forced migration inside the country. Egypt's vision to launch a national strategy for climate change by 2050 receives special attention in this context. The chapter concludes with some reflections on the dynamics behind social mobilization around environmental issues as well as state responses to environmental activism.

Socioeconomic and Environmental Problems

Domestically, as one study demonstrates, given that the agriculture sector is the largest consumer of water in Egypt, it is vitally important to assess the impact of climate change and temperature increase on water demands for irrigation. Clearly, the surge in water demands owing to climate change will affect water security in Egypt, as the available water resources are likely to be limited.[10] It is estimated that by 2050, climate change will increase water demand by an average of 5 percent.[11] Egypt is among the countries that will certainly face food crisis during the coming forty years. The burden of Egypt's increasing water crisis will present a colossal challenge for its government and population.[12]

Population growth presents yet another challenge that requires proper management of water resources. Egypt's large population has made the country extremely vulnerable to climate change. Moreover, its densely populated Nile Delta is seriously threatened by sea level rise.[13] As Africa's most populous country, Egypt has a population of nearly 115 million people (2024) and is projected to reach 120.8 million people by 2030 and 159.9 million by 2050. An estimated 43 percent of the current population resides in urban areas, which is expected to reach 56 percent in 2050.[14]

Egypt's environmental problems include, but are not limited to, water scarcity, air pollution, damage to historic monuments, animal welfare issues, and deficiencies in its waste management system. In light of the anticipated climate change risks, more attention should be given to water-saving methods to grow crops in the Nile Delta and Valley. Cotton, rice, clover, and sugarcane are all major crops of Egypt. Farmers also grow different kinds of beans to eat, as well as citrus fruits like oranges, tomatoes, and potatoes. The crops are easily irrigated if needed, thanks to the plentiful water supply of the Nile River. Egypt is highly vulnerable to water scarcity, droughts, rising sea levels, and other adverse impacts of climate change.

Without adaptation, agriculture, tourism, and coastal communities will be at great risk. Climate change effects can have various impacts on the agricultural sector and stability of food security in Egypt. Crop production will be negatively affected due to the expected increases in temperature, extreme weather events, drought, plant diseases, and pests. "As the agriculture sector consumes the largest amount of water resources," as one study finds, "this sector must apply all agricultural practices that will rationalize water and raise the efficiency of irrigation at the field level; this will maximize the utilization, maintenance, and sustainability of every drop of water."[15]

The crop reduction has had significant economic repercussions. With many farmers forced out of work, Egypt's agricultural sector's employment rates plunged from 44 percent to 27 percent from 1991 to 2019. Not only did this impair the livelihood of individual families, but it also debilitated the country's $28 billion agricultural sector.[16] The question is: What is Egypt doing to be sustainable? Egypt increased its capacity to produce electricity by at least 15 gigawatt of electricity using very advanced, sustainable, and efficient technologies. Indeed, the energy strategy aims at increasing the share of the renewable energy to reach 42 percent by 2035.[17]

The Dispute over the Nile River

The Nile Delta shoreline stretches from Alexandria in the west to Port Said in the east, approximately 285 kilometers long. The Nile Delta region in the Mediterranean coastal zone embraces the country's major industrial, agricultural, and economic resources. The region is known to have a relatively low land elevation, which renders it acutely susceptible to rising sea levels. In addition, it suffers from local drainage issues and sudden sinking of soil, magnifying the effects of rising seas. This area is subject to shoreline changes stemming from erosion, land subsidence, and rising sea levels resulting from climate change.[18]

Egypt is enormously vulnerable to climate change due in large part to its paramount dependence on the Nile River, which is crucial to potable water, agriculture, industry, fish cultivation, power generation, inland water transport, mining, oil and gas exploration, cooling of machinery, and power generation. The dependency upon the Nile River's water renders the country vulnerable to soaring temperatures, declining rainfall for both the upper Nile Basins and the rest of the east Mediterranean coastal zone.[19] Egypt has submitted its NDC and Third National Communication (NC3) to the UNFCCC in 2016, demonstrating its efforts to adapt to climate change, while focusing on the sustainability of the environment, water resources, energy, agriculture, and health.[20]

The Nile River has been at the center of such disputes between Egypt, Sudan, Uganda, and Ethiopia. The river runs from Uganda and Ethiopia through Sudan into Egypt. The lives of more than half of Africa's total population—that is, nearly 600 million people—are affected by the Nile Basin. Caught in an endless internal political strife and mounting pressures to feed a growing population, many Africans residing in the region have been traumatized by these conflicts.

Even though the Nile's origin lies outside Egypt's borders, the latter has controlled the lion's share of its waters. This exclusive ownership of the Nile waters has been granted to the Egyptians based on a 1929 treaty between Egypt and Britain's East African colonies: Burundi, Kenya, Rwanda, Tanzania, and Uganda—colonies that gained independence from Britain in the 1960s. The treaty awarded 57 percent of the waters to Egypt while still giving Cairo an upper hand vis-à-vis other neighboring countries in terms of launching any major water project on the river. A subsequent treaty, signed in 1959 between Egypt and Sudan, raised Egypt's share to 66 percent. The two signatories distributed virtually all the Nile waters, without even consulting Ethiopia. After the 1959 accord, both Egypt and Sudan built mega-dams to exploit the water for irrigation.[21]

If unresolved, the conflicts between down- and upstream countries will negatively affect Egypt's share of the Nile's flow, especially considering the fact that upstream countries face growing populations and development agendas. Ethiopia and Sudan's populations are increasing rapidly yet Egypt continues to control most of the water in the region. Over the years, disagreements have deepened and arisen over the Nile Waters Agreements. Meanwhile, various Egyptian leaders have threatened to go to war to preserve these agreements. Upstream riparian states such as Kenya, Tanzania, Uganda, and Ethiopia have questioned the legitimacy and legality of these agreements due to the fact that they were never parties to them. In fact, shortly after independence from Great Britain in 1961, Tanzania's new leader, Julius Nyerere, argued that the Nile Waters Agreements ran counter to the sovereignty of his country and other upstream riparian states. In fact, he asked for a new, more inclusive legal framework for governing the Nile River Basin.[22]

The 2015 agreement between Egypt, Ethiopia, and Sudan—with Sudan acting as an intermediary—represented a new reality based on the realization that the colonial agreement was no longer sustainable. Given that approximately 85 percent of the water that flows into the Nile River comes from the Ethiopian highlands through the Blue Nile, it was unrealistic to prevent Ethiopia from using water resources located within its boundaries.[23] As Africa becomes more and more vulnerable to climate change, these riparian states must seek a diplomatic solution to the water issue that could be sustainable in the coming years. The region is prone to prolonged drought, with a humanitarian crisis likely to ensue soon. All three countries—Egypt, Sudan, and Ethiopia—need to focus on finding a sustainable solution for the region.

As noted previously, the agricultural activities in Egypt consume approximately 80–85 percent of the water supply of Egypt. Groundwater contributes a negligible amount to the country's agriculture. Crop production will most likely be negatively affected by the effects of climate change, including increases in temperature, extreme weather events, drought, sea level rise, deluge, and declining precipitation. Climate change will have negative impacts on food security to the extent that it affects water availability, soil erosion, and crop productivity.[24] Climate change adaptation strategies in the Nile Delta region should include altering cropping patterns, growing salt-tolerant land crops in areas near the Mediterranean Sea, and developing policies for coastal protection.[25]

Agriculture and Food Security

Drought and water scarcity have a substantial adverse impact on crop yields and hence the agriculture sector more broadly. The risk of food insecurity will dramatically increase, as the issues of water shortage and provision of food are inextricably intertwined. There are two cultivation systems in Egypt: irrigated agriculture and rain-fed agriculture. The largest irrigation area is in the Nile Valley and Delta. The rain-fed areas in the North Coast of Egypt depend on rainfalls in the eastern Mediterranean from El-Saloum (west of Libya) to Rafah (East of Palestine).[26] It should be noted that few field crops and fruit trees are cultivated in the rain-fed areas, as farming in these areas uses old, traditional, and primitive soil and crop management practices.[27]

Food security has become a national concern in Egypt given the rapid growth of population, a changing climate, and the anticipated water scarcity. The climate change effects, along with the construction of Grand Ethiopian Renaissance Dam (GERD) in 2021, will affect the nature of food security in Egypt. The main purpose of the dam, Ethiopian authorities insist, is electricity production to alleviate Ethiopia's pressing energy shortage and for electricity export to downstream countries of Sudan and

Egypt.[28] The GERD could adversely affect the flow of the annual share of Nile water, disrupting the Nile's sediment transport and impeding the natural process of silt soil in Egypt. Consequently, the water in the Nile Delta is likely to lack the nutrient elements for agricultural productivity.[29]

In Egypt, most croplands depend largely on irrigation water from the Nile. Very few regions, such as Mersa Matrouh, are characterized as maintaining rain-fed agriculture. The Nile Delta region is known for its high-production irrigated smallholder agriculture, rising urban water demand, and rapidly growing population. Water for fertilization depends wholly on the Nile River and varies due to changes in freshwater availability and to regional rivalries among water users.[30] The adaptation measures include, among others, changing cropping patterns, utilizing new drought-tolerant plants, reducing the farming of high-water consuming crops, altering crop rotation method, changing irrigation practices, and improving the efficacy of water use.[31]

Numerous studies have suggested that Egypt would be a food importing country in 2050. Approximately 2 percent of current Egyptian croplands are projected to be more suitable by 2100. By 2030, more than one-third of current Egyptian croplands are projected to become less suitable for cultivation because of climate change.[32] A receding Nile would jeopardize Egypt's food security. To tackle water scarcity, the Egyptian government reduced the production of many water-intensive crops. Rice water consumption is about 2–3 times higher than what is required for producing other cereals, such as maize or wheat.[33] Producing 1 kilogram of rice requires around 2,672 liters of water. Rice production has negative impacts on the environment given that its growth demands heavy pesticide usage. Furthermore, when rice field is flooded, it undergoes complex microbial processes, resulting in the development and release of massive methane into the atmosphere.[34]

The total area of rice cultivation, according to one study, was reduced by 31 percent in January 2018.[35] Additionally, the water contamination of the Nile will have grave implications for the fish industry. Most importantly, the Nile's pollution has prevented water from being physically distributed to surrounding farms and cities. The pollution has cost Egypt 15 billion cubic meters of water annually. Pollutants have also contaminated marine life and agricultural lands that reach the Nile's waters. Environmental experts have regarded half of the Nile's fish as ill-suited for human consumption due to infected water. Similarly, much of Egypt's homegrown produce is infected with sewage water and chemical residue. Health conditions, including neurological and reproductive abnormalities or terminal diseases such as cancer, dementia; advanced lung, heart, kidney, and liver diseases, have been attributed to these pollutants.[36]

It is estimated that the country's population will reach 140 million inhabitants by 2050, and Egypt's per capita share of water will dramatically reduce by half, even excluding the effects of climate change. Agricultural production should increase by 70 percent by 2050 to keep pace with this rapid population growth, while also reducing agricultural vulnerability to climate change.[37] Agriculture policies, according to one study, should aim at increasing food production, especially the essential crops like wheat, maize, sugar, lentils and broad bean, oilseed, and meat and dairy products. Moreover, the Egyptian government should make serious efforts to find alternative sources of water to harness water shortage, which may include desalination of seawater, treated sewage and industry water, and recovered agricultural drainage water, and also develop and revamp the entire agricultural system.[38]

The same study indicates that one in five Egyptians is currently experiencing food insecurity due to structural flaws in the food supply and distribution system. The limited water resources and arable land are key contributing factor to food insecurity. Egypt imports about three-fifths of its basic food, which costs more than $10 billion annually. Egypt was among the five biggest buyers of wheat, namely, Indonesia, Nigeria, mainland China, Turkey, and Egypt, purchasing $2.47 billion worth of wheat in 2021.[39] Egypt imports 32 percent of its sugar at a rate of 1–1.2 million tons/year, about 70 percent of broad fava bean, and the whole of lentil and edible oil.[40]

Climate Change Impact

Egypt is particularly vulnerable to the impacts of climate variability and change, especially with respect to water security, agriculture, and livestock, as well as increasingly adverse conditions to health, human settlements, and energy demand and supply. Projected climate change effects on food security and agricultural livelihood are inextricably intertwined with the future projected water supply concerns and temperature rise. The latter renders Egypt's agriculture sector vulnerable, as the rise in temperatures adversely affect irrigated yields across the country, leading to a reduction in the productivity of staple foods (wheat, maize, rice, and tomatoes), sugarcane, and milk.[41]

Average temperatures are expected to rise in the coming years, reaching more than 2 degrees Celsius higher than present day by the end of the century. Extreme temperatures will continue to increase in Egypt over the coming decades and century. By the end of the twenty-first century, the hottest day per year is likely to be 3–8 degrees Celsius warmer that it is today, indicating a persistent climate change.[42]

On a national level, sea level rise could have serious implications for food security and economic growth. The Nile Delta in Egypt, which is a breadbasket for all neighboring countries, serves as such example. Egypt's Nile Delta is one of the most densely populated areas of the world and is extremely vulnerable to sea level rise. A rise of just 1 meter would displace atleast 6 million people and flood 4,500 square kilometers of farmland.[43] Air pollution has been a problem for Egyptians for decades, particularly in large cities such as Cairo. In Greater Cairo, levels of the fine particulate matter PM10 and PM2.5, which pose the greatest risk to people's health, are several times higher than the levels recommended by the WHO.[44]

In addition to potential damages to the ecosystem in general, climate change is expected to have a significant negative impact on tourism since it requires suitable weather conditions and a clean environment.[45] Droughts and sea level rise will certainly undermine the Egyptian tourism industry, which is one of the country's key sectors that generate major foreign currency reserves. Nearly 20 percent of Egypt's foreign currency earnings stem from tourism and 12.6 percent of the workforce depends upon this industry.[46]

Another consequence of climate change is energy scarcity and rising energy prices that increase poverty, strain national budgets, and jeopardize Egypt's competitiveness for the future. Aware of the fact that global warming results in sea level rise due to the melting of glaciers and Arctic ice, Egypt submitted its NDC and NC3 to the UNFCCC in 2016, in support of its efforts to realize its development and economic goals and increase its adaptive capacity to climate change. The country's vulnerability to the impacts of climate variability, particularly with respect to energy demand and supply, has intensely captured the attention of its leaders. Egypt's NDC is consistent with the country's overall goals of reducing vulnerability and poverty and achieving long-term sustainable, economic development. Key areas of focus include the sustainability of the environment, water resources, energy, sustainable land management, agriculture, and health.[47]

Climate Governance

Opinions vary on climate governance in Egypt. Some experts hold that the Egyptian sustainable agricultural strategy has adopted several successful measures to adapt to climate change and increase water use efficiency. These include

> the reduction of area cultivated with rice; improving farming systems for animal husbandry; introducing the cropping pattern with improved agricultural extension service; improving on-farm water management in the old lands; concentrating on the agro-industrial complexes in the new land reclamation projects; improving the capacity building for human resources working in agricultural sector; and increasing the public awareness on climate change through integrated programs including private and public sectors.[48]

Others point out that Egypt has shown a deep commitment to the climate agenda as it has finalized the National Climate Change Strategy to support its 2030 SDG development agenda. Most relevantly, Egypt hosted COP27—the 27th Session of the United Nations Climate Change Conference of the Parties. Egypt has pioneered the first sovereign green bond in the MENA region (worth $750 million) by tapping investors interested in financial and environmental programs. The World Bank has supported a $200 million project aimed at cleaning up the air that has improved traffic. The project seeks to reduce emissions, including GHG emission. The project plans to introduce electric buses in the public transport fleet of Cairo, while also supporting the monitoring of climate pollutants and developing an integrated climate and air quality management plan.[49]

Still others underscore the importance of Egypt's long-term plan to deal with climate change. Egyptian minister of environment Yasmine Fouad stated on August 11, 2021, that Egypt was prepared to combat climate change because it was one of the most affected countries in the region. In December 2020, Egypt's Supreme Committee of the National Council for Climate Change held its initial meeting led by Egyptian prime minister Mostafa Madbouly, who asked the council to prepare a general strategy to confront climate change. On August 1, 2021, in the face of a heat wave that swept through Egypt, Minister Fouad laid down the goals of the national strategy for climate change: "The plan ensures the achievement of low emissions development in various sectors, the increase of renewable and alternative energy sources, and preservation of natural resources and green spaces."[50]

Relying on hydropower has also gained much traction in Egypt in recent years. Part of the explanation for why hydropower has received significant attention lies in the fact that it can be adequately stored, is less vulnerable to climate variability, and is more reliable than other renewable energy sources. In recent years, hydropower has become the third largest energy source of Egypt.[51] The country's vast majority of the hydropower generation stems from the Aswan High Dam and Aswan Reservoir Dams along the Nile River. Increasingly, however, Egypt's New and Renewable Energy Authority has shifted its focus from hydropower to seeking other renewable energy alternatives (REA) such as solar and wind power as the Nile's hydropower capacity has already been largely utilized.[52]

Critics, however, point out that Egypt has failed to effectively utilize its fossil fuel resources to develop alternative energy supplies in sync with the country's alleged broader goal of sustainable development. State-owned energy firms control the production of fossil fuels, relying on partnerships with multinational oil companies to provide know-how and competitive technologies. Energy exports constitute an important source of hard currency for the state. The Egyptian state has attempted restructuring fuel subsidies on several occasions only to retreat from implementing such reforms.[53]

Egypt's rising population, increased carbon emissions, and greater reliance on imported foodstuffs, as other critics note, pose threats both to the country's and the region's well-being. Egypt's failure to square the energy-environment-growth triangle poses grave threats to itself and to others. Egypt, like other countries in the MENA region, has long suffered from the Dutch Disease affliction (the so-called oil curse) of an overvalued currency, driven not only by the impact of hydrocarbon exports but also by a large and relatively overpaid and bloated civil bureaucracy, and by prevailing wages in the region's labor-importing countries, most notably in the Arab Gulf countries and Libya.[54] One of the grave consequences of Egypt's Dutch Disease has been due to the sluggish growth of manufacturing industry, the only sector of the economy capable of absorbing large numbers of newcomers into the labor market.[55]

The oil revenue was largely and disproportionately used in several respects, including to beef up the oversized state and its sprawling civilian and military bureaucracies; to provide food and energy subsidies to most of the population; to be channeled into illegal capital flight, amounting to nearly 2 percent of GDP between 2003 and 2011, used primarily to cultivate crony capitalists and their patronage. Allocated to regime survival and stability, hydrocarbon rents reinforced the grip of the Dutch Disease over Egypt, as was typical of the MENA region economies.[56]

Meanwhile, the el-Sisi government's approach of expanding energy production and consumption has been costly for an ailing economy and a population, 50 percent of whom now live on less than $2 daily. Much of the population cannot afford electricity and fossil fuels at current market prices. The military's command economy, along with a submissive civilian community that it has supported, will most likely prioritize economic growth over the environment. Regardless of the outcome, Egypt will contribute to global carbon emissions while failing to take sufficient advantage of its potential to wind and solar energy.[57] It is important to note, as one study demonstrates, different parts of Egypt are expected to experience diverse trends of changes in the annual wind speed. Similarly, the northern parts of Egypt, as compared to upper Egypt, are expected to have the highest solar power potential due to the mutual effect of both solar radiation values and relatively low temperatures increase.[58]

Environmental Change and Migration

Some analysts have denied a direct link between climate change and migration, arguing that migration from environmental degradation has been relatively confined in Egypt due to extreme population density and the fact that urban areas are already strained by a growing youth bulge. Technological and market solutions can arguably prevent migration. Egypt, however, has engaged in neither; yet migration has remained minimal. Despite the dispute over the Nile River with Ethiopia, regional tensions have been generally unrelated to environmental issues.[59] Thus, in Egypt, it is concluded that due to limited places to which migrate internally, migration cannot be an effective adaptation technique.[60]

Other researchers point to a surging trend that links environmental problems to forced migration. In Egypt, they argue, a combination of the slow-onset hazards of desertification and sea level rise is likely to contribute to migration. Sea level rise of 1 meter would displace up to 10 percent of the population of Egypt due to flooding in the Nile Delta. Similarly, if the sea level rises by half a meter, 67 percent of residents, 65.9 percent of the industrial sector, and 75.9 percent of the service sector will submerge.[61]

Some Egyptian researchers have warned that if no action is taken to combat climate change, seawater will rise by 50 centimeters by 2050 and displace approximately 1.5 million people in the governorate of Alexandria alone. The report has also predicted that surging temperatures and water scarcity caused by rising demand from the farming industry will escalate the pressure on agriculture and minimize crop yields. As these areas become barren because crops can no longer endure the climate, farmers will most likely migrate to marginal lands. Farming in those lands—whose soils contain high salts and lack rich organic matter—further expedites their environmental degradation.[62]

In Egypt, according to one study, environmental degradation caused by air and water pollution has had negative economic and health consequences for society, resulting in soil degradation. A World Bank report, using 2016–17 data, estimated that air pollution in Greater Cairo alone has likely cost Egypt an estimated 47 billion Egyptian pounds in 2017, which has been equivalent to 1.35 percent of the country's then GDP.[63]

Desertification and soil degradation have affected large areas of land on the Eastern and Western Nile Delta, driving some people to migrate internally in search of subsistence. In fact, large swaths of land are likely made inaccessible by the dual climate change-related forces of desertification and sea level rise. Future increases in sea levels will intensify the concentration of a rapidly increasing population into certain areas. In the future, according to one study, sea level rise could affect an additional 16 percent of the population.[64]

Those who were resettled in the Eastern Delta were largely unemployed young men from urban ghettos. The people who moved to the Western Delta, by contrast, were mainly farmers affected by a law that favored landowners who could push away sharecroppers from profitable agricultural areas. Both the Western and Eastern Deltas lack access to potable water, proper infrastructure, public and educational facilities, health care, and operative sewage systems. Many migrants have left these areas, either going to other regions or returning to their original regions.[65]

Absent appropriate measures, notes Egypt's former irrigation minister, Mohamed Abdul Ati, the climate change effects on the country's northern coast are likely to lead to the migration of at least 5 million people from the Nile Delta. Climate change studies, according to Abdul Ati, predict a reduction in the productivity of two key crops—wheat and maize—by 15 percent and 19 percent, respectively, by 2050. Also, approximately 15 percent of the most arable land in the Nile Delta is anticipated to be salty.[66] While these factors significantly contribute to human movement, other variables such as poverty, high population growth and density, and the lack of economic development have worsened the situation even further.[67] Consequently, internal migration has been a constant element in the lives of many Egyptian farmers and fishermen who seek livelihoods against such a background.[68]

The city of Alexandria has been highly vulnerable to sea level rise, leaving its communities at risk of storm surges and flash floods. One preference of vulnerable people in Alexandria, according to one study, is preventive resettlement in response to sea level rise. In cases where displacement and environmentally related migration offer complicated adaptation strategies due to the impact of climate change, resettlement of the exposed population may provide the only viable strategy. Preventive resettlement, this study demonstrates, is an appropriate adaptation pathway in which some or all parts of a community are relocated due to the fact that these communities face the risk of an impending disaster, and mitigating all the risk factors appears impossible.[69] If no plans are put in place to adapt to a sea level rise of 0.5 meter, about 30 percent of the Alexandria will be exposed to deluge, over 1.5 million people will have to be relocated and approximately 195,000 jobs will be lost, and $3.5 billion economic loss has to be endured—not to mention that water and soil salinity and coastline erosion will jeopardize the livelihoods of local fishermen.[70]

Displacement and Migration

Over the past century, Egypt has seen waves of migrants, refugees, and displaced people. During the first half of the twentieth century, especially during both world wars, Egypt hosted waves of European refugees. The early waves were the Armenian refugees who escaped the Turkish massacres in 1915. This was followed by a larger wave (1941–5) with the Greeks, Poles, and Yugoslav refugees who reached Egyptian shore because of the Nazi occupation of many European countries.[71]

Throughout the twentieth century, Egypt saw several waves of internal displacement: the Suez War—Fleeing Port Said (1956), the Nubian resettlement (1964), and the aftermath of the Six-Day War (1967). The Second Gulf War (1990), following the Iraqi invasion of Kuwait, resulted in a wave of Kuwaitis coming to Egypt. The US invasion of Iraq and its occupation (2003–7) caused many Iraqis to seek refuge in Egypt. Roughly 150,000 Iraqis resettled in Egypt.[72]

With the advent of the 2011 Arab Spring uprisings and the ensuing civil war in Syria, many Syrian civilians started to flee the conflict zones and sought refuge in neighboring countries and Europe. Egypt has become one of the main destinations for the Syrian migrants and refugees.[73] Egypt hosts more than 270,000 registered asylum seekers and refugees from sixty-five countries.[74] The majority of these displaced people are from Syria, Sudan, South Sudan, Eritrea, Ethiopia, Libya, Yemen, and Somalia. The ongoing conflicts and political instability in East Africa, and the Horn of Africa, as well as the turmoil in Iraq and Yemen, have compelled thousands of Sudanese, South Sudanese, Ethiopian, Iraqi, and Yemeni individuals to seek refuge in Egypt. As of June 30, 2021, according to UNHCR, refugees registered in that organization comprised 133,568 Syrians, 50,665 Sudanese, 20,240 South Sudanese, 20,174 Eritreans, 15,671 Ethiopians, 9,404 Yemenis, 6,805 Iraqis. 6,771 Somalis, and more than fifty other nationalities.[75]

Most refugees and asylum seekers, according to UNHCR, live in urban areas in Egypt and are largely concentrated in Greater Cairo, Alexandria, Damietta, and several towns on the North Coast. With the rise of sectarian strife in the immediate aftermath of the 2011 uprisings, Egypt saw an increasing desire for migration among Christian Egyptians to regular destinations in Europe and North America, in addition to Georgia and some Eastern European countries. One expert even suggested that some 75 percent of Christians considered migrating from Egypt.[76]

Many Libyan asylum seekers wound up in the Salloum camp, which is located between the Egyptian border post of Salloum and the eastern Libyan frontier. This refugee camp hosts hundreds of migrants who have fled the civil war in Libya yet are unable to legally enter Egypt. Shortly after the violence in Libya, approximately 800 migrants, mostly of African origin, who fled civil war and discrimination in Libya, remained in limbo in Salloum.[77] If these asylum seekers lack the proper papers to enter Egypt, they will remain at the border. International organizations' generous care for these individuals notwithstanding, this care is unlikely to translate into a dignified form of active participation in everyday life activities where the asylum seekers will be afforded a choice, and they feel a sense of belonging and empowerment.[78]

In recent years, however, the country's growing difficult economic conditions have noticeably increased the vulnerability of refugees and nationals alike. Lacking a steady source of income, coupled with rampant inflation, these refugees' basic needs are hardly ever covered. Other challenges include limited means of support and subsistence and the language barrier facing non-Arabic-speaking refugees. Some also lack access to basic education that could support their livelihoods and development. A great number of refugees and asylum seekers must rely on humanitarian assistance to meet their basic needs and to provide medical or mental support.[79]

Unlike South Africa and Israel, where active media and civil society organizations (CSOs) tend to publicize and criticize the incidence of anti-migrant violence, in Egypt, despite the fact that anti-migrant xenophobic attitudes and behaviors toward refugees are widespread, there has been far less civil society opposition expressed in this way.[80]

Environmental Activism

Labor activism, including wildcat strikes, protests, and attempts to create independent unions, soared significantly in the second half of the 1990s. With the declining public services and safety nets, private voluntary organizations emerged to fill the gap. The growth in the number of Islamic and civil society organizations has also indicated emerging neoliberal economic changes in the regional political economy, as ideas, labor, migrants, and remittances flowed to and from the Persian Gulf to the rest of the MENA region. Population growth, especially among those aged 15–29, surpassed job opportunities, causing high unemployment rates for women, youth, and college students. It was from these educated cohorts that networks of activists emerged.[81]

In 2008, the Egyptian port city of Damietta saw rising environmental protests against several fertilizer plans by the Canadian firm Agrium near the popular middle-class tourist resort area of Ras al-Barr. Thousands of protesters regularly took to the streets, closing workplaces and bringing transport to a halt. They hung black banners with slogans in Arabic urging: "For the sake of our children's health, no to the project," "No to pollution," and "Respect the will of the people of Damietta!"[82]

The Damietta campaign was one of the most discernible and effective environmental campaigns to emerge in Egypt and in the region more generally. The movement included members from among residents, civil society, the parliament, and the public sector in a coalition that crossed class and religious lines. A new set of Egyptian independent newspapers and private TV stations, in addition to Al Jazeera, covered the protest extensively. In the face of widespread and continued popular protest, the Mubarak regime abandoned the investment agreement with Agrium and granted the Canadian firm shares in another state-owned fertilizer plant with guarantees that production would be increased.[83]

With the growing popular mobilization of the 2000s, and especially the mass protests of the 2011 uprisings in Egypt, environmental politics has attained renewed life. Activist and expert networks have broadened their momentum and diversified their mechanisms and strategic objectives, giving rise to a dynamic phase in Egyptian environmentalism. In the post-2011 Arab Spring uprisings, the Egyptian state has exhibited repressive measures to renewed protests against environmental campaigns. Marginalized communities, including the Bedouin of South Sinai and the Nubians in the south, which have organized protest movements aimed at gaining access to water resources, have failed to advance their agenda in the face of state repression.[84] Many environmental campaigns have diminished given the widespread crackdown on protests that ensued the military coup of June 2013. The urban activists who contributed to the 2011 uprisings turned their attention to grassroots initiatives and local economy, while moving activism off the streets and back into the press and online.[85]

Several groups, connected with the Tahrir uprisings, spearheaded a campaign against the government's attempt to import coal to power a variety of industries in the face of natural gas scarcity. The environmental activists who campaigned against the public health hazards caused by air pollution from such policy failed to mobilize strong local constituencies and advance their climate agenda.[86]

The CSOs and the landscape around climate change in Egypt for the most part appear fragmented and small. Few organizations have the funding and connections required to make a massive difference in dealing with climate action, enjoying significant support from local mobilizations and engaging national CSOs on the international stage. Many groups working on the climate agenda avoid describing their activities in that language, or they may not even be seen as traditional NGOs, with some seeking to register as social enterprises. Yet the fact remains that these organizations are part of the mosaic of nonstate actors shaping the climate dialogue and the nature of work at the grassroots level.[87]

National Strategy for Climate Change 2050

On May 19, 2022, Egypt launched a National Strategy for Climate Change 2050 to improve its citizens' quality of life, promote its sustainable economic growth, and safeguard its natural resources. "Climate change," noted Egyptian prime minister Moustafa Madbouly in his speech at the launch of the national strategy for this plan, "is one of the most important issues facing people as its impacts pose threats to sustainable development plans, food security and water availability, and thus will affect the national security."[88] Although Egypt's share of global CO_2 emissions does not exceed 0.6 percent, Prime Minister Madbouly added that Egypt continues to be most vulnerable to the negative impacts of climate change on many fronts, including its coasts, agriculture, water resources, health, population, and infrastructure.[89]

In the same meeting, Egyptian minister of environment Yasmine Fouad emphasized the need for financial support for this vision, noting that the strategy originated from Egypt's Sustainable Development Strategy 2030. Ms. Fouad stated that the government has allocated a financial package that includes energy, transportation, agriculture, water, irrigation, and carbon reduction in the petroleum sector at a cost of about $211 billion for mitigation and $113 billion (a total of $324 billion) for adaptation to such programs to meet its targeted date. The five-point strategy includes: "reducing CO2 emissions; adapting to climate changes with the country's ability to address this issue; governance framework of the state and society; improving the financial infrastructure; and improving scientific research and technology management."[90]

Conclusion

Egypt, as noted above, is surrounded by a whole host of problems and complicated policy decisions regarding a changing climate, forced labor migration, IDPs, and environmental activists who call into question the government's lack of a transparent, firm, and long-term climate change agenda. Clearly, the country faces emerging challenges, if not dire conditions, posed by internal displacement and forced migration that have been exacerbated by drought, hunger, flooding, earthquakes, war, and economic circumstances.

At the same time, Egypt has become a destination for many Arab and African immigrants and a major host of hundreds of thousands of Palestinians, Sudanese, and—especially in the aftermath of the 2011 Arab Spring uprisings—Syrian refugees. Egypt has also served as a transit country in migrant routes used by sub-Saharan Africans who have recently attempted to cross the Mediterranean toward Europe.[91]

As has been shown in this chapter, Egypt has been particularly affected by climate change as its location and topography render it more susceptible to dwindling precipitation levels, extreme weather patterns, rising sea levels in the Nile Delta, and resulting water shortages and food insecurity. With millions of Egyptians living below the poverty line, the country's lack of financial resources and technological capabilities has intensified growing worries over climate variability and change. A confluence of factors, including climate change, urbanization, population growth, and economic development, are likely to affect the sustainability of agriculture sector and water resources in Egypt.

Water scarcity poses the greatest threat to the country's food security. The adaptation regarding irrigational strategies requires putting in place new hydrological regimes. Since the most vulnerable sector of the economy is agriculture, it is important to put in place effective adaptation measures, including growing new strains of rice that are resistant to drought and water scarcity. While adaptation plans, including vulnerability analyses, have been drawn for urban areas such as Alexandria, Rosetta, and Port Said, as one study concludes, there is also an urgent need to assess the vulnerability of several other economic areas in the Nile Delta—coastal agriculture, lakes and wetlands, and major coastal archeological sites—to sea level rise.[92]

The consistent review of drought responses and ways to improve long-term forecasting will most likely promote Egypt's ability to cope with prolonged drought.[93] Climate change will mount added pressure on the riparian states along the Nile River to compete for accessibility of water. The Nile water crisis has dictated cooperation with the Nile basin countries, especially between Egypt, Sudan, and Ethiopia. That's why it is vitally important to simultaneously determine concrete actions to limit the negative effects of climate change and develop adaptation strategies for helping migrants displaced by it.

On balance, it is important to integrate climate change programs into policy and operational frameworks. Some experts suggest that Egypt should upgrade its capacity for monitoring and forecasting climate variability. "Egypt should prepare for climate change by developing crops that can tolerate environmental stresses such as drought and salty soil," says Waheed Mahmoud Emam, a professor of environmental sciences at Ain Shams University, Cairo, and director of the Egyptian Academic Society for Environmental Development.[94] Egyptian officials should seek both technical and political cooperation of all the Nile Basin countries if they wish to effectively promote the country's development and manage its water needs, while also curbing the adverse effects of the climate change and drought.

CHAPTER 7
SYRIA

CLIMATE CHANGE AND SYRIAN CIVIL WAR REEXAMINED

Located on the east coast of the Mediterranean Sea in southwestern Asia, Syria is bordered by Turkey to the north, Iraq to the east and southeast, Jordan to the south, and Israel and Lebanon to the southwest. Cyprus is west of Syria and Lebanon across the Mediterranean Sea. It also includes territory in the Golan Heights that has been occupied by Israel since 1967. The present area is different from ancient Syria, which connected the eastern Mediterranean coast and the desert of northern Arabia.[1]

The declining level of precipitation, along with rising temperatures, has long resulted in desertification and devastation of agricultural land in Syria, as in the rest of the MENA region. In Syria, this has been particularly the case in eastern parts of the country. The evidence indicating how a significant aridity correlates with the upending of agricultural activities is well documented.[2] So too is the evidence of how past climatic shifts or shocks have reduced the supply of resources that sustained the cohesion and stability of different societies.[3]

Today, the production of wheat crops has suffered from low quality and sells for much less than before the current drought. In fact, nearly 70 percent of the farmers left their fields dormant in 2022 because it would cost more to grow crops than to profit from selling them. The impacts of the drought are vividly tangible in the small city of Al Shaddadi, 50 miles south of Hasaka in northeastern Syria. The Khabur River, which flows through the town and which many farmers depend on to irrigate their fields, has reduced to ponds of murky water. The low level of the Khabur River has forced farmers to operate their diesel-powered pumps longer than usual to get the same amount of water. As a result, the cost of diesel fuel has spiked, and so too have the prices of other essentials.[4]

Additionally, the war against ISIS has left many parts of Al Shaddadi dilapidated. US-led air strikes have destroyed a large residential complex, including water pumping stations, schools, and bakeries used by ISIS. Across the country and beyond, acute poverty and lack of employment opportunities have contributed to young men joining ISIS. Such grievances are likely to multiple by climate change. The ongoing drought continues to drive farmers and their families to the urban areas where there are even less opportunities to make a living.[5]

Drawing on both international political economy and human rights-based theories, this chapter attempts to achieve two primary objectives: (1) to identify critical gaps in the existing regime of the refugee convention and (2) to contribute to policy analysis and prescription with a view toward improving strategies and operational practices aimed at opening up opportunities for the legal employment and technical education of refugees. In the sections that follow, I first explore the impacts of droughts and climate-induced effects on Syria's social and political stability. I next turn to the refugee crisis in Syria and the way in which it has presented difficult trade-offs for the international community as well as the region.

The flaws of the current refugee regime are examined in the ensuing section of this chapter. The terrorist attacks in the West and their impact on the fundamental norms set in the refugee regime are dissected in the next section. The war against ISIS, including the dilemmas of military intervention, receives considerable attention in the following section. The chapter concludes by looking into the prospect of change in the refugee regime and the much-needed drive for new norms and standards.

Droughts and Climate-Induced Displacement

The debate over whether climate change directly causes conflict remains unsettled. Experts have refuted a direct connection between climate change and conflict, although they have argued that the former may indirectly enhance the risk of the latter by intensifying existing socioeconomic and environmental problems. Arguably, the civil war in Syria was a result of complex interrelated factors associated with climatic conditions that significantly contributed to the deterioration of the country's socioeconomic and political conditions. Ultimately, however, the specific conditions of each war make it impossible to establish universally acceptable precepts linking climate change to conflict.[6]

Some studies have suggested clear causal links between drought and climate change in Syria and Sudan. As elsewhere in the world, the case of Syria is not typical in that it involves major conflicts and migration, rendering it immensely difficult to distinguish the role of climate change from those of socioeconomic and political crises.[7] Factors related to drought, including agricultural failure, as well as water scarcity and mismanagement, greatly contributed to the decay of social structures and the spread of violence. Specifically, a combination of exceedingly severe drought, tenacious multiyear crop failures, and economic slums led to drastic dislocation and migration of rural communities to the cities. These factors, experts note, further contributed to urban unemployment, economic disarray, and social unrest.[8]

While experts concur that climate change alone cannot be blamed for the eruption of civil war in Syria, neither do they dismiss it outright as an entirely irrelevant factor.[9] Between 2006 and 2010, according to one study, nearly 1.3 million dwellers of eastern Syria were afflicted by agricultural failures. An estimated 800,000 people lost their livelihoods and essential food supplies. Yields of wheat and barley plummeted 47 percent and 67 percent, respectively, and so too did livestock populations. The ensuing drought in 2011 drove between 2 and 3 million people into food insecurity. More than 1.5 million Syrians—largely agricultural workers and farmers—moved from rural areas to cities and camps on the outskirts of the country's major cities of Aleppo, Damascus, Dara'a, Deir ez-Zour, Hama, and Homs.[10] The regime's failure to effectively bring under control the adverse impacts of drought prompted massive mobilization of dissent against the Bashar old regime.

It should be noted that poverty in the Arab region more generally, but in Syria in particular, mostly exists in rural areas and is growing alongside population growth and unemployment. Nearly 62 percent of poor people in Syria live in rural areas.[11] The ongoing drought has deepened poverty in the country's breadbasket. Many of the fields in rural areas have been left fallow by farmers who cannot afford to buy seeds, fertilizer, or diesel to run water pumps to replace the shortage of rainfall the country has experienced in the past years.[12]

In 2009, the UN and the International Federation of Red Cross and Red Crescent Societies reported that more than 800,000 Syrians had lost their entire livelihoods as a result of the droughts, and herders in the northeast lost around 85 percent of their livestock, adversely impacting 1.3 million people.[13] This led to a massive exodus of farmers, herders, and their families from the rural areas to the cities. The resulting failure of crop yields (particularly the Halaby pepper) around the city of Aleppo led "200,000 rural villagers to leave for the cities."[14] Those farmers and herders who opted for staying in their own villages were mainly destitute; and they subsequently became easily tricked, manipulated, and even recruited by terrorist groups such as the so-called Islamic State in Iraq and Syria (ISIS, also known as ISIL and by its Arabic acronym as Daesh).

In October 2010, a UN report estimated that 50,000 families had migrated from rural areas, in addition to hundreds of thousands of people who had fled in earlier years. Given that Syrian cities were already coping with influxes of Iraqi refugees since the US invasion in 2003, this had placed additional strains on a troubled economy and disenfranchised population.[15] The climate disruption amplified the existing problems, creating a political crisis that was further exacerbated by the Bashar old regime's decisions to remove fuel, water, and food subsidies in the ensuing years. These policies brought to surface, and even intensified, existing inequalities and tensions between Kurds, Arabs, Alawites, and Sunnis.[16]

The 2011 civil war in Syria created a dire humanitarian crisis as the mass influx of refugees overwhelmed the neighboring countries of Jordan, Lebanon, Iraq, and Turkey. These countries hosting Syrian refugees have almost reached a stage where they can no longer absorb or accept any more refugees. With the increasing influx of refugees, as one study demonstrates, local communities have come under enormous pressure to sustain their livelihoods in terms of food, water, education, health services, and employment. The presence of these refugees poses serious challenges to the already fragile stability of host countries and their local communities. Tensions arising from economic difficulties, social problems, and altering ethnic and sectarian balances have led to conflicts between these host countries' nationals and the Syrian refugees. This has resulted in implementing further border restrictions and controls, as well as security measures, around areas bordering Syria.[17]

Jingoism and callous living conditions have initially characterized the Jordanian refugee camps as well as those urban areas hosting Syrian refugees in Lebanon.[18] Jordan has all but closed its border, and the KRI, which is home to a significant number of Syrian refugees, lacks the necessary infrastructure to meet the challenge. This situation has forced hundreds of thousands of displaced Syrian people to seek shelter in Europe, even as the latter has accounted for less than 10 percent of the total Syrian refugees.[19]

Meanwhile, the traditional way of relying on purely humanitarian remedies has proven counterproductive. There is a need for a paradigm shift that envisions relief and humanitarian aid reinforced by the granting of work permits to the refugees. Helping refugees should begin with an effort to pair "technical education" and "vocational training," with a strategy focused on creating jobs in development zones, from which both host societies and displaced Syrians can benefit. The Syrian crisis has underlined the need to construct a new refugee regime—one that guarantees the rights of asylum seekers and refugees, imposes sanctions on states that cause displacement, and reduces the need for long-distance asylum seeking. Similarly, a growing sense has emerged among the refugee recipient countries that new human rights norms are needed, which should be equally based on the self-interest of refugee-hosting states.

The Refugee Crisis

Coming to grips with the refugee crisis—emanating mainly from Syria's civil war, but also more generally from the MENA's political environment in the aftermath of the 2011 Arab uprisings—has never been more urgent.[20] The war has killed hundreds of thousands of people since it began in 2011. More than half of Syria's prewar population of 22 million has fled their homes in search of safety and opportunity. Approximately 83 percent of Syrian people live below the poverty line and some 60 percent of the population suffer from undernourishment.[21] About 6.8 million Syrians are refugees and asylum seekers, and another 6.7 million people are internally displaced within Syria. This brings up the number of Syrian people who are forcibly displaced to 13.5 million—that is, more than half of the country's population. Over 11 million people in Syria need humanitarian assistance. And about half of the people affected by the Syrian refugee crisis are children.[22]

The majority of Syrian refugees, roughly 5.6 million, have fled—either by land or by sea—across borders to neighboring countries, including Turkey, Lebanon, Jordan, and Iraq. As of early summer 2022, nearly 3.7 million Syrian refugees were in Turkey, the largest refugee population worldwide. About 90 percent of Syrians in Turkey lived outside of refugee camps and had limited access to basic services. Some 855,000 Syrian refugees made up about one-eighth of Lebanon's population. Many of these refugees lived in poor conditions in informal tent settlements, which fell far short of official refugee camps. With rarely any legal opportunities to earn money, they have continued to struggle to afford residency fees, rent, utilities, and food.[23]

Of the nearly 668,000 Syrian refugees who lived in Jordan, some 120,000 people lived in Za'atari and Azraq refugee camps. The impact of a massive refugee influx into Jordan, according to one study, has created one of the most thorny and strenuous humanitarian issues in the world today, as substantial political, socioeconomic, and environmental tensions have undermined the notion of solidarity and hospitality between Syrian refugees and Jordan citizens. This situation has clearly posed threats to balancing the needs of the citizens with those of refugees.[24]

The hosting of Syrian refugees in Jordan has provoked tension with the old government in Syria, creating intense friction between the political regimes in Amman and Damascus over the border issues. The presence of refugees has led to competition over limited resources while placing a mounting burden on the quality of public services. All this has unleashed tensions between refugees, citizens, and institutions over questions of identity, territory, security, and the general understanding of what constitutes fair and/or unfair aid distribution.[25] Economically, employment problems added yet another layer of complexity to the current situation on the ground. Following the Syria crisis, Jordanians experienced rampant inflation, surging food prices, decreased government subsidies, and declines in regional investment and tourism.[26]

Approximately 247,000 Syrian refugees lived in the northern parts of Iraq, in the Kurdistan region where more than a million Iraqis fled to escape ISIS. Most refugees have been integrated into communities, placing great demands on services. Some 132,000 Syrian refugees lived in Egypt.[27] In 2015, 1.3 million Syrians requested asylum in Europe. The number of new asylum seekers, however, has significantly dwindled since then. The United States, by contrast, admitted 18,000 Syrian refugees between October 2011 and December 31, 2016.[28] Between FY 2016 and July 31, 2021, according to one source, 1,534 Syrian refugee arrived in the United States.[29]

As a result of the civil war, millions of Syria's young people and children have grown up amid trauma. The one constant danger facing children, according to one study, has been recruitment by soldiers. ISIS has routinely enlisted children to carry out some of its most atrocious crimes, including suicide attacks, while also boasting of training what it has called "caliphate cubs."[30] Jordanian officials, who viewed the Syrian crisis with much apprehension, fearing reprisals and refugee floods, publicly refrained from demanding Assad's ouster and persistently called for a negotiated political resolution to the crisis. Yet the greatest regional spillover from the Syrian crisis was felt not in Lebanon or Jordan, however, but in Iraq.[31] It was in Iraq where the Islamic State emerged and expanded its domain.

Defeating ISIS, while at the same time fulfilling the R2P in the Syrian situation, has presented complex and profound challenges to the international community. The growing threat of ISIS as shown in attacks in Paris (2015), Brussels (2016), and subsequently in Spain (2017) magnified by the increased threat of individual terrorists—the so-called lone wolves in San Bernardino, the United States—has experienced a wave of nationalist, right-wing alarm, reinforcing a general concern about the influx of immigrants and asylum seekers, while underlining the shift from a regional to the international nature of both threat and risk.

These security issues also illustrated the willingness of great powers to support military actions against ISIS in order to stem the atrocities perpetrated by this group, while putting the departure of the old regime on the back burner—at least for the near future. For all intents and purposes, ISIS posed—and continues to present—a much greater threat to Europe than does the current regime in Damascus.

An Unavoidable Trade-off

Confronting and dismantling ISIS need not have been achieved by stigmatizing refugees and subjecting them to religious litmus tests. Closing Europe's porous borders and politicizing the admission of refugees at a time when the growing humanitarian crisis poses mounting human rights challenges to the international community is fundamentally misguided. After all, fortifying European borders, while effective in the short term, strengthens the hands of ISIS and other terrorist groups that portray such policies and practices largely in terms of apocalyptic visions and arcane Islamic prophecies of great battles against Western imperialists.

Defeating ISIS requires strategic endurance and long-term, prudent political decision-making involving internal and external actors in the MENA region. While doing so, it is important to avoid the enemy's repressive, atavistic, and brutal methods, eschewing certain tactics that could potentially play into ISIS's hands. It is important to bear in mind that the tactics that terrorist groups like ISIS employ pose mostly political and ideological challenges to the West and that the real fight will be in defeating and destroying the claims and values that these groups assert. In the end, defeating ISIS requires that its demonic ideology and tactics be confronted and exposed.[32]

Preventing further refugee crises in the future requires that fighting ISIS be at least temporarily prioritized over the overthrow of the old regime. Seeking a political solution in conjunction with harnessing a multipronged strategy may in fact be among the most effective tools and processes for dealing with the current humanitarian crises that the world faces. The possibility of working with the Russians and the Iranians in order to seek a political solution in Syria also raises questions about whether this trade-off is justified. If the intent is to defeat ISIS, Al-Qaeda, and the Levant Conquest Front (formerly known as al-Nusra Front), then this trade-off is inevitable even as it buttresses the old regime in Syria for the foreseeable future. This solution is not problem-free or without significant challenges, however.

Meanwhile, the spread of the Covid-19 pandemic has ironically prolonged the reign of authoritarian regimes. In Syria, for example, the old regime has in some ways been a beneficiary of this unfolding crisis, as other Arab countries have offered medical assistance and remedies to help Damascus better contain the spread of the virus. The UAE has frequently called on the UN to ensure a ceasefire in Syria in order to assist in efforts aimed at preventing the spread of the coronavirus pandemic. On March 27, 2020, for example, the UAE offered to help Syria battle the coronavirus pandemic during a phone call between Sheikh Mohamed bin Zayed, crown prince of Abu Dhabi and deputy supreme commander of the UAE Armed Forces, and Syrian president Bashar Assad. Sheikh Zayed told the Syrian president that their countries needed to "place humanitarian solidarity over political issues during this common challenge we are all facing."[33]

Paradoxically, but understandably, when priority is given to combating a global pandemic and other emerging threats, such as the campaign against terrorism, authoritarian regimes will wind up consolidating their rule, thus exploiting such crises not just for grabbing further power at home but also for seeking geopolitical advantages in the region. It is safe to conclude that the pandemic has spurred no dramatic shift in the regional power and security dynamics. Protecting and promoting human rights amid the global pandemic has become more difficult, if not chaotic, than ever. Under such circumstances, demands for economic stability and security have come to supersede demands for liberty. Yet stability cannot be sustained for long if it is achieved to the detriment of the most vulnerable sectors of the population. The current authoritarian trend should not obscure the reality that spontaneous uprisings may pose unpredictable but destabilizing threats to autocratic regimes over the longer term.

Competing Views

While the military actions against ISIS were and are warranted and can be coordinated by both inside and outside actors, the nature of such military missions and their consequences has also been subject to further debate and negotiations. The inclusion of Russia and Iran, allies of Assad, in the fight against ISIS raised and continue to raise concerns that their stated goal of curbing ISIS is merely a pretext for prolonging

Assad's rule. Likewise, Turkish participation in air strikes in northern Syria in the war against ISIS has raised the possibility that Ankara will target the Kurds, who have successfully fought against ISIS since the beginning of the conflagration. Turkey's interest in settling political scores with the Kurds, an interest that it believes is vital to its security, imperils whatever impartiality one might have hoped for in a fight solely against ISIS.

The massive movement of migrants and refugees to Europe, coupled with the ISIS-led attacks on soft targets in Europe, has created a new urgency among Western leaders to fully confront this new global threat and seriously contemplate the possibility of cooperating with Russia in a coordinated effort. Compromises must at times be made when a multifaceted campaign that includes both countering ISIS and precipitating the removal of the old regime is waged. The collapse of the old regime would create a significant security void that ISIS and other terrorist groups could easily exploit.

There is no denying the fact that the Paris, Brussels, and San Bernardino terrorist attacks have heightened the securitization of the refugee threats, as the Islamic State has been using the wave of migrant influx to infiltrate Europe and North America. The number of terrorists hiding among the refugees is small. ISIS has exploited the flood of refugees to smuggle jihadis into Europe by distributing fake passports in Greek refugee camps to allow its terrorists to travel within Europe.[34] Western media reported that more than three dozen suspected militants who had posed as migrants have been arrested or died while planning or carrying out acts of terrorism. They included at least several individuals who were directly linked to the bloody attacks in Paris and Brussels.[35]

The Flaws of the Current Refugee Regime

Today, most refugees tend to flee from violence and flagrant human rights violations—not necessarily from the threat of persecution, which is a key requirement of the 1951 Refugee Convention. The issue remains controversial within the core conception of the mandate of the UNHCR. This has resulted in confusion and a lack of clarity regarding who qualifies for refugee status and what are the rights to which all refugees are entitled—issues left up entirely to states to interpret. To compound matters further, in the case of Syrian refugee crisis, the neighboring countries of Lebanon, Jordan, and Iraq have never ratified the 1951 Refugee Convention. Now, more than ever, a robust and comprehensive discussion about the future of refugees, whose numbers will only increase with the worsening effects of climate change, is not only timely but necessary, given that the global environmental politics faces extreme turbulence.

It is noteworthy that refugees can be a destabilizing factor, especially when displaced, alienated, and bitter persons among them are recruited into armed extremist factions. Many studies have shown that the absence of a protective and enabling environment is likely to render more young people vulnerable to racist ideologies and movements and ease the process of their recruitment into the ranks of radical groups like the Islamic State or Al-Qaeda.

The all-too-familiar way of relying on purely humanitarian remedies is no longer tenable. The focus of the 1951 refugee protection regime should shift to new ways of dealing with displaced persons that recognize the self-interested reasoning of host countries and the concerns of their citizens regarding competition over jobs. This shift will help to eliminate risks to refugees' personal security by reducing human smuggling and trafficking by land and sea. Some experts, such as Alexander Betts and Paul Collier (2016), have offered solutions along the lines of creating "spaces of opportunity" for the refugees through "special economic zones" that provide jobs, training, and education.[36]

Helping refugees should begin with technical education and vocational training, coupled with a strategy focused on creating jobs immediately in economic zones from which both host societies and displaced Syrians can benefit. The need to work is inseparable from human security and thus crucial to preserving human dignity. The refugees' right to legal employment makes good ethical and logical sense. Designing, for example, a vocational skills training program tailored to the needs of women refugees can significantly reduce the incidence of sexual trafficking and abuse. These projects offer a more plausible solution in the long term, not only because they will develop transferable skills that refugees can use in their countries of origin upon return, but also because they create monetary disincentives for refugees to emigrate to Europe in the first place.

Paris and Brussels Attacks

Global terror reached Paris and Brussels with a vengeance on November 13, 2015, when a group in Syria directed French people in Belgium to commit attacks in Paris. France and its intelligence officers and agencies were clearly taken off guard. Why, one might ask in hindsight, given that by late 2015, ISIS had claimed responsibility for the killing of 425 people, were the French so ill-prepared? The January 2015 attacks at the satirical newspaper *Charlie Hebdo* and kosher supermarket massacres, which left seventeen dead, were in fact the harbinger of things to come.[37]

The attacks at the Brussels airport and subway that killed 31 and left 316 injured on March 22, 2016, pointed to the links between Paris and Brussels attacks. A clear connection emerged between the terrorists who carried out the Brussels airport and metro bombings and the Islamist cell behind the Paris attacks in that assailants in both cases were part of the same Islamic State network.

These new attacks have once again underscored not only the vulnerability of the European security context but also the persistence and increasingly dangerous prospect of what these sophisticated attacks could mean for the rise of further terrorist attacks in Europe in coming years. Western politicians and experts have increasingly noted that the Brussels attacks served as yet another reminder that the Syrian crisis laid at the heart of the new waves of terrorist attacks in Europe. The then US secretary of state John Kerry reflected on this reality: "Events in Brussels underscored to all of us the urgency of every country that has the ability to make a difference to end this evil scourge that comes from the ISIS and violent insurgency."[38]

The migrant issue has divided Europe, posing a more serious threat to its unity and future than anything in recent memory. Virtually all far-right movements, including the anti-immigrant National Front Party in France, Britain's UK Independence Party, and Germany's Alternative for Deutschland, are likely to exploit this situation to close the country's border. Such reactions have been even more marked in Eastern Europe.[39] This is likely to further stigmatize Muslims and aggravate the already existing state of Islamophobia throughout Europe, despite the courageous and dignified call by the then French president François Hollande, as well as the strong stance taken by German chancellor Angela Merkel, for still agreeing to accept Syrian refugees and refraining from the detaining of Muslims.

Some experts remind us that American Muslims are much less attracted to ISIS and its ideology than European Muslims seem to be. European Muslims are poorer, ghettoized, and marginalized. These crises should compel us to dig deeper for the reasons and motivations behind these attacks without perhaps rationalizing such heinous crimes against humanity. Surrounded by privation and destitution, as well as living on a continent with porous borders that would allow for a cache of arms and machine guns, and the likes of AK-47s to move freely, these underprivileged youth search for both a religious justification and identity with an ideology that sees secular infidels as primary foe and knows no limit on the loss of civilian lives.[40]

Despite the growing fears and frustration about the presence of potential terrorists in the West, attacking Muslim minorities and discriminating against them is unlikely to provide long-term solace—much less a solution. To make matters worse, some politicians in the West have shown their apprehensions and qualms about Syrian refugees and the challenges in vetting them.[41] The intuitive panic after the Paris attacks, manifested primarily in ideological, not logical, backlashes such as closing mosques or registering Muslims, would contradict the very ideals that the West has come to defend and represent.

A deeper analysis of the rise of ultranationalism and xenophobia in the West demonstrates that economic instability, populist politicians, and a refugee wave have combined to change the face of politics in Europe. In part, the rise of the far right coincides with fears aroused across Europe by a tidal wave of hopeless migrants. At least 550,000 refugees have arrived in Italy and Greece in 2015 alone. It is estimated that Germany could receive upwards of 1.5 million migrants by 2017. In 2013, Germany spent roughly $14,340 to house and feed each refugee in its care—a bill that amounted to roughly $21.5 billion at a time of economic stagnation in much of Europe.[42]

The migration crisis has given ultranationalists—National Front in France and the Alternative for Germany, among others—necessary ammunition to frequently couch their criticism of migrants in terms of economic pragmatism, explaining there is simply not enough money and jobs to go around in their welfare states. Despite the fact that these parties often have known histories of racist tendencies and policy orientations, their economic message now resonates across a continent exhausted by multiple crises in recent years.[43]

Ultranationalism and xenophobia have become embodied in the electoral success of right-wing political parties and the reassertion of nationhood in European nation-states confronted by both immigration and European integration.[44] Increasingly, nationalist parties across Europe are tapping into voters' economic and cultural anxieties by calling for ending immigration and denouncing the EU, the IMF, and international financiers. Leaders across Eastern Europe, but most notably in the Czech Republic, Poland, and Slovakia, are rejecting the EU's call to impose immigrant resettlement quotas in their countries. Citizens across Europe, but largely in Poland, have protested plans to accept migrants from Syria and Africa to accommodate requests from the EU commission.[45]

Similarly, in a swift and strong response to the terrorist attacks in Paris on November 19, 2015, the US Congress overwhelmingly—the vote was 289–137—passed legislation aimed at tightening controls on refugees from Syria and Iraq.[46] Moreover, the Republican presidential candidate Donald J. Trump's call to temporarily bar Muslims from entering the United States in the aftermath of San Bernardino's massacre has been very divisive and unconstitutional, but perhaps would likely aid ISIS and other terrorist groups by playing into their narrative.[47]

Critics have pointed out that these xenophobic reactions against Syrian refugees, which defy both morality and logic, seem reminiscent of misguided policies of the past, including internment during the Second World War of Japanese-Americans for fear that they were not loyal citizens, or turning away Jewish refugees.[48] Arguing that refugees are not the foe, former US secretary of state Madeleine Albright argues that the United States has in place "a rigorous process of refugee resettlement that balances generosity with security. It works, and it should not be stopped or paused."[49] By portraying Syrian refugees as the enemy, Albright contends, the United States will play into the terrorists' hands, undermining both its moral authority and international credibility.[50] It is in this context that the Pentagon has asserted that statements by particular politicians are likely to do disservice to US position in the Middle East. Given the supposedly nonpolitical nature of the US military, this clarification is extremely significant.

Under Article 1 of the 1951 Convention relating to the Status of Refugees, legal scholars assert, the relationship between refugee protection and human rights is self-evident: "refugees are individuals whose human rights are in serious danger of being violated."[51] There are few more assertive means of safeguarding human rights than granting refugee status to an individual who has a "well-founded fear" of persecution. Individuals who are no longer granted human rights protection from their country of origin are entitled to receive this protection from another country. As the most vulnerable groups in society, asylum seeker populations and refugees should continue to enjoy protection under all other international human rights treaties, as if they were still residing in their country of origin.[52]

Furthermore, according to the *non-refoulement* doctrine, as provided in Article 33(1) of the Refugee Convention, a refugee shall not be expelled or returned "in any manner whatsoever to the frontier of the territories where his life or freedom will be threatened on account of his race, religion, nationality, membership of particular social group or political opinion."[53] However, according to Article 33(2), a refugee cannot enjoy the benefit of *non-refoulement* if there are "reasonable grounds" that they pose a danger to the security of the country of asylum. It should be noted that this principle is now a part of customary international law—that is, even states that are not party to the various treaties must uphold it. That said, no state can be compelled to grant entry to any person seeking refuge. It is entirely at the discretion of the state to determine who shall enter its country in accordance with its application of the criteria for the eligibility of refugees.[54]

Balancing the War against ISIS with the Fundamental Norms

Even if there is no global consensus about how to broadly enforce laws and monitor them in state practice, we still should turn our attention to the strict observance of existing fundamental norms, such as the principle of civilian immunity from attack, safe asylum from both persecution and war violence, and, most importantly, R2P, which should not be bent toward the interest of the stronger powers. "We lose the war against ISIS," Frank Bruni writes, "if the bloodshed in Paris—or in Beirut—becomes fertile soil for bad ideas like religious litmus tests for refugees or religious litmus tests in general."[55] Fanaticism cannot be determined by the outfit someone wears or the text one carries; but rather, it resides in the heart of the individual, something that is beyond plain view and detection.[56]

Moreover, vilifying refugees is far from the solution, as it represents the worst kind of demagoguery for political reasons. Confusing refugees with terrorists is both morally and politically disgraceful. Legally, it also raises many questions regarding the international community's obligations to protect civilians fleeing war and persecution. The notion of accepting only "Christian refugees," as reflected in the voices frequently heard on airwaves and social media, is not only immoral but just an egregious violation of universal human rights.[57]

In both policy and practice, such measures are likely to drive young Muslims toward joining ISIS, further reinforcing the commonly held view propagated by such terrorist groups that the West and Islam are on a civilizational collision course, a tired and weary misapplication of an outmoded paradigm. Instead, the West and the rest of the world should contemplate a myriad of alternatives.

Defeating ISIS has been increasingly prioritized over removing Syria's old regime. The fight against ISIS has also left several unanswered questions, the most obvious of which is how to wage a war against ISIS while at the same time adhering to international norms. Critics of the military intervention against ISIS have noted that the best defense against such Islamic extremist groups is improved governance throughout the region. The US interventionist policies and operations, so goes the argument, are likely to provide less incentive for local actors to get their own houses in order. It is important to bear in mind, critics warn, that not every foreign tragedy is a menace to US interests and not every conflict in the globe needs to be disentangled by American power.[58]

Given that Russia and Iran share many common goals and concerns in Syria, including being concerned about the growth of Salafi and Wahhabi movements in Syria, their opposition to regime change in Damascus is understood.[59] At the same time, however, Moscow has reiterated its message that the Russian SU-30SM Flankers that were recently sent to Syria are intended to reinforce and protect the Syrian air force, impeding the possibility of a US air intervention against the Syrian government forces. This situation made it difficult for Washington to have a free hand in Syria and must treat Moscow as a key player on the political front.[60] Under such circumstances, the possibility of cooperating with Russia, which had seemingly placed itself in the deliberate position of supporting the old regime, with the strategic intent to fight against ISIS appeared inevitable given that nearly 30,000 foreign fighters traveled to Iraq and Syria from more than 100 countries since.

Clearly, migratory pressures have caused deep rifts within the EU, as the actions of various leaders navigate between a humanitarian obligation, a populist backlash, and a broader concern about unwittingly letting in economic migrants and asylum seekers.[61] Many of these refugees, who were caught between the violence in Syria and oppression in areas under the control of ISIS, sought to escape and take refuge somewhere else. The EU agreed to spend 1 billion euros, approximately $1.1 billion, on Syrian refugees in Turkey, Jordan, and Lebanon and other countries.[62]

Turkey's president Recep Tayyip Erdoğan claimed that his country had spent nearly $8 billion managing migrants from Syria and Iraq. More than 3 million Syrians were already in Turkey, with restricted working rights and educational opportunities—factors contributing to viewing migration to Europe as a better alternative—not to mention the fact that many Turks also considered them an economic and cultural menace.[63]

The Dilemmas of Military Intervention in Syria

On August 21, 2013, when reports emerged that the Syrian government had used chemical weapons on its citizens, the Obama administration explicitly expressed its willingness to use air strikes against certain targets in Syria. This willingness was qualified in terms such as "limited, tailored, surgical attacks" and not full-scale war aimed at regime change. President Obama sought congressional approval to do so, especially after the UK lower house halted British participation in military strikes.

The US Congress, however, was divided on the possibility of an intervention. Some of its members were driven by isolationist impulses, while others seemed to prefer a "fight to the finish," not merely a few air strikes against the old regime in Syria. In the meantime, the American people revealed that they had lost their appetite for military intervention, as the shadow of the continuing war in Afghanistan and the Second Iraq War (2003) weighed upon their minds and collective conscience.

The Obama administration endeavored to show the American people that any military action in Syria would not involve the presence of American soldiers on the ground and that full-scale war, of the kind that transpired in both Afghanistan and Iraq, was not an option. Yet, the Obama administration's case was made poorly, as there was a legitimate fear on the part of the public that the willingness of the American people to intervene not only would prolong and escalate Syria's civil war but would almost necessarily result in further civilian suffering. Public opinion polls constantly showed that approximately 55 percent of Americans believed that the US government should take no major military action against Syria, and 78 percent believed there was a chance such weapons would eventually be used against the United States.[64]

These fears—whether justified or overstated—were amplified when one considered the fact that the potential collapse of the old regime would have had profound implications for the regional as well as the trans-regional balance of power. One legitimate concern, however, was that the fear of a Sunni Islamist takeover (ISIS) after the collapse of the old regime in Syria loomed large. A military attack on Syria by NATO—a la military intervention in Libya—was risky, as more evidence in Libya demonstrated that the NATO operation, in the aftermath of the UN Security Council's decision in 2011 to ensure the protection of civilians by air strikes, created a Libya marred by ethnic and sectarian violence, awash with weapons, besieged by armed bands wandering freely about, and tribal warfare. In addition to these uncertainties, a growing threat from the rise of radical Islamist groups and militia rivalries contributed to poor security conditions and an unruly political system.

Human rights lawyers, legal scholars, and activists criticized NATO operations in Libya, questioning the shift in the operation from initially a humanitarian objective to one of regime change. The concern at the heart of this critique was that once the line between preventing atrocities and precipitating regime change became blurred, air strikes became patently inconsistent with the R2P as it had originally been adopted at a 2005 meeting of heads of state and government. The mission of NATO, which was to prevent genocide in Libya, morphed into regime change over the course of the intervention, hence violating the requirement of right intention under R2P.[65]

In addition to the failure to fulfill the "right intention" requirement, it was evident that it failed to fulfill the requirement of armed intervention as a last resort, for Qaddafi's acceptance of the ceasefire was never taken seriously. Furthermore, in Libya's post-conflict period, the responsibility to rebuild—another key component of the R2P—represented the most blatant failure of R2P in this case. The international community clearly failed to provide adequate financial, military, institutional, and political support for Libya's postwar transition.[66]

Under such circumstances, a military intervention was likely to make things worse for Syrian civilians, while plunging Syria and its neighboring countries into a sectarian conflagration and the West and the region into a protracted civil war with no visible endgame in sight. "What happens in Cairo and Damascus," writes John Mearsheimer, "has little effect on American security [and that] . . . no vital American interests are at stake in either Egypt or Syria."[67] For this and other geopolitical reasons, the case for military intervention in Syria was considered unappealing.

The problem was further compounded by the fact that the Syrian opposition (the Syrian National Council) was utterly fragmented, dominated by extremist elements, and it was neither broad-based nor organized. Whereas in Egypt and Libya, all the major and regional powers supported—either explicitly or tacitly—intervention, in the case of Syria, no such obvious consensus existed. A military intervention could turn the situation into a proxy war (Russia and Iran supporting the old regime vs. Saudi Arabia and Turkey supporting rebels) with considerable regional spillover impacts, subsequently resulting in a protracted civil war with huge civilian casualties, reminiscent of the decades-long civil war in Lebanon (1975–91), which resulted in thousands dead and millions displaced.

There was neither the political will nor reasonable hopes of combining regime change with sustainable post-conflict stability until normalcy could be established by new national leadership. To make matters worse, the military option would likely increase the magnitude of internal violence without generating better hopes for the future in a society conflicted and divided by deeply rooted enmities, bitter historical memories, and pervasive fears of uncertainty. Such anxiety was particularly prevalent among the Syrian minorities (Alawites, Christians, and Druze, roughly 30 percent of the population) that seemed to fear the collapse of the present regime more than they disapprove of its oppressive and inhuman conduct.[68] This somber conclusion is best captured by one observer: "Just as doing nothing is unacceptable, mounting a military intervention is unrealistic, and perhaps undesirable, and for now politically impossible."[69]

There is a better than even chance that if the old regime falls, one observer points out, Syria will be divided up into ethnic and sectarian-based mini-states in constant conflict with each other. It is becoming abundantly clear that the strife in Syria has had enormous negative consequences in the region, especially in the arena of inter-sectarian relations, for the neighboring states of Lebanon and Iraq that are themselves fractured societies. The war in Syria is likely to turn the Fertile Crescent into a perpetual conflict zone.[70]

Conclusion

As discussed in this chapter, views vary as to what contributed to the long-lasting civil war in Syria. One view holds that environmental problems brought on by poor policies and climate change fueled the civil war that was already precipitated, among other things, by discrimination, inequality, and political coercion in Syria as we know it today. The civil war in Syria created a harrowing refugee and humanitarian crisis, as the mass influxes of displaced people have overwhelmed the neighboring countries of Jordan, Lebanon, Iraq, and Turkey, rendering these host countries and their local communities vulnerable as a result.[71] Regardless of what factors have contributed to the Syrian civil war, habitat destruction and soil erosion are the two key factors that caused the war. Approximately 11.9 million people have been forcibly displaced within Syria and across its borders, of which 5.7 million are refugees and 6.2 million are internally displaced, including 2.5 million children within Syria.[72]

A contrasting view underlined the importance of the country's economic liberalization policies that discarded agricultural fuel subsidies, driving farmers into massive debt.[73] While the climate change explanation is compelling, economic policies merit equal attention. After all, full-blown social conflict, some experts insist, probably cannot be attributed to climate change alone. As complex social occurrences, civil conflicts cannot be broken down into single causes.[74]

Regardless, the spillover of the Syrian crisis into Lebanon has caused growing apprehension about these refugees, some of whom were involved in suicide bombing networks that perpetrated several acts of terrorism against Hezbollah and other Shia groups. Sadly, as one observer notes, whereas most refugees have been peaceful residents, the activities of a few of them, in conjunction with ISIS and al-Nusra, have turned them in the eyes of many Lebanese into suspects or (real or imagined) culprits of terrorism and hardship.[75] Conditions in many refugee camps are inferior, and many of the recipient countries lack the necessary infrastructure to effectively cope with the influx of refugees without the assistance of relief agencies and donor countries.[76]

This chapter has also argued that the traditional way of relying on purely humanitarian remedies has proven ineffective. There is a need for rethinking that connects relief and humanitarian aid with the granting of work permits to the refugees. Helping refugees should begin with an effort to pair "technical education" and "vocational training," with a strategy focused on creating jobs in development zones, from which both host societies and displaced Syrians can benefit. A new refugee regime must be constructed that guarantees the rights of asylum seekers and refugees, imposes sanctions on states that cause displacement, and reduces the need for long-distance asylum seeking—and new human rights norms that are equally based on the self-interest of refugee-hosting states—if not their mendacious propaganda regarding protecting the refugees.

Yet, the reality is that Syria's neighboring countries themselves face the upsurge of poverty, unemployment, disruption of social services, and the overall sinking socioeconomic fortunes. In this context, several nagging questions persist: How, then, can these countries provide educational and vocational opportunities for the Syrian refugees? Who decides what should be given to the refugees? Sovereign states or the United Nations?

The new refugee regime should not be predicated solely on an ethical basis; it needs to take the self-interests of the refugee-hosting countries into account. The fact remains that keeping refugees in the refugee camps permanently has not gone well over time. Vocational and technical training provides the refugees with necessary skills to be deployed when/if they are returned to their countries of origin. This also acts as a monetary disincentive to travel to other parts of Europe in the hope of seeking employment there. The solutions for the refugee crisis in Syria are multiple, complex, and perhaps more importantly, regional. Such solutions, as discussed in this chapter, fall outside the purview of the military realms. As Antonio Guterres, the then UNHCR, and currently UN secretary general, has noted, the solutions are ultimately diplomatic not military.[77]

CHAPTER 8
YEMEN

THE WAR, HUMAN SECURITY, AND THE ENVIRONMENT

Yemen is located at the southern tip of the Arabian Peninsula. The country's mountainous terrain is surrounded by lean coastal plains to the west, south, and east and by highland desert to the north bordering Saudi Arabia. The country is located on the Bab Al Mandab Strait at the southern entrance of the Red Sea and Aden Gulf, which are crucial for international trade. The country has three climatic zones. It is mostly hot and humid along the west coast. Its western mountains receive seasonal rain and monsoon. And its desert in the east is extremely hot, dry, and harsh.[1]

Less than 3 percent of the land is cultivable. Yemen has limited freshwater resources, and its groundwater is swiftly being depleted.[2] "Yemen," according to one study, "may offer the most serious case of groundwater overuse in the world."[3] The increase in the total area of irrigated land has led to further use of water supplied by deep groundwater aquifers—a development strongly fostered by government policy. The government also safeguarded the domestic fruit and vegetable market, as well as the boom in the production of qat (a flowering plant comparable to a mild narcotic), whose cultivation consumed roughly 37 percent of all agricultural water in the country. In the past, merchants and wealthy tribesmen have managed to capture most of the groundwater by investing in pumps.[4]

Yemen's oil reserves are also limited and have been largely misused in a corrupt system that has been heavily reliant upon patron-client networks, with millions of Yemenis being severely food insecure. The Yemeni population presents a complex mixture. Unlike the densely populated Zaydi Shia in the north, those who live in the south identify with Sunni Islam of the Shafii school. The relations between north and south remain tense as evidenced by sporadic civil uprisings and tribal agitation between the two.[5] The Zaydi-Shafi division has been disruptive and even turbulent, but the broader barriers to inclusive development have been regional, tribal, externally charged, or economic rather than sectarian.[6]

Yemen's population of just over 33.7 million (2022) has a median age of 20.2 years and is growing at a rate of 2.28 percent per year. The Yemen Unemployment Rate was projected to be around 17.5 percent in 2023.[7] Yemen is the second poorest country in the MENA region. The United Nations Development Program estimates that 55 percent of Yemen's population lives below the poverty line, and 18.4 percent lives in extreme poverty.[8]

Mean annual temperatures are expected to rise 2.1–5.9 degrees Fahrenheit (1.2–3.3 degrees Celsius) by 2060, with higher increases in the non-coastal regions of the country. It is estimated that runoff will increase by 13 percent, the number of hot days will soar, and the number of cold days will decrease. The risks posed by these environmental and climate changes include "floods, droughts, sea level rise, landslides, and increasing desertification."[9] The country's geography and conflict dynamics restrict the options of people trying to flee abroad. People are penned in by the ocean and the desert, with only Saudi Arabia and Oman as direct neighbors. Yemen is surrounded by extensive and highly insecure terrain to the east. Its neighboring countries (Saudi Arabia and Oman) are reachable only via a treacherous sea journey to the south. Yemen's dwindling rainfall over the past decade has exacerbated the civil war in that country.

This chapter underlines the many complex and interlinked factors contributing to the current humanitarian crisis in Yemen, including poverty, social inequities, tribalism, and sectarian tensions. These factors render the country's battle against climate change impacts excruciatingly difficult. In the sections that follow, I first zero in on specific causes of the civil war in the country. I then trace patterns of social displacement and mass migration brought on by the civil war, water scarcity, and food insecurity in the context of climatic variability and environmental changes. I next turn to the refugee situation in Yemen, specifically focusing on the Markazi camp. I conclude by offering ways to combat elite infighting, clientelism, and internal power struggles in the wake of the serious negative consequences of climate change.

Humanitarian Crisis

Yemen faces one of the worst humanitarian crises in the world today. The country's infrastructure has been bombed, as hospitals, schools, bridges, and roads have been destroyed. Millions of people across the country are now chronically malnourished. The refugee crisis in Yemen was sparked by the Yemen Civil War of 2014. The history of the civil war dates back to the 1960s, when a military coup dominated the political scene in which outside actors (Saudi Arabia and Egypt) supported the opposite sides of the Yemeni politics. In 2004, with the formation of the armed movement of Zaidi Shia, who looked for greater autonomy over the Northern Province of Sa'ada, the civil war gained a renewed intensity.[10]

After the 2011 Arab Spring uprisings, the country disintegrated into tribal strife, religious factionalism, and civil conflict (more on this later). The Ansar Allah militias (better known as "the Houthis") took over the Yemeni government upon the claim that the Yemeni government was corrupt. The Houthis are supported by Iran, although the extent to which their movement is externally fueled by the latter is unclear, or perhaps even overstated. In Yemen, as in Libya, Syria, and Bahrain, social disparities have intersected with tribal or sectarian prerogatives and helped turn the uprisings into bloody civil wars.[11]

The Houthi rebels captured Sana'a in September 2014, placing President Abd Rabbuh Mansur Hadi under house arrest and forcing him and his government to relocate to the southern city of Aden. There they lost control of much of the city while fighting with Ali Abdullah Saleh supporters. President Hadi fled to Saudi Arabia, which prompted "Operation Decisive Storm" by the Saudi-led coalition backed by some of GCC countries. The operation was aimed at destroying the rebel stronghold in the region and reinstating the government of Mansur Hadi. The conflict had far-reaching effects on sectarian violence in Yemen, while also intensifying the competition between Saudi Arabia and Iran—two key regional rivals.

Furthermore, the conflict led to the arrival of ISIS and Al-Qaeda in Yemen. This civil war has ratcheted up more pressure on a country that is already tormented by war, poverty, water scarcity, malnutrition, and bacterial diseases such as cholera, among others. All these developments have amounted to one of the world's most severe humanitarian crises. The fear that the dire situation may descend into famine is frequently expressed by international organizations and human rights groups. An ICRC official once said that "after five months of war Yemen looked like Syria after five years of conflict."[12]

In the absence of lifesaving assistance, thousands of lives have been lost owing to ongoing violence and lack of access to food and medicine. Known as the Arab world's poorest country, Yemen suffers from a poverty rate of nearly 75 percent. Two out of three Yemenis need humanitarian assistance, and there are nearly 4 million IDPs in the country, according to the UN refugee agency.[13] A combination of several factors have caused a great deal of pain to the people of Yemen, creating the world's worst humanitarian crisis, including civil war, flood, diseases, famine, economic blockade, and economic disasters, which have caused pain to the people of Yemen.

Many NGOs, including the ICRC, have provided lifesaving assistance and emergency aid. The situation in Yemen became even more urgent than ever in the wake of deaths resulting from the Covid-19 pandemic. The outbreak of this pandemic, along with the fastest-growing cholera epidemic ever recorded and widespread malnutrition, continues to put millions of children and families at risk. The gravity of lingering humanitarian crisis has accelerated population movement, with far-reaching implications for the region.[14] In an extremely poor country like Yemen, it has been very difficult to provide health care and fight the Covid-19 pandemic and other devastating diseases simultaneously.

The Civil War

In the 1960s, a military coup ushered in a long civil war (1962–70) in Yemen that became a manifestation of the Arab Cold War between pro-monarchy elements in Saudi Arabia and Arab socialist republics in Egypt. Whereas Jordan, Saudi Arabia, and Britain supported the Royalist side, the Egyptian president Gamal Abdel Nasser supported the Republican government by deploying 70,000 Egyptian troops. The war led to the establishment of the Yemen Arab Republic in 1970, when an agreement was reached to have a Republican government that incorporated members from the Royalist factions.[15] In 1967, the British left southern Yemen, and the People's Republic of Southern Yemen was subsequently created. By 1970, the People's Republic became the Marxist People's Democratic Republic of Yemen, known as South Yemen, which turned into a client state of Moscow.

Throughout the 1980s, approximately 1 million Yemeni workers were employed in Saudi Arabia, whose remittances made up over 40 percent of the GDP of both North and South Yemen.[16] In 1990, the two parts of the Yemen unified into the Republic of Yemen. The socialist economy of the south was to merge with the free market of the north. The Iraqi invasion of Kuwait on August 2, 1990, forced Yemeni officials to assume a political position vis-à-vis this conflict consistent with the dominant actor in the Arabian Peninsula—Saudi Arabia. President Saleh's decision to oppose US efforts in the region, including the 1991 Gulf War, while also refusing to denounce the Iraqi invasion of Kuwait as well as supporting the idea of sending Arab forces to join the allied coalition, prompted Saudi Arabia to expel Yemeni workers in that country.

As a result, some 800,000 of them returned to Yemen. The other Gulf states not only expelled an additional 50,000 Yemeni workers, but they canceled their financial aid programs to Yemen.[17] This led to the fall in remittances and caused the government to initiate structural adjustment programs. Consequently, the government raised the price of diesel fuel used by pickup trucks and in water pumps, while also ending credit subsidies and the lifting of import bans on fruits and vegetables intended to incentivize greater water conservation and more efficient agricultural use of water. These policies, which led to increase in fuel and food prices, have on numerous occasions triggered deadly riots.[18] In line with Saudi reactions, Washington also cut off all US aid to Yemen, ratcheting up the pressure on Saleh as his country faced a serious economic blow to its newly united republic.[19] In 1994, boiling north-south tensions erupted, with President Saleh sending armed forces to defeat a southern independence civil war.[20]

In the years following the 9/11 terrorist attacks on the United States, however, Saleh cooperated with Washington on the latter's "war on terror," a tactical alliance that put him at odds with his domestic Islamist elements and some Afghan Arab jihadists who had helped him defeat southern separatists in the 1994 civil war.[21] By the late 1990s, the civil-military relations were mired in the question of succession and exclusive patronage system.[22] The Yemen's internal power struggles and civil war have been intensified by outside military intervention, generating a complex humanitarian and political crisis. Since 2014, outside intervention in Yemen has been a persistent feature of the country. Using the pretext of civil war in Yemen, international and transnational actors have frequently shaped domestic state institutions.

The non-state actors, such as Al-Qaeda and a southern separatist group, as well as foreign powers, such as Saudi Arabia and the United States, respectively, have at one point or another intervened to steer the political process toward state reconstruction.[23] The UAE was instrumental in the creation of militia forces known as the Southern Transitional Council—a development that subsequently led to the growing conflict between these militia forces and the Yemeni government.

For several years, Washington cooperated with Saudi Arabia on their war in Yemen by sharing intelligence, refueling their aircraft, selling them weapons, and providing them with diplomatic cover. Six years after the Saudi intervention, the war in Yemen proved to be a disaster. This war further destabilized the Middle East region, empowered Iran, and tarnished the US global reputation. Crucially, however, the war devastated the Yemeni people, who now face the world's worst ongoing humanitarian catastrophe. Approximately 250,000 Yemenis have died because of the conflict, with roughly half from other causes such as malnutrition and disease—not to mention many millions more are starving or becoming homeless. Given the fragmented and decentralized sources of power among Yemenis, the conflict has calcified and become less prone to political solution.[24]

Following the US invasion of Iraq in 2003 and the emergence of Al-Qaeda in the Arabian Peninsula (AQAP), the United States and other regional actors demanded that President Saleh commit to fighting AQAP. In response, Saleh launched a series of attacks, backed by Saudi Arabia, against northern Yemeni Zaydi Shia fighters known as Houthis, whom he accused of pursuing separatist goals and of trying to impose their religious ideology on the state. The Houthis—making up roughly 40–45 percent of Yemen's population—had grown vocal in their complaints of discrimination and disenfranchisement under Saleh's autocratic rule.[25] While the Houthi movement initially emerged as a religious opposition to the state-adopted Shawkani Salafism and Saudi Wahabi propaganda, it slowly evolved into a full-scale movement embraced by those in the Sa'ada region and beyond who felt economically marginalized and politically disadvantaged by the republican government in Sana'a.[26]

"Over the course of his 33-year presidency," wrote Asher Orkaby, "Saleh crafted the modern Yemeni state, which was built on precarious foundations of tribal agreements, temporary truces, negotiated boundaries, political nepotism and patronage, and widespread corruption."[27] Saleh was in fact the only person who understood the intricacies of his own craft. As he stepped down from the presidency in 2012, the country's political, economic, and social life grew chaotic. Furthermore, his mismanagement and exploitation of Yemeni resources for his own personal gain proved detrimental to the future stability of the country.[28]

During this time, the ongoing politics of patronage has given the Houthis movement particularly favorable conditions to grow. The government patronage—favoring one tribal group or shaykh at the expense of another—had driven a wedge between some influential tribes. Without accomplishing much in the way of state-building, which could have allowed Yemeni state to push its agenda forward, the patronage system, both internally and exerted by the Saudi government, has led to the socioeconomic imbalances and inequities even prior to the emergence of the Houthis movement. In this volatile context, as Marieke Brandt rightly notes, the surging socioeconomic discontent and the struggle over resources and political participation brought to the fore a sectarian force that fueled the emergence of a multidimensional resistance movement—a movement that ultimately morphed into the Houthis rebellion.[29]

Thus, well before the 2011 Arab Spring uprisings, growing disillusionment and anger at Yemen's political economy had made the country a most likely target for violent social unrest aimed at removing a regime that appeared to be in fatal fall.[30] The 2011 Arab Spring uprisings prompted protests in Sana'a, initially over the issues of corruption and economic hardships, but subsequently turning to widespread protests that included several opposition parties, including the Houthis in the north, CSOs, and various tribes. President Ali Abdullah Saleh, who had been in power since 1978 and had presided over the unification of North and South Yemen in 1990, had proven to be particularly adept at manipulating internal politics of a fractured tribal society only to enhance his personalized rule. These protests, however, forced him to hand over formal power to his vice president, Abrabbuh Mansur Hadi, in February 2012.[31]

As was the case across the broad swath of the MENA, the 2011 Arab Spring uprisings, as Asher Orkaby notes, "brought neither democracy nor freedom. Rather, the fall of a dictator opened the gates to a multitude of militant groups, previously on the radical fringe of mainstream politics, and led to the beginning of Yemen's current civil war."[32] Fearful of political instability in the region, Yemen's GCC members, while drawing on US support, persuaded President Saleh to resign in favor of Hadi. As part of the GCC deal, Saleh was granted immunity from local prosecution and Hadi was to become the next president in a two-year term transitional period. Hadi became the president of Yemen but lived in exile due to political instability and tension. In 2013, the UN-brokered National Dialogue Conference (NDC) secured a degree of compromise between political groups from south and north.[33]

The following year, Houthi-Sunni clashes further complicated the implementation of the NDC outcome. Spawned by a reduction in fuel subsidies, popular protests erupted against the Hadi government in September 2014, as the Houthi rebels broke the NDC deal and prevailed.[34] With Houthi insurgents taking control of Yemen's capital and largest city—Sana'a—while demanding lower fuel prices and a new government, the country's new civil war began. The rebels seized the presidential palace in January 2015, forcing President Abd Rabbu Mansour Hadi and his government to resign.

In the ensuing months (March 2015), a coalition of Gulf states, spearheaded by Saudi Arabia, began a campaign of economic isolation and air strikes against the Houthi rebels, with US operational and intelligence support. Having revoked his resignation, Hadi returned to Aden in September 2015—a development that has intensified and prolonged the civil war to this very day, even as Hadi has continued to reside in exile in Saudi Arabia. In December 2017, Saleh was killed by the Houthi insurgents, as he sought a compromise solution that betrayed his commitment to the Houthi. The subsequent intervention of regional powers in Yemen's conflict, by Iran, Saudi Arabia, and the UAE, further intensified the larger Sunni-Shia divide.[35]

Significantly, the Houthis' attempt to overthrow Hadi's transitional government won widespread support, as the latter lost political legitimacy. Most Yemenis, according to one expert, supported the Houthis' accusation that Washington and Riyadh were pursuing their imperial interests in the country.[36] While the Saudi-led invasion of Yemen has to some extent hampered the advancement by the Houthis, it has failed to reinstate the Hadi government.

The fighting, now eight years old, has clearly turned into a proxy war. On one side are Iran-backed Houthi rebels who overthrew the Yemeni government and have proved astonishingly resilient. Their mark on the capital Sana'a today is unquestionable.[37] Houthis are pitted against a multinational coalition led by Saudi Arabia on the other side. Over the years, the Saudis have exported a toxic mixture of sectarianism and corruption to Yemen.[38] The involvement of other fighters, including militant Islamist groups and separatists backed by the UAE, during the earlier phases of the civil war complicated the picture.[39] While the Saudi military operations to reinstate Hadi as the president of the country failed to yield any results, they intensified the sectarian tensions between Iran and Saudi Arabia by turning the conflict into a new cold war between the two countries nonetheless.

Throughout the war, both Saudi and Emirati forces conducted aerial bombing campaigns that destroyed Yemen's water and sanitation infrastructure. These assaults further intensified the water scarcity that Yemenis were facing well prior to the war, leaving over 19 million without access to potable water.[40] Moreover, this water insecurity was exacerbated by the Saudi-led blockade of two major ports, al-Hudaydah and Salif, where 80 percent of food imports enter the country. As such, external powers placed Yemenis at great risk for communicable diseases such as cholera.[41]

Meanwhile, the UN estimated as of May 2020 that 131,000 of the estimated 233,000 deaths in Yemen since 2015 were the result of indirect causes like food insecurity and lack of access to health services. Today, approximately 25 million Yemenis remain in desperate need of aid, some 5 million face famine, and a cholera outbreak has affected more than 1 million people. The war has turned Yemen into a breeding ground for cholera spread by contaminated water. Heavy rains in some parts of Yemen have also contaminated wells.[42]

Aside from the ongoing civil war, Washington continues counterterrorism operations in Yemen, striking at Al-Qaeda targets there.[43] The war has left more than 5 million people on the verge of famine, and despite having plenty of high-yielding land, Yemen imports some 90 percent of its food.[44] Food insecurity, which predates the war but certainly exacerbated by it, has led to the further deterioration of climate change effects on fertile lands, escalating the cultivation and use of a flowering plant called qat. According to one study, over 90 percent of the Yemenis developed the habit of chewing qat. Many fields are devoted to growing qat instead of other daily necessary nutrient crops such as wheat and rice. Qat is traded and sold for high prices throughout the country, in part due to grave privation and the demand for the product.[45]

Yemen's economy has been severely damaged by the widespread social habit of chewing qat. As a cash crop, the qat bush is easy to grow, allows multiple cropping, and provides quick cash. As a result of increasing demand, some fields that previously grew wholesome and exportable crops have been converted into qat fields, rendering Yemen into an import-dependent country.[46] Given that qat is a water-intensive crop, it is easy to understand why growing qat would present severe problems to a water-poor country like Yemen. Significantly, however, international donors, especially the World Bank and the IMF, pushed the Yemeni government to shift from the subsistence agriculture that had sustained its population for centuries to a model that placed a high premium on farming more water-intensive crops, such as oranges and grapes, for export in return for international loans. The amount of water necessary to sustain these crops in a water-stressed region has an immensely detrimental impact on the country's water basin, for which qat was never exclusively culpable.[47]

Since June 2022, Saudi Arabia and Yemen's Houthi movement have resumed direct talks to discuss the ongoing conflict in Yemen. Iran's support for and its extensive military training of Houthis have enabled the latter to regularly attack Saudi soil. Estimates from 2015 indicated the Yemeni War was costing Saudi Arabia up to $6 billion a month. The ongoing talks between Saudi Arabia and Iran, shortly after a negotiated ceasefire in Yemen, have significantly contributed to ending the war there.[48]

Water Scarcity

During the 1970s and 1980s, according to one study, unchecked and unsustainable private and public well drilling irrigated Yemen's coastal lowlands and highlands to support groundwater-dependent agricultural projects. Meanwhile, the Yemeni government invested in large public and private enterprises espoused by USAID and other international organizations. The World Bank, for example, supported water diversion projects administered by foreign contractors, accelerating the production of opulent market-oriented crops rather than sustainable and necessary produce for domestic consumption.[49] During the 1990s, a demographic explosion, along with the return of 800,000 migrant workers from the GCC member states following the Iraqi invasion of Kuwait, as well as growing consumer demands for cash crops such as qat, placed untenable demands on Yemen's agricultural economy and water resources of its key urban areas.[50]

The overutilization of Yemen's water resources while relying on modern irrigation technology has gravely depleted ground aquifers and jeopardized household water supply. Sana'a, for example, is likely to become the first major global capital to run short of water.[51] The extent of this menace is best captured by Asher Orkaby: "Yemen's water crisis poses a far greater threat to the country's continued existence than the current civil war."[52] Seasonal flooding has been mismanaged owing to woefully inadequate state infrastructure that has more often than not failed to garner the rainstorms for the country's depleted aquifers, water reservoirs, and groundwater restoration. The monsoon season often leads to destruction and water-borne diseases.[53]

The postwar reconstruction, some suggest, should focus on diversifying the country's economic projects and resettling internally displaced refugees in areas with more bountiful water resources.[54] The long-term solution, Orkaby notes, lies in "a realignment of water rights and sharing arrangements that more clearly reflects the needs and resources of local population."[55] One plausible solution, Orkaby goes on to argue, is to follow the model of the advanced desalination technology that has been deployed at a new plant in the city of Ashdod, Israel, which can produce low-cost desalinated water for mass consumption.[56]

Yemen's future, as Helen Lackner rightly argues, depends on effective water management. "Without effective water governance and management," Lackner insists, "Yemen's long-term perspective is one of worsening poverty, and therefore of further deterioration of security and safety."[57] If unchecked, Lackner warns, water shortage is likely to lead to civil strife on an unprecedented scale. Several strategies must be put in place to conserve groundwater and reduce saline intrusion in the coastal plains. Agricultural policy must be directed at improving the production of rain-fed crops and such drought-resistance varieties. It will be also essential to enact policies focused on guaranteeing the accessibility of water-saving irrigation technology to poor farmers with shallow wells. Yemen's future development, and indeed its very existence, hinges upon an effective water management, hoisted by nationwide actionable policy options.[58]

Climate Disasters

Yemen is likely to experience significant changes in its average temperatures, precipitation, and sea level, all of which will exacerbate the effects of the socioeconomic and political issues millions of citizens already face. Yemen's average temperature is expected to increase by 1.2–3.3 degrees Celsius. The expected temperature spikes as well as rise in number of extremely hot days is likely to lead to water scarcity, a dramatic decline in agricultural productivity, and enormous human health risks.[59] Sea level is expected to increase by 0.3–0.54 meters by 2100.[60] This rise in sea level would put more than half of Yemen's coastal areas at risk and badly affect more than 55 percent of the country's coastal population.[61] Sea level rise will most likely lead to the movement of saline water into freshwater aquifers, aggravating the country's already severe water scarcity.[62]

New studies have shown that climatic disasters resulting in water and resource scarcity tend to enhance the risk of conflict breaking out in ethnically fractionalized countries, spurring intercommunal violence. Several developments have triggered large-scale migration from rural to urban areas in the developing world. These include, among other things, widespread famine, population displacement, the precipitous decline of agricultural productivity across the developing world, and food security.[63]

The issue of food security in Yemen is highly vulnerable to the vagaries of food price surges and climate change.[64] It is estimated that by 2050, between 80,000 and 270,000 people could go hungry due to climate change. The number of hungry people is projected to rise, even with perfect mitigation, largely due to rising global food prices brought on by global increases in demand. The negative consequences of climate change on food security are largely confined to rural households. Within rural areas, it is chiefly nonfarm households who suffer, as they profit from rising prices for agricultural goods and simultaneously have the highest food cost shares of all household groups, rendering them particularly susceptible to food prices changes.[65]

It is equally important to remember that the historical, deeply rooted, and complicated causes of violence in Syria and Yemen cannot be solely blamed on the impacts of climate change and natural disasters. Underlying the blatant political and societal tensions, however, lies the issue of water scarcity. Yemen is ranked among the most water-stressed countries in the world due to being constantly exposed to drought, desert climate, and mismanagement.

Amid rising population, spiking unemployment, political instability, terrorist operations, outside military interventions, and plummeting water and food resources, climate change is likely to make matters worse.[66] Water shortages have at times caused real humanitarian concerns during the civil war. Several reports have shown that both the Houthi and Saudi forces have blocked deliveries of humanitarian assistance, including food and water. In a water-stressed country like Yemen, competing sides tend to treat water as a resource to be kept back from the enemy.[67]

Other environmental stressors, such as deforestation, poaching, flooding, and air pollution, have exacerbated previous socioeconomic and political grievances in Yemen. It is believed that local disputes over religious endowment (*waqf*) lands and the rivalry over water resources in Sa'ada governorate have led to the outbreak of the first Sa'ada war in 2004.[68]

It is worth noting that in Yemen, as elsewhere in the region, land and water disputes are interconnected. Given the country's water scarcity and inequitable water distribution, maintaining control over land is the key to accessing water. For this reason, private land ownership determines who gains access to water.[69] The war has rendered large parts of agricultural land inoperable, further accentuating food insecurity. This is particularly alarming as nearly half of the population faces hunger, with over 16 million people suffering from food insecurity.[70]

In recent years, reliance on firewood has dramatically increased as a way to cope with fuel shortages brought on by the conflict. This spike in demand for firewood has led to widespread deforestation, with people illegally cutting down trees to meet household energy needs. Absent proper government regulations, Yemen's flora and fauna will likely be threatened in the coming years, further intensifying the ecological and environmental crisis. Additionally, a progressive deterioration of livelihoods is also taking place, with Yemen's agricultural and fishing sectors being severely impacted.

In short, climate change-related impacts have wide-ranging implications for the effective enjoyment of human rights. Increasingly, climate-induced displacement and migration affect the basic rights of inhabitants. Climate change effects in Yemen have presented existential threats to several groups in the country. The number of refugees and IDPs has increased in Yemen, where open-ended civil wars have driven many people from their homes.[71]

Internal Displacement and Refugees

Internal displacement in Yemen predates the 2011 Arab Spring uprisings. Political problems in the country during the 1960s, 1970s, and 1980s had already led to noticeable displacements from south to north. Additionally, natural disasters, such as the floods of 2008, which affected the governorates of Hadhramaut, Lahej, al-Mahra, and Ta'iz, have displaced more than 25,000 people.[72] The Sa'ada war (2004–10) forced a massive displacement, as inhabitants of the northern governorates escaped the conflict. The Saudi air strikes in 2009 caused a new wave of displaced people.[73] Despite international support by several local and global NGOs and other organizations such as the Organization of the Islamic Cooperation, both IDPs and refugees remain vulnerable to internal conflict and a climate of instability.[74]

Since a new war erupted in 2014, approximately 4 million people have been compelled to flee their homes, many of whom were farmers.[75] These displaced people have turned to unsustainable subsistence ways and means, such as the heavy reliance on humanitarian aid. According to the IOM, some 73 percent of the displaced population in Al Dhale'e governorate were farmers driven from their lands and unable to find work.[76]

This has further undermined the country's food production capacity. Similarly, another report has also highlighted the war's impact on yet another vulnerable community: Yemen's fishermen. The Saudi-UAE—led coalition, according to this report, issued restrictions on fishing in certain areas, forcing Yemeni fishermen into the shallows where their jobs became fruitless. This act deprived fishermen of their sole source of livelihood and threatened their basic right to survive and maintain a decent sustenance.[77] Additionally, the UAE exploited the conflict as an opportunity to occupy and control Socotra—an archipelago off the southern coast of Yemen in the Arabian Sea—in 2020. The UAE officials sought to make the archipelago an Emirati territory as part of a broader attempt to enhance the UAE's influence over the region's vital waterways.[78]

It is worth remembering that Yemen is among the few countries in the Arabian Peninsula that has signed the 1951 Refugee Convention and its 1967 Protocol.[79] It has been a destination for refugees from the Horn of Africa, particularly from Eritreans in the 1970s and 1980s, and Somalia since 1991.[80] Since the 9/11 terrorist attacks on the United States, international refugees have come to be seen as potential jihadi fighters or arms smugglers, fortifying the illegal arms-trade networks between the Horn of Africa and terrorist groups in Yemen.[81] In recent decades, most Africans arriving in Yemen prefer "spontaneous settlement" or undocumented livelihood in the key urban centers, in the hope to cross over to the Arab Gulf states, and hence avoiding the Yemeni authorities and regulations.[82]

As noted in the previous chapters, the explosive aftermath of the 2011 Arab Spring has led to a disastrous refugee and IDPs crisis in Syria, Libya, and Yemen, which has culminated in unprecedented hardships, difficulties, and challenges for those and neighboring states, as well as extending to Europe. Following the 2015 Saudi-led coalition's air strikes against the Houthis, Yemen became the worst global displacement disaster in terms of the number displaced, which that year amounted to 2.5 million—nearly 9 percent of its entire population at the time.[83] As a result of seven years of conflict, according to one report, some 70 percent of the population depend on aid for their survival and millions have fled their homes.[84]

As of July 14, 2022 (or after seven years of war), according to the UN Refugee Agency, nearly 6 million Yemenis have been uprooted from their homes and more than 23.4 million are in desperate need of humanitarian assistance. The country faces an increasingly gloomy and uncertain risk of a large-scale famine. Tens of thousands already live in starvation, with a bewildering 5 million more on the brink of famine.[85] By the end of 2021, according to the same report, Yemen had the fifth largest number of IDPs in the world—after Syria, Colombia, the Democratic Republic of Congo, and Afghanistan.[86]

Many IDPs in Yemen live in perilous locations, wrestling with rampant food insecurity and water scarcity, as well as poor healthcare and sanitation services. The advent of the Covid-19 pandemic and the threat of an impending famine in the country have even further exacerbated the living conditions of these IDPs. Despite its massive internal difficulties and an ongoing civil war, Yemen hosts more than 97,000 refugees and asylum seekers, largely from Somalia and Ethiopia, rendering it the world's third-largest host of Somali refugees.[87]

Migrants and refugees, one study suggests, continue to be among the most vulnerable people in Yemen. They are easily stigmatized by local militia, who accuse them of carrying viruses or planning to join enemy forces as mercenaries. If arrested, migrants and refugees run the risk of becoming detainees. Well prior to the 2015 conflict, human rights organizations reported systematic and widespread rape and other sexual abuses, overcrowding, and lack of access to the bare minimum medical aid in both men's and women's detention centers and prisons. Kidnapping and rape by armed or criminal groups represent the major protection concerns.[88]

The outbreak of the Saudi-led war redrew the migratory flows given that Yemen is surrounded by the sea or the desert. Some Yemeni elites sought refuge in Saudi Arabia, the UAE, and Jordan. Their persistent involvement in patronage networks and their association with wealthy businessmen of Yemeni descent facilitated the process of obtaining temporary residence permits.[89]

Other elites turned to Beirut, where leftist political networks and supporters of Hezbollah resided. Islah party leaders chose Saudi Arabia, Qatar, or Turkey, given that Istanbul had become a home for many of Egypt's Muslim Brotherhood escaping the wrath of the el-Sisi government. Oman has also accepted thousands of displaced Yemenis. Several thousand less well-off Yemenis fled to Somalia in 2015. In the ensuing years, some 20,000 Yemeni refugees settled in the UN-managed camp Markazi in Obock, Djibouti.[90]

A combination of the war, climate change, and the anticipated decline of water levels in Yemen has triggered more and different migratory flows.[91] In terms of the gender breakdown of those Yemeni refugees who ended up in the Markazi camp, 36 percent were men above eighteen years old, 25 percent were women, and 38 percent were children. Some men explained they were escaping for security reasons, avoiding forced recruitment; others lost their jobs in Yemen, hoping to find new opportunity in Djibouti as a way to support their family in Yemen.[92] The cost of living in Djibouti city, however, was exorbitant for many Yemenis, forcing them to go back to the Markazi camp in Obock.

The Markazi Camp

Markazi camp (Camp de Markazi) is located 4 kilometers southwest of Obock, a desert land in Djibouti, separated from Yemen by the 32-kilometer Bab al-Mandeb Strait, where temperatures exceed 35 degrees Celsius (95 degrees Fahrenheit) for almost half of the year, with sandstorms prevalent during the summer months. Djibouti's authorities have made significant efforts to adapt the country's French education system to Arabic and English-speaking refugee students. Yemeni children, however, are taught in Arabic in the camp by refugee teachers.[93]

Whereas Ethiopians are granted asylum seeker status, those fleeing Yemen and south-central Somalia are granted refugee status automatically (prima facie) without any need for screening beyond affirmation of nationality.[94] Migrants and refugees in Djibouti are heavily reliant on international humanitarian aid and are highly susceptible to economic manipulation and abuse, grievous bodily harm, gender-based violence, and detention in miserable conditions, exposing them to heightened risks of disease transmission and even loss of life. Moreover, many migrants and refugees transitioning out of Djibouti are held against their consent by smugglers and brokers until they pay the requested ransom.[95]

The post-2015 Yemeni migrants who seek refugee status in Djibouti differ significantly from those who arrived much earlier through chain migration that has sustained generational continuity of the sort. The difference clearly lies in migration—that is, forced versus voluntary migration. Today's refugees are granted special legal status and often have fewer resources and opportunities to integrate socially and professionally. Consequently, they have lower social standing and are mostly physically confined to a refugee camp. While the Djiboutian government has attempted to provide these refugees with some means of social mobility, such policies have for the most part yielded no desirable results. For many refugees, the key obstacle to socioeconomic integration is their exclusion from the well-established Yemeni trading communities. These Yemenis prefer to call themselves "foreign workers," even as they have settled there for such a long time.[96]

According to the 2019 UNHCR figures, there were 4,916 Yemeni refugees in Djibouti, with 2,264 registered in the Markazi camp and 2,652 in Djibouti city.[97] From 2015 to 2018, Yemeni migrants from the coastal Tihama region constituted the majority (60 percent) of the camp's population. In 2019, individuals from the Tihama made up 90 percent of the Yemeni population there, with the remaining 10 percent coming from such urban areas as Aden, Taiz, and Sana'a. Many of the Yemeni families in the camp are Somali-Yemeni, Ethiopian-Yemeni, or even Djibouti-Yemeni. This diversity between and among locals, urban Yemenis and biracial Yemenis determines their social position and spatial mobility.[98]

Some 2,800 individuals are living in Markazi camp (Djibouti). It is anticipated that the number of camp residents will continue to increase, not only with Yemeni refugees coming from Yemen but also with Yemeni refugees presently living in the city of Obock, who cannot sustain themselves in the city. The Government of Djibouti recently adopted a new policy to allow Yemenis to reside in urban areas. The cost of living in Djibouti city, however, is prohibitive for many Yemenis, and as such, they may choose to stay or go back to the Markazi camp in Obock. Yemeni refugees living in the Markazi refugee camp cannot access the capital unless they give up their right to receive governmental and nongovernmental aid (housing, food, etc.), and are expected to remain in the camp.

A Gender-Based Approach to the Humanitarian Crisis

Three quarters of the 4.5 million displaced people in Yemen, according to the UN Sustainable Development Group, are women and children, with some 26 percent of displaced households led by women. "Displaced people in this region live in very poor conditions, struggling to find ways to earn a living. Many go through entire days without earning anything; some women may work for a full day on the farm for only $2.50."[99]

IOM's community center in Al Khukhah, located in Yemen's Hadidah Governorate, was originally created as a platform for the internally displaced Yemenis to voice their concerns and needs. Additionally, it served as a refuge resort sanctuary shelter for women across thirteen IOM-managed displacement camps in southern Ta'iz to express their basic needs and demands. The women's participation projects were launched as a place for them to actively engage in camp activities. While encouraging women to contribute with various suggestions, IOM also fosters such platforms for interacting and sharing information. Within such a collaborative context, women can increase their participation in camp activities, as well as acquiring useful skills that contribute to their personal growth.[100]

This center actively enhances the well-being of displaced families by organizing skill-building activities for women, such as sewing, hairdressing, henna tattooing, and straw weaving. At the center, many women now play a central role in sharing their knowledge and skills with other displaced women. The ripple effect of this process fosters a sense of community among IDPs, all the while providing more women with the means to generate income and become a contributing member of the society.[101]

A gender-based approach to the humanitarian crisis can also affect the way in which funding priorities can be determined. This could offer an opportunity for enhancing access to household revenue for vulnerable families. Currently, women business owners face a myriad of difficulties in raising capital and are often prevented from deploying their assets as collateral. Absent collateral conditions, as in the case of microfinance services, women make up nearly 90 percent of the customers. Many studies have demonstrated that entrepreneurship, including home-based entrepreneurship, can afford unique economic opportunities for women in Yemen and could likely be congruent with traditional and local conservative norms as well as the eruptive security context.[102]

Conclusion

The civil war in Yemen has conflated different yet interrelated issues such as governance, socioeconomic development, human rights issues, humanitarian crisis, and the environment. All of these issues have been and continue to be impacted by climate change. Many years of civil war in Yemen have resulted in one of the world's worst humanitarian crises, in which almost three-fifths of the population rely on aid on a daily basis for their subsistence, and millions have fled their homes and country. Since the 2014 civil war, political instability has impeded the government's ability to manage and direct the country's limited resources.

Both local CSOs and international nongovernmental organizations working in Yemen face myriads of challenges in the face of poverty and civil wars. Additionally, Yemen faces multiple challenges in its attempts to combat climate change. Yemen is immensely poor and thus vulnerable to climate change—but it has also been embroiled in conflict since the early 1960s. As a result of deteriorating political conditions, the distinction between, on the one hand, criminal and political networks of migration, typified by human trafficking, violence, and terrorism, and on the other, asylum and economic-migratory networks, is becoming increasingly blurred, in large part because the former precipitously use the same routes and agents as the latter.[103]

With the unraveling of the Yemeni state, the spread of conflict on both sides of the Red Sea and the Gulf of Aden, and burgeoning sources of financial and ecological distress, massive and urgent human migration is likely to intensify in both Yemen and the Horn of Africa.[104] Many Yemeni displaced persons and refugees have been exposed to exploitative practices by smugglers and traffickers, including blackmail. Many have also been subject to abduction by trafficking networks and criminal gangs, with some even at risk of being recruited by armed groups to fight in conflict.[105]

Yemen and Djibouti, according to one study, exhibit the highest extent of food, human habitat, ecosystem services, and health and infrastructure vulnerability.[106] These countries also have the lowest agricultural capacity and the highest rate of child malnutrition due to reginal conflicts, political instability, and high levels of poverty.[107] Additionally, the risk of climate-related mortality is higher in these two countries than it is in the rest of the MENA region.[108]

For these reasons, Yemenis need to see an end to the international and regional blockade. All ports and airports must be reopened, public services restored, and a nationwide ceasefire implemented so that talks can bring an end to the conflict. There is no other way to break the vicious cycle of human suffering and tragedy. The UN should get involved in ending the civil war and work toward a ceasefire. Regional cooperation between Iran and Saudi Arabia has opened the door for a ceasefire. UNHCR's role has not been used properly for the rehabilitation of the refugees. Selective rescue targeted at several refugees is woefully inadequate.

The needs of urban and camp refugees are massive, ranging from health to nutrition, shelter, self-reliance, nonfood items, and protection. To avoid more deaths and suffering, it is essential to continue to scale up humanitarian assistance, and donors should give more money. The use of blockades and restrictions on food, fuel, and medicines must be lifted. Putting an end to the civil war and outside intervention is urgent. Formidable adaptive and institutional capacity is needed to combat climate change effects in Yemen. Women empowerment is indeed the best solution to malnutrition. Policymakers should facilitate private sector-led growth with a view toward addressing climate-induced impacts. Better investment should target the most food insecure households.[109]

From the logistical standpoint, it is important to guarantee the right of movement to refugees in the Markazi camp by allowing them to travel to Djibouti city to integrate socially and professionally and network with the existing Yemeni community. It is equally important to enable access to fishing licenses for Yemeni fishermen to avoid pushing them into illicit activities, the underground economy, or to return to Yemen to bring goods back to Djibouti.[110]

Insofar as international organizations are concerned, they can do much in the way of reinforcing socioeconomic integration and cooperation between locals and refugees by involving refugees and locals in the process of developing projects and reducing institutionalized exclusion by including them in the decision-making process (especially women). Particularly relevant is the reduction of aid dependency (such as food rations and temporary jobs) by supporting local businesses and entrepreneurship, such as innovative agricultural methods for farming in arid areas. Most importantly, however, is the promotion of foreign direct investment in the form of microcredits and professional development counseling. International organizations should provide the necessary tools to render individuals independent, such as sewing machines, and implement Technical and Vocational Education and Training for youth and adults on various potential trades, so that refugees can choose training for trades they are eager to pursue.[111]

PART III

ENVIRONMENTAL DECLINE, POLITICAL ECONOMY, AND ADAPTATION PLANS

CHAPTER 9
IRAN

ENVIRONMENTAL DECLINE, CLIMATE DISPLACEMENT, AND GROWING PROTESTS

Formerly known as Persia, Iran is situated at the crossroads of Central Asia, South Asia, and the Arab states of the Middle East. Iran is bordered in the north by the Caspian Sea and in the south by the Persian Gulf and the Gulf of Oman, both of which shape its strategic location in the region. A range of high, rugged mountains in the west separates Iran from Iraq and Turkey, while the eastern region, which is bordered by Afghanistan and Pakistan, is covered by an arid, salty desert. Most people in Iran live along the edges of a high plateau that runs through the middle of the country. A mountainous, barren, and ethnically diverse country of southwestern Asia, Iran maintains a rich and distinctive cultural and social continuity.[1]

With an area exceeding 1.6 million kilometers, Iran's topography presents a mixed picture of comprising lowlands at sea level and mountainous areas of several thousand meters of elevation. The Alborz Mountain range is located in the north and the Zagros mountain range is in southwestern Iran, extending the border areas of eastern Turkey and northern Iraq to the Strait of Hormuz. Because of its vast space, topography, and distance from the sea, Iran has a diverse climate, but greater parts of it are arid (65 percent), with remaining parts of it semiarid (20 percent).[2] The annual rainfall over the country ranges from less than 100 to 2,000 millimeters and drought is a chronic event there.[3]

Historically, the occurrence of drought can be traced back to the late nineteenth century. In the early and latter parts of the twentieth century, the country experienced widespread and several serious droughts. The debate about climate change has gained urgency in Iran amid the country's multitude environment difficulties ranging from more extended periods of extreme high temperatures to air pollution, flooding, droughts, disappearing lakes, and inefficient water and natural resource use. In April 2021, the Iranian Meteorological Organization warned of an "unprecedented drought," as well as the precipitation levels that were substantially below long-term averages. The amount of rainfall in Iran's main river basins from September 2020 to July 2021 was, in most parts of the country, significantly lower compared with the recent past.[4]

In this chapter, I first examine internal migration caused by environmental degradation. I then explain the reasons behind the increasing vulnerability to climate and environmental changes that have given rise to forced migration. Of particular relevance to this chapter is the nature of Iran's rentier state and its broader impact on neglecting the role of climate factors. In the ensuing section, I will tease out potential impacts of water mismanagement and its political consequences for migration flows. The desertification of the rural areas, and its resultant climate-induced migration, receives closer scrutiny in the ensuing section. This section will be followed by an explanation of the surge in protests against water shortage in some parts of the country as well as some backlashes from local NGOs and civil society. The chapter concludes by examining Iran's Achilles' heel and the implications of climatic changes for migration.

Land Degradation and Internal Migration

Iran's geographic, economic, and climatic characteristics have rendered it one of the most vulnerable countries to climate change in the region. A large area of the country is prone to floods, while shifting patterns in precipitation have also led to droughts and forest fires.[5] Over the past two decades, however, particularly in the first decade of the twenty-first century, the country has seen devastating droughts in part because of the mismanagement or nonmanagement of water resources, and partly due to the precipitation deficits than to the decline in the other elements of hydrological cycles such as streams and wells.[6] Global-scale studies of climate change have predicted the rise in the frequency and duration of droughts over Iran, showing also increasing severity in various climates of Iran. These studies indicate that droughts with extreme severity and duration would increase in the future.[7]

Environmental and land degradation, dust and sandstorms, as well as floods and droughts, have driven patterns of internal migration within Iran. These developments have contributed greatly to mounting pressures on urban settings, namely the nation's capital Tehran, which has hosted an average influx of a quarter million people annually for the past two decades. Today, more than 70 percent of Iran's population resides in cities, placing increasing ecosystem stress on urban areas. Furthermore, population growth has intensified the country's already troubled labor market, exploiting water resources, and contributing to excessive energy use. These challenges illustrate how Iran's environmental issues have further exacerbated the nation's socioeconomic vulnerabilities and inequalities.[8]

While Iranian leaders have repeatedly underlined the gravity of the crisis facing the country in terms of climate change, and while they have signed the 2015 Paris Climate Agreement, the Iranian parliament (*Majles*) has yet to ratify it. In recent years, the government policy of increasing the price of fuel has caused nationwide protests, compelling the government to continue to subsidize the price of oil and gas. There has been vast urban and industrial development and deforestation, with soil erosion and grazing and salination visible around the Alborz Mountains.[9]

Since the 1990s, massive droughts have hit Iran. The Iranian government ratified the Vienna Convention in 1990 and participated in the UNFCCC signed in Rio de Janeiro in May 1992. Iran accepted the Kyoto Protocol to the UNFCCC in August 2005 and sent a delegation to the subsequent international conferences on climate change, including the UN Climate Change Conference in Bali in December 2007.[10]

Despite many policies and programs directed at tackling climate change, several problems, including widespread corruption in the country, especially in the fossil-fuel sector, alongside sanctions, institutional capacities, access to energy-efficient technologies, and availability of capital, have presented significant barriers to Iran's mitigation measures. One of the most contentious issues facing the country has been its economic isolation, rendering it extremely difficult to gain access to foreign investment and technology transfer required to meet its renewable energy goals.[11]

Damage from Climate Change

Iran's physical geography features, according to one study, are particularly amenable to the consequences of climate change. More than 80 percent of the country is arid or semiarid, with nearly 20 percent desert coverage and only 9 percent of the country's 1.648 million kilometer area is forest land.[12] The shares of arable and agricultural areas, respectively, at 10.1 percent and 39.2 percent in 1996, had noticeably reduced to 9.3 percent and 28.5 percent by 2011.[13]

Iran has also been subject to the increasing rise in temperatures, with the national high of 53.0 degrees Celsius (127.4 degrees Fahrenheit) in 2016 followed by nearly 1 degree Celsius in 2017 at 53.7 degrees Celsius (128.7 degrees Fahrenheit). It is predicted that Iran will see a further increase of 2.6 degrees Celsius (4.68 degrees Fahrenheit) in mean temperatures and a 35 percent reduction in rainfall in the coming decades. These trends paint an unsettling future, one in which an increase in drought periods will be coupled with extreme flood events.[14] Several experts have predicted that in the next three decades, Iran would likely experience more extended periods of extreme temperatures in the southern part of the country, more extended periods of dry and precipitation, along with higher intensity and frequency of floods. The projected spiral in both exceedingly hot and wet periods simultaneously, as this study demonstrated, indicated long dry periods sporadically disrupted with precipitations of high intensity and brief duration, dramatically enhancing the chances of drought and flood conditions.[15]

Predictably, Iran has been hit by a myriad of floods, some resulting in the displacement and death of thousands of people. In May 2021, floods struck fourteen provinces and killed ten people. These floods vandalized villages, livestock, and water supplies.[16] Between March and April 2019, flash flooding overwhelmed approximately 90 percent of Iran, causing the displacement of over half a million people while also resulting in hundreds of millions of dollars of damage done to infrastructure.[17] In the aftermath of the disaster, Zahra Falahat, the Iranian Red Crescent's under-secretary-general for International Affairs and IHL, reflected on the unfortunate outcome: "Entire villages [were] washed away in a matter of minutes, countless homes and buildings [were] damaged and completely destroyed."[18]

Known as one of the most water-stressed countries in the world, Iran has in recent years suffered from a lack of rainfall. During the 1996–2011 period, the country's annual average precipitation was 203 millimeters, compared to 1,121 millimeters globally.[19] Rising temperatures and declining precipitation are likely to significantly reduce Iran's crop yield in the coming years. Climate change will substantially reduce Iran's four major crops—wheat, barley, rice, and corn—in the next three decades.[20]

The fact remains that traditional irrigation systems that require a great deal of water are no longer sustainable. The following items and products require more water: alfalfa (to feed domesticated livestock), rice, almonds, pastures, vineyard, cotton, sugar beets, and apples. Given the projected impacts of climate and environmental changes, the government should support the farmers to change the crops they are cultivating.[21] It is important to know that the following crops require less water: eggplants, squash, watermelons, pineapples, strawberries, onions, spinach, carrots.

Among the Middle Eastern countries, according to one study, Iran will undergo an increase of 2.6 degrees Celsius in mean temperatures and a 35 percent reduction in precipitation in the coming decades. Iran's increasing level of contributing to emissions of GHG continues to depend on a significant production of oil, gas, and rapid urbanization.[22] The regularity with which extreme precipitation occur will decline in Iran but the rising risk of droughts in the future will likely threaten water and food security, especially for people who live in the densely populated cities of the country.[23]

As noted above, Iranian leaders have shown on numerous occasions that they are aware of the kinds of dangers facing the country as a result of climate and environmental changes. They have actively participated in the twenty-first COP of the UNFCCC in Paris 2015 (known as COP21), adopted its agreement on December 12, 2015, in Paris, and signed its treaty on April 22, 2016, in New York, even as the Iranian parliament has yet to ratify the Paris Climate Accord. Clearly, Iran has shown a willingness to espouse the international efforts to mitigate GHG emissions and to adapt to the impacts of climate change, consistent with the principle of CBDR.[24]

The Rentier State Problematic

Iran has an economy that is largely reliant on oil and cheap fuel. Iran holds some of the world's largest deposits of proven oil and natural gas reserves, ranking as the world's third-largest and second-largest reserve holder of oil and natural gas, respectively, in 2020. At the end of 2020, Iran accounted for 25 percent of oil reserves in the Middle East and 12 percent in the world. Iran's large reserves notwithstanding, its crude oil production has fallen since 2017 because the oil sector has been subject to underfunding and international sanctions for several years.[25] Iran was the fifth-largest crude oil producer in OPEC in 2020 and the third-largest natural gas producer in the world in 2019.[26] In the same year, Iran earned $30 billion in net oil export revenues, down from $66 billion in 2018. Iran's export revenues dramatically fell in 2019 due to US sanctions on Iran's oil exports, resulting in a decrease in both crude oil production and exports.[27]

The impacts of climate change, along with mismanaged water resources, have in recent decades posed a profound environmental challenge to the Islamic Republic. Slowly but steadily, there has been a growing realization at the highest levels of the Iranian government that climate change needs to be taken seriously as it poses a realistic threat to the regime's stability. In 2015, following the signing of the Paris climate accord, Iran agreed to reduce its emissions by 12 percent and to spend $5 billion on conservation efforts.[28]

The central government's dependency on oil revenues is likely to slow down any progression toward tackling the impacts of climate variation. Climate change has undoubtedly damaged Iran's economy more generally, but its agricultural sector in particular.[29] Water scarcity and floods pose the most serious dangers to the latter. The Iranian government has attempted to tackle this issue by relying on new farming techniques, specifically updated irrigation techniques and alterations in cropping patterns, to avoid utter agricultural disintegration in Iran. These techniques notwithstanding, the agriculture sector is likely to stall, and Iran could stand to lose nearly 70 percent of its farmland in the coming decades.[30] Iran's turn to nonproliferation nuclear energy can significantly reduce the amount of CO_2 generated. Iran has constructed a nuclear power plant, Bushehr 1, and has planned to complete its second plant by 2026.[31]

Ecological Changes and Water Mismanagement

Less rainfall for several years in a row in Iran has led to an increase in the number of climate migrants and displaced people, as many Iranians have headed toward wetlands. Facing enormous pressure during the last decade or so, the wetlands themselves are currently in a state of ecological crisis with major impacts on biodiversity and socioeconomic conditions. While contributing to dust and sandstorms, reduced precipitation has also caused new waves of climate migrants as never before.[32] The last decade has seen an influx of migration from the southeastern province to Khorasan Razavi, Tehran, and Golestan. "A total of 20% of the people have moved to Golestan in the northeast."[33] Alireza Khosravi, director of Iran's Cultural Heritage, Handicrafts and Tourism Office in Zabol notes, "In the past people would use the 120-days wind to power wind-mills and cool the air; however, the dried wetlands and river beds have consigned the region to perdition."[34]

A report by the Statistical Center of Iran (SCI) states that during the five-year period from 2007 to 2011 more than 5 million people have become displaced, relocating within the country. From among the thirty-one provinces, eighteen have become "senders" of migrants while thirteen remained "recipients." The SCI report further added that Alborz, Isfahan, Gilan, Yazd, and Mazandaran are the top five provinces receiving the most numbers of migrants owing to their better climatic and environmental conditions.[35] Sistan and Baluchestan, Khuzestan, Kermanshah, Lorestan, and Fars provinces, by contrast, have been identified as the largest senders of migrants largely because of ecological disaster and other socioeconomic reasons. Thousands of villagers in various parts of the country have left their habitats as they have faced consecutive years of drought in the past fifteen years, causing a large influx of people migrating to big cities in search of jobs. Calling our attention to these worrisome trajectories and a "code red" for the movement of population from countryside to cities, experts have warned that if the current trend continues, the rural population will further sink to 22 percent by the year 2031.[36]

The Case of Khuzestan Province

The climate and environmental changes have contributed to the shrinking cities of Khuzestan province, which is located in the southwest of the country, bordering Iraq and the Persian Gulf. The province, both strategically and economically, is crucial to Iran's national security due to its oil and gas reserves—not to mention that it is also notable for its agriculture, animal husbandry, and fishing industry. It has a population of 4.7 million (as of 2018), making it the fifth most populated province in Iran, with twenty-seven townships.[37] The temperature of the province has spiked from 0.25 to 1.87 degrees in the past thirty years. The large-scale activities in oil and gas industry, along with the vast rice farms, can cause a faster rise in temperature.[38]

While these climate and environmental problems have been long known to government officials in Khuzestan, they have stirred attention in light of the recent Covid-19 pandemic and its enduring impact on the region. Khuzestan was one of the hardest hit by Covid-19, as a wave of infections forced several counties in that province to reimpose strict lockdown measures. At one point, infections in Khuzestan accounted for about a quarter of Iran's total daily cases. By late summer 2020, according to Johndi Shapour University of Medical Sciences, some 600 people had died and more than 13,000 were infected.[39]

The regression of precipitation in the province presents yet another challenge to this region. During the past thirty years, more than 270 millimeters of precipitation has been lost. Since 2008, drought has been a frequent occurrence and has significantly diminished agricultural water sources, negatively affecting income of rural residents of this region.[40] Immigration has also been the key reason for the negative growth of rural population and consequently the rural evacuation. Over the past two decades, the number of vacant villages has risen from 1,217 to 2,398. During the same period, the rural population has undergone a swift negative growth from (−0.02) to (−4.6).[41] Clearly, environmental mismanagement that led to protests in southwest parts of the country in July 2021 spread to several other Iran provinces.

At the heart of this unrest was the issue of water shortage in a province that used to be water rich. Critics were quick to put the onus on the government's mismanagement. For example, Esmaeil Kahrom, a well-known Iranian ecologist, has said in an interview with Khabaronline news agency: "Khuzestan is one of the water-rich provinces of Iran, but unfortunately some unqualified individuals have been in charge, and the result is the current situation we that officials' environmental mismanagement has significantly contributed to Khuzestan's water crisis."[42]

Similarly, Somayeh Rafiei, the head of environmental affairs in the Iranian parliament, views the causes of the water crisis in Khuzestan along the same lines:

> The water problem in Khuzestan has much deeper and more hidden roots than a multi-year management period. We see a clear example of this in the management of the country's wetlands. . . . In fact, the main problem of water scarcity in Khuzestan and its related environmental issues, as well as the problems facing many other areas of the country, are the officials for whose past failures and inadequacies the ordinary Iranians must pay a huge price today.[43]

The impact of climate change, Rafiei adds, needs to be understood alongside several other factors, including improper cultivation pattern, construction of a dam by nonprofessionals, irregular transfer of water to other provinces, and the existence of destructive environmental industries such as oil, petrochemicals, and sugarcane.

These ecological changes have a significant influence on out-migration. Ahwaz, the provincial capital, is one of the country's most polluted cities. The significant reduction of precipitation and severe droughts, coupled with the destruction of palm groves in southern Iran, have caused atmospheric instability leading to dust storms.[44] There has been, according to one source, an average of eighty-five dusty days in the shrinking city of Abadan over the past years. Dust storms and extreme temperatures in summer have had a negative influence on public health, agriculture, and education (e.g., school closure) and airports (e.g., flight cancellations).

The cumulative result has been the growing economic downturn and unemployment in the shrinking cities of Khuzestan province.[45] Although the indirect impact of climate and environmental changes on other migration drivers, including socioeconomic and institutional factors, is far more significant and complex than their direct impact, they have compromised the human habitability of this region and have compelled more people to migrate.[46]

While drought, flood, and sandstorms have caused displacement in an already overcrowded city, it is important to remember that Iran's ambition to become a regional power has led the government to invest in rapid development and infrastructure, with still less attention to environmental standards. Additionally, water has become a convenient political tool, which is strategically used by Iran's ruling elites, particularly its Islamic Revolutionary Guard Corps, to manipulate ethnic conflicts. To that end, Iran's authorities have built many dams over the rivers across the country to divert water from minority regions to central parts of the country. The diversion of Karoon River from Arab ethnic minority illustrates the point here—a strategy that has led to ongoing unrest and demonstrations in the province of Khuzestan.[47]

To further demonstrate the failures of government planning, as one expert asserts, since the 1970s, Iran has built several dams in Khuzestan to contain seasonal flooding, generate hydroelectric power, boost crop irrigation, and provide other provinces with water. Many of these dams, however, have degenerated and become unsustainable. Isa Kalantari, the head of Iran's Environment Department, referenced such a planning failure in 2018: "We made these mistakes in the 1980s. Then we came to realize that in places that we'd built dams, we shouldn't have built any, and in places where we should have built dams, we didn't build any."[48] Some of these dam reservoirs are barren and others have precipitated environmental degradation. Consider, for example, the Gotvand Dam, which was completed in 2012 and was constructed on salt beds that made the water in the reservoir unfit for irrigation.[49]

The collapse of the ten-story Metropol building under construction in the Khuzestan province city of Abadan on May 23, which led to the death of twenty-four people and injured dozens more, has added yet another trouble for a regime that is already facing many challenges. The following day, Khuzestan governor Sadegh Khalilian noted that noncompliance with laws, regulations, and technical principles in the construction was the main cause of the crumbling of the building. Similarly, Siamak Sadeghi, the head of the Association of Housing and Building in Khuzestan, told the Iranian Student News Agency that the building was originally designed to be seven stories but that subsequently three more stories were added, escalating the stress on the foundation and most probably contributing to its collapse.[50]

Following this incident, on May 27, 2022, several protests were held in Abadan and other cities, including Khorramshahr and Isfahan, where citizens expressed support for the people of Abadan and chanted anti-regime and anti-authority slogans. The collapse resurrected the memory of the 2017 fire and collapse of the archetypical Plasco building in Tehran that killed twenty-six people. Most crucially, however, the incident served as yet another grim reminder of a pervasive lack of transparency—both nationally and at the local level—and flaws in state regulatory policies throughout the country.

Climate Migrants

Water shortage has driven many villagers and farmers from their homes in search of safety. Over the past two decades, according to one study, climate-driven migration has increased ten-fold.[51] The provinces most severely affected by climate migration include Khuzestan, Lorestan, Esfahan, Hamedan, Kerman, Yazd, and Sistan and Baluchistan. This has resulted in widely known migration routes in some parts of the country: in Kohgiluyeh and Boyer-Ahmad Province in southwestern Iran, for example, these routes stretch to southwest (e.g., Khuzestan province) and southeast parts of the country.[52]

Iran's environmental problems have further intensified internal displacement. Some 75 percent of Iran's nearly 88 million people now live on less than 40 percent of its land, mostly in the water-rich areas of the west and in the north. It is worth noting that half a century ago, 60 percent of Iran's 30 million people lived in those same areas. Increasing desertification and deforestation in these regions, together with ill-advised environmental practices and high water consumption patterns, have badly affected underground water resources and caused significant water scarcity at the local level.[53] Migration trends have made some regions depopulated, a development fraught with serious implications for national security. Border areas, which are particularly integral to maintaining Iran's territorial integrity, have seen large-scale migration propelled by climate and economic factors. This has left many of these regions devoid of inhabitants.[54]

The vast majority of climate migrants are farmers, workers, and fishermen who have moved with their families from the countryside to major urban areas seeking alternative livelihoods. In the meantime, Iranian officials have blamed deteriorating water scarcity and rising desertification on climate change—not their mismanagement and poor planning. While the precise number of climate migrants is unknown, Iranian media have estimated that in 2021 some 41,000 people were forced to migrate due to the effects of climate change, including drought, sand and dust storms, deluge, and natural disasters. That figure increased to 42,000 people in 2022.[55]

In July 2023, officials warned that more than 1 million hectares of the country's territory—approximately equivalent to the size of Qom Province or Lebanon—is at risk of becoming essentially uninhabitable annually. The central government's failed attempts to rectify water scarcity, including dam building and water-intensive irrigation projects, have precipitated the drying up of rivers and underground water reservoirs.[56]

Under the Islamic Republic, some observers note, dams have been built to divert water to politically powerful areas, resulting in the drying up of lakes. Faced with diminishing water levels, Iran has resorted to new technical solutions, like transferring water from one region to another to support industrial plants in the bigger cities, as well as desalinating seawater, a costly, energy-intensive, and polluting practice.[57] Searching for a remedy, the government is constructing a 620-mile pipeline to bring desalinated water from the Sea of Oman to Sistan and Baluchistan province and other parts of the country. Such measures, however, are unlikely to provide sustainable solutions to Iran's rapid descent into water depletion.[58]

Protests against Water Shortage

Since 2018, Iran has been witnessing protests against water shortages in both southwestern provinces and central provinces such as Isfahan. The Islamic Republic authorities have been transferring water from the southwestern region to the central plateau of Iran for years. The impetus for the transfer scheme is most notably for the development of a single city: Isfahan. Throughout June 2018 and beyond, Arab farmers of southwestern Iran have frequently protested in front of the Ministry of Water and Energy.

Ethnic strife between Khuzestan's Arabs and the central government has dramatically increased. Many human rights NGOs have frequently pointed to very serious tensions in this area. "Concentrated in poor urban outskirts lacking in basic facilities," Amnesty International reported in 2015, "many Iranian Arabs have alleged that the government systematically discriminates against them, particularly in employment, housing, access to political office, and the exercise of cultural, civil and political rights."[59] In 2017, one local human rights organization asserted that Arabs held less than 1 percent of public and private sectors' managerial positions in Khuzestan. Since the 1979 Iranian Revolution, this source went on to note, none of Khuzestan's governors, with the exception of one, have been Arab.[60]

Arab separatists—even before recent climate and environmental difficulties—have also launched sporadic attacks against the central government. Several Ahwazi separatist groups continue to operate inside Khuzestan province and abroad. Some of these groups have called for an independent Arab state, the most prominent of which is the Arab Struggle Movement for the Liberation of Ahwaz (ASMLA), a group that consists of two rival branches that are based in Copenhagen and The Hague, as well as the Mohiuddin Nasser Martyrs Brigade, a local Arab militia that operates in Khuzestan province. The ASMLA has organized several terrorist attacks on petroleum and gas installations, government agencies, and banks since 2005.[61]

In addition to affecting the ethnic Arab population of this region, the water transfer has also put the lives and livelihoods of many Bakhtyaris at risk. Several thousand villagers and farmers in Zagros mountains, according to one report, were forced to sell their lands at low prices to the government.[62] Drinking water shortages have since the summer 2018 sparked demonstrations across the province, especially in Abadan and Khorramshahr, over the quality of drinking water from the Bahmanshir river. Many were arrested, and four demonstrators were reportedly killed. Following these events, Abadan's local government dealt with the situation by rationing water usage through various kinds of restrictions.[63]

On November 9, 2021, thousands of protesters gathered in Isfahan province in central Iran to demand the revival of a major river: Zayandeh Rood. The city of Isfahan is Iran's third largest, with a population of nearly 2 million. Its tourist attraction is due to its heritage sites, including a historic bridge that crosses the Zayandeh Rood River. As this river has dried up, the livelihoods of hundreds of thousands of farmers in the province have been jeopardized. While climate and environmental changes have been partly responsible for this situation, farmers have blamed the authorities' diversion of the river water to neighboring Yazd province. Energy Minister Ali-Akbar Mehrabian apologized to farmers for this problem and noted: "We hope to fill these gaps in the coming months."[64]

Exposing Iran's Achilles' Heel

To be sure, climate change appears to be a new and enduring threat hovering over the Islamic Republic of Iran as the country has become politically divided, ideologically fragmented, and economically insecure.[65] Iran's litany of environmental problems cannot be effectively tackled unless local and national plans undergo significant shift. If current trends are any guide of what is to come, and absent necessary adjustments by the government, Iran is likely to experience its frequent political unrest in the coming years.

Iran's then president, Seyyed Ebrahim Raisi, was invited to attend the 26th UN Climate Change Conference of the Parties (COP26) in Glasgow, Scotland. He chose not to attend. Perhaps, according to one source, the decision for Raisi not to travel to Glasgow has been due to a formal complaint filed with Scottish authorities calling for his arrest if he ever sets foot in Scotland. The plaintiffs behind the complaint included the former Scottish MEP Struan Stevenson, human rights activist Tahar Boumedra, as well as many witnesses of the 1988 massacre in Iran, in which Raisi has been implicated, and the families of the victims.[66] The summit brought leaders from more than seventy countries together to expedite action toward accomplishing the goals set in the Paris Agreement and the UNFCCC signed in 2016 by 195 signatories. Iran's former president, Hassan Rouhani, participated in the COP15 held in Copenhagen in 2009 and the 2014 UN climate summit in New York. The 2015 United Nations climate change conference in Paris, however, was attended by Iran's vice president.[67]

In general, the issue of the environment has been a nonpolitical issue for Iranian presidents. The environment has become important only when it has taken on a "security" form and threatened the security of the ruling regime.[68] In the 2021 presidential debates in Iran, the issue of the environment was not on the top of candidates' list of priority. In these debates, environmental issues were by and large placed on the backburner.[69] During these debates, the main topics of discussion revolved around economic issues and people's livelihoods. In a move that proved to be nothing but a publicity gesture, Raisi's campaign unveiled its "Environment Charter." This charter outlined in detail his government's general policies regarding wide-ranging environmental concerns, including sustainable development, water management, natural resource management, waste management, public environmental education, watershed management and development of aquifer feeding plans, underground dams as well as rehabilitation and reconstruction of wetlands and lakes, and many other environmentally friendly policies.

Some observers have pointed to saving Iran's precious Lake Urmia, which is located in northwestern Iran and is one of the largest natural permanent saltwater lakes in the world, arguing that if the current land use and water consumption patterns continue, then the lake will vanish in the near future. Barring any unexpected, great amount of rainfall, the only way to restore the lake is to take emergency actions to prevent further retreatment of the lake, while also reducing water consumption in a carefully administrated plan.[70]

The lake's increasing salinity, according to a UNEP report in February 2012, has reached near saturation and threatens to decimate the lake's brine shrimp population, which is essential to the ecology of the lake and surrounding wetlands. Finding the volume of water necessary to rejuvenate the lake, without going outside the basin, would probably require diverting water away from other key areas of irrigated agriculture.[71] Water transfer from the Caspian Sea would be costly and labor-intensive and may come too late to avert damage to the ecosystem given the historically low water levels and high salinity of the lake. Diverting water from neighboring basins would be cost-effective and swift but it also presents some serious challenges. A far-reaching and well-managed water plan would take all facets of the basin's water budget into account, balancing demands for irrigation, ecosystem protection, socioeconomic and human impact and water quality, as well as functioning within the national and regional political contexts.[72]

The Unfulfilled Promises

It remains to be seen if the state can actually fulfill his campaign promises of protecting the environment from climate change-related vulnerabilities.[73] Clearly, the environment has not been a top priority of the Iranian government. To begin with, the appointment of the head of the Environment Organization in the then Raisi administration was blatantly delayed, an indication that environmental issues will be relegated to the lower priority in his government. Furthermore, ex-president Raisi lacked a sound plan and strategy to deal with the country's environmental changes and problems.

The fact remains that the head of Iran's Environmental Protection Organization (IEPO), however theoretically powerful and important the position that one holds as vice president, must compete with other potent and important government ministries, such as the Ministry of Oil, the Ministry of Energy, the Ministry of Roads, and the Ministry of Industry that have much larger budgets and superior lobbying influence. These ministries, whose activities and programs are detrimental to the environment, will more than likely shape the government agenda absent a charismatic and dynamic head of IEPO.

Not ironically, the reason that Raisi turned his attention to issues such as water scarcity and drought was that precisely fifteen days prior to taking office as president, Iran witnessed protests in Khuzestan that swiftly spread to other parts of the country. A report by the Ministry of Agriculture showed that drought has had a dire impact on the entire country, but more specifically nearly 200 towns and villages in Khuzestan province have been adversely affected by water shortages, with their lifelines on the verge of becoming degraded or even extinct. Meanwhile, the Minister of Energy under the Raisi administration advised his subordinates to encourage people to pray for "rainfall."[74]

The magnitude, gravity, and complexity of the environmental problems at the national level cannot be effectively dealt with absent a transformational strategy and plan. Iran is headed for a multidimensional environmental crisis in which economic, political, cultural, and social issues are inextricably intertwined with the environment. Environmental policies and decisions will have significant implications for other areas. The Yale University "Environmental Performance Index" report shows that Iran lacks a fundamentally sound plan for meeting its environmental challenges. The last report published in this field was in June 2020, in which 32 indicators were measured in 11 fields in 180 countries, and the Islamic Republic of Iran was ranked 67th with a score of 48.[75]

Among the serious environmental problems that the country faces are the drying of rivers and wetlands, severe groundwater decline, desertification, biodiversity reduction, air and soil pollution, waste management, soil erosion, destruction of pastures and forests, and dust storms that threaten the life and health of Iran—not to mention the serious socioeconomic consequences of such human-caused environmental debacle. These difficulties are a direct result of several decades of the accumulation of problems ignored by governments and Iran's water and environmental governance system. It is not possible, therefore, to solve these problems immediately and overnight by making emotional appeals and setting unrealistic goals.[76]

Another important issue in relation to environmental issues is the Raisi administration's attitude toward the Paris Climate Agreement. The attempt to join the Paris Climate Agreement was approved in the previous term of the Islamic Consultative Assembly but was not approved by the Guardian Council, which raised two key objections to it. In order for the resolutions of the Iranian parliament to become law, they must be approved by the Guardian Council. The Guardian Council's key objections were: (1) Iran's obligations were not fully specified in the government proposal and (2) Article 4, section 8 of the Paris Climate Agreement ("In communicating their nationally determined contributions, all Parties shall provide the information necessary for clarity, transparency, and understanding in accordance with decision 1/CP.21 and any relevant decisions of the Conference of the Parties serving as the meeting of the Parties to this Agreement") will considerably limit the operations of Iran's national industries in regard to GHG emissions.

In both Iran's parliament and the Guardian Council, "the process of joining and even implementing the commitment in this regard remained unresolved and vague," said Massoud Tajrishi, deputy director of the human environment in the Rouhani administration. At the Paris summit, Iran pledged to reduce its GHG emissions by 4 percent unconditionally and 8 percent on the condition that sanctions be lifted by 2030, but his attempt to do so was opposed in the implementation phase: "Iran has not officially and legally joined the Paris Agreement and therefore we have not taken any action in this regard so far," Massoud Tajrishi told IRNA.[77]

The fact remains that Iran will be one of the most vulnerable countries in the world more generally and in the Middle East in particular due to the hazardous effects of climate change. Iran is currently one of the ten most polluting countries that still relies on fossil fuels. Despite these realities, the Raisi administration showed no sign of joining and enforcing the Paris Climate Agreement.[78]

It is safe to assume that joining the Paris Agreement by the Raisi administration became inextricably linked to the outcome of the nuclear deal (JCPOA). The Raisi administration insisted that if no progress was made in the JCPOA talks and if the past sanctions remained in place, it was highly unlikely that Iran would abide by the Paris Climate Agreement.[79] One key question remains: Who is to be blamed for such conditions: Iran's government, climate change effects, or the ongoing US sanctions? The answer may lie in all three. But government mismanagement of water resources was said to be the main contributor to this crisis. Over 90 percent of Iran's water is used for agriculture. Due to international sanctions, the government wishes to render the country self-sufficient, thus encouraging farmers to grow water-intensive crops like wheat, rice, and sugarcane.[80]

This approach has proven costly and ineffective as the Iranian government has failed to come up with novel methods to explore renewable energy applications and to mitigate GHG emissions in order to mitigate the various risks of climate change vulnerabilities. As long as the government prioritizes ideology over practical solutions to address the effects of climate change, the country will find itself in the constant and tight grip of protests and riots in the coming years. The nonideological nature of this threat, along with its all-embracing scope, renders it virtually impossible to ignore. If not properly tackled, the impacts of climate change are likely to spell disaster for the government. Any leader who writes off the effects of climate change is setting herself/himself up for failure.

NGOs and Civil Society Reactions

The activities of Iran's nongovernmental organizations in the environmental arena began with the UNCED in 1992, also known as the "Rio Summit," when Iranian NGOs' activities in the field of environment flourished. Activists and their attendant NGOs sought to address many issues, including air pollution, forest and rangeland protection, and the protection of endangered animal species. By 2010, water shortages in different parts of the country had raised the concerns of the public and environmentalists alike. The water shortage became an environmental inflection point for civic activism in the country's history. In recent years, the participation of NGOs has gained prominence not only in the governmental sphere but also in the parliament and the judiciary.

At the same time, the development of digital media and social networks has played a key role in bringing to the fore lingering environmental concerns while also conferring legitimacy and attention on the activities of environmental NGOs. Today, many Iranians have become acutely aware of the adverse impacts of climate change as the country faces many challenges, especially land subsidence in Neyshabur, the drying of Lake Urmia and the Gavkhoni Wetland, and the ongoing dust storms in Khuzestan and Sistan and Baluchistan. Since the early twenty-first century, various international organizations have provided financial assistance to these NGOs in order to cope with global warming and climate change, desertification, and the protection of endangered species.

Consequently, many environmental NGOs have been formed with the task of confronting climate and environmental changes in Iran. NGOs' activities, however, have failed to sustain their efforts over time, in large part due to budgetary problems, but also because many of these NGOs risk becoming a contract company rather than holding onto their key mission, which is operating as independent agencies, mission-driven advocacy, or service organizations in the nonprofit sector.[81]

Making matters worse, recently the government has arrested and imprisoned many environmental activists. In February 2018, eight environmental activists were arrested.[82] Many of these detainees were the members of an NGO called the Persian Wildlife Heritage Foundation, a private nonprofit organization dedicated to the conservation and preservation of wildlife in Iran, as well as others who collaborated with the Lavardin Nature Conservation Institute. These individuals were sentenced by the Iranian judiciary to several years in prison on charges of espionage and separatism, among other accusations, despite the fact that Iran's Department of Environment claimed that there was no evidence to prove such charges.[83]

In February 2018, another detainee, the Iranian-Canadian professor, Kavous Seyed-Emami, died under suspicious circumstances in the notorious Evin Prison in Tehran two days after his arrest.[84] Prison officials claimed that the cause of death of this pro-environmental activist was suicide.[85] Issa Kalantari, the head of Iran's Department of Environment under Rouhani's government, was banned by the Iranian judiciary from playing any role in environmental cases. In fact, the judiciary took full control of the case of detainees of environmental activists, even though the Rouhani administration vehemently opposed such arrests.[86] The activities of environmental activists have been regularly restricted by the Iranian intelligence agencies. In 2018, Kaveh Madani, the former government deputy minister of the Department of Environment, who returned to Iran at the invitation of the Rouhani government, spoke out about the criminalization of environmental defenders in Iran. He was later accused of espionage and was forced to depart the country.[87]

Despite all these problems, challenges, and difficulties that activists and NGOs have encountered, social networks have raised widespread awareness and informed ordinary citizens, while raising objections to projects that are harmful to the environment. These networks have sparked outrage and mobilized several nationwide protests, forcing the Iranian government to either back down or respond effectively. Consider, for example, the case of building a petrochemical plant in the Miankaleh Wildlife Sanctuary and Wetland, the construction of which had already begun. Pressure from environmental activists halted the project, and as a result, the Supreme Council of Environment, Iran's highest environmental body, terminated the construction contract of Miankaleh Petrochemical.[88]

Using official information from the World Economic Forum and the IMF, Reporters Without Borders, and other sources, the Legatum Institute publishes an annual global ranking of more than 170 countries based on its Welfare Index. This index also includes individual freedoms and other markers of prosperity. Iran's ranking dropped from 107th in 2014 to 117th in 2016. Iran's downgrade in individual freedoms was even more discouraging, indicating a drop from 128th in 2014 to 144th in 2017.[89]

Today, environmental activists in Iran face various security charges, including espionage, separatism, and ethnic provocations. For years, environmental activists in East and West Azerbaijan have been accused of separatist and ethnic activities in support of Lake Urmia. This oppression has also happened to some Khuzestani Arab protesters in recent years. A similar charge imposed severe restrictions on environmental groups and campaigns, particularly in Kurdistan and Sanandaj. Given that environmental problems are often significantly more pervasive in the border provinces, and considering the fact that a higher percentage of the population of Iranian ethnic minorities live in these provinces, the judicial and security system have almost always applied separatist label to these areas' environmental activities.

Likewise, religious minorities face similar restrictions on participating in environmental activities. These activists are often accused of religious propaganda or simply belonging to a religious group, regardless of their environmental activities. The environmental activism of a group of Baha'is in Shiraz, for example, and their subsequent detention, as well as the restrictions imposed on a group of Sunni activists in Sistan and Baluchistan province, or the Yarsan group in Kermanshah, are all cases that show a clear religious persecution angle with environmental activism as the true source of the conflict.[90]

Following the 2018 protests, security, judicial, and executive organizations have become cognizant of the magnitude and acuteness of the environmental concerns of the protesters. These protests also broadly reflect other issues related to social injustice, lack of access to natural resources, and other problems in the daily lives of residents of villages and small- and medium-sized cities. Environmental NGOs developed a special mechanism and method of coordination and information to create empathy in achieving their goals of using all legal and judicial channels to oppose antienvironmental actions, and this model could have been successful in the absence of an independent union. Both the location of rallies and protests and environmental activities are geopolitically sensitive to the security and military system, and most military and governmental institutions and even subdivisions of ministries are somehow implicated in the abuse of natural resources.

The long-term consequences of these protests could spell disaster for some government officials. As Iranian citizens continue to experience water crises, air pollution, drought, deforestation, global warming, and climate-induced impacts—problems that make them restless and prone to taking to the streets in an attempt to protest against the government—the Islamic Republic will most likely gravitate toward viewing such environmental activities as posing a direct threat to its security and the country's political stability.

The Dispute over the Helmand River

The Helmand River is the longest river in Afghanistan and the primary drainage divide for the Sistan Basin. A historic waterway, the river flows from southwestern Afghanistan to eastern Iran. It holds significant ecological importance as it nourishes the Hamoun wetland and the Goud-e Zareh, a salt lake situated in the Sistan endorheic basin. Additionally, both Iran and Afghanistan heavily depend on the Helmand River for drinking water supply and agricultural irrigation.[91]

Some observers even argue that water disputes between Iran and Afghanistan date back to as early as the 1870s when Afghanistan was under British control. British officials drew the Iran-Afghan border along the main branch of the Helmand River.

In 1939, Iran's Reza Shah Pahlavi and Mohammad Zahir Shah of Afghanistan signed a treaty to share the river's waters, but the Afghans refused to ratify it.[92]

In 1948, another attempt to resolve the dispute was initiated by Washington. Based on a US suggestion, a three-person commission was selected by Iran and Afghanistan to investigate the issue and offer a settlement. On February 28, 1951, the Helmand River Delta Commission presented its report, recommending that Iran's share of the Helmand waters amount to 22 cubic meters per second. Iran, however, rejected that recommendation, asking for a larger share. Afghanistan had offered to provide more water only if Iran would give Afghanistan better access to the Iranian ports at Chabahar and Bandar Abbas, as well as development assistance.[93]

Although a treaty was concluded in 1973 to address the sharing of this cross-border resource, the dispute over the water was not fully resolved. Each side continues to accuse the other of violating the provisions delineated in the agreement.[94] The 1973 Treaty falls far short of addressing the realities of evolving climate conditions. Although it incorporates provisions to consider climatic events, droughts, floods, and natural disasters, it lacks specific provisions regarding the development of mitigation and adaptation measures to tackle the adverse effects of climate change on the Helmand River. To reduce conflict and foster sustainable collaboration over transboundary water resources, it is important to turn to international water law while also striving toward establishing a coherent international climate change regime.[95]

The Helmand River is vitally significant for farmers in both Afghanistan and Iran's southeastern Sistan and Baluchistan province. A number of hydroelectric dams have created artificial reservoirs on some of the Afghanistan's rivers, including the Kajaki Dam on the Helmand River. In recent years, the water scarcity in Afghanistan has become an acute issue for the country and its leaders. The drought has been exacerbated by climate change, further intensifying pressure on water resources. The Global Climate Risk Index indicates that Afghanistan is the world's sixth most affected country by climate-related threats.[96]

With the US withdrawal from Afghanistan and the Taliban's return to power in Kabul since 2021, however, the Helmand River has become a main point of contention between the neighboring countries.[97] The 621-mile-long river, which flows across Afghanistan and empties into lakes in southeastern Iran, is an essential source of water for both countries. It has been critical to agriculture, fishing, and human consumption for millions of people in Afghanistan and Iran's southeastern Sistan and Baluchistan province.[98] More recently, the Taliban regime has created several hydroelectric dams that have increased artificial reservoirs on some of the Afghanistan's rivers, including the Kajaki Dam on the Helmand River.[99]

This development has had far-reaching implications for the Iranian side, as dry soil has also left Sistan and Baluchistan vulnerable to flash floods. The Indian monsoon, which largely affects the Indian subcontinent and its surroundings, has also adversely affected this region. In January 2020, three days of heavy rain led to the displacement of thousands of people in Sistan and Baluchistan region, killing at least three and destroying crops, homes, and infrastructure. The floods caused at least $53 million in damage, while also exposing the Iranian government's limited capacity to provide relief to this region's people.[100]

Despite the fact that both countries have taken steps to revitalize the Hamoun wetland on the Iran-Afghanistan border during the 2015–16 period, recent tensions point to further conflicts over the water rights between the two countries. Some Iranian experts have noted that the Islamic Republic has failed to invest in improving the water management system by missing numerous opportunities. Over the past four decades, Iranian authorities have not only ignored experts' warnings but also have arrested and imprisoned environmental activists who have raised legitimate concerns about the water crisis. Without a lasting agreement between Iran and Afghanistan, we are likely to see mass migration and further conflicts between the two countries in the future due to water and climate change.[101]

Conclusion

Counteracting global warming and climate change in Iran is a more complex and daunting task than the government admits or even realizes. In the coming years, Iran is likely to experience more extended periods of acute high temperatures as well as more frequent dry weather and deluges, adversely affecting climate change there. Many factors have contributed to Iran's existing environmental crisis, including, but not limited to, drought, air pollution, dust storms, sanctions, mismanagement, and the lack of enforcement of existing environmental regulations.

Furthermore, socioeconomic, political, and institutional complications are likely to render meeting the challenge of climate and environmental changes all the more difficult. Civil unrest and street protests in reaction to government inaction and poor planning—especially with regard to water shortages—are likely to grow. Given the rise in temperatures and the decline in precipitation, it is easy to predict that climate change will pose serious challenges to the country's ruling elites. The ongoing mismanagement and corruption would most likely complicate tackling climate changes effectively. A core challenge remains to be how to strike a sensible balance between policies aimed at installing climate mitigation measures on the one hand, and relying on oil and natural gas industry to finance its developmental goals.

Thus far, the Iranian authorities' efforts to cope with water shortages have been ineffective to say the least, illustrating a lack of a formidable water strategy. This failure is likely to further undermine the legitimacy of the current governing system in the eyes of many ordinary Iranians. Without an effective strategy, social tensions and unrest will spike, putting additional pressure on large cities and thus the central government to manage the impacts of climate change.[102] Iran would do well by cooperating with its neighbors to seek long-term solutions for its ecological variation, especially as it relates to water scarcity.

On a broader level, assigning a higher priority in Iran's foreign policy to geopolitical considerations and regional security issues has eclipsed the country's environmental issues. If current trends continue, perpetuated by poor governance in terms of water management, it is not unreasonable to assume that water shortages and ensuing food insecurity will most likely threaten Iran's future control over its resource structure. This mismanagement further heightens the existing socioeconomic and ecological pressures stemming from both a changing climate and woefully inadequate planning. Poor governance under such circumstances will ultimately lead to the waves of displaced people and climate refugees.[103]

CHAPTER 10
GCC
CLIMATE CHANGE AND THE POLITICAL ECONOMY OF ADAPTATION

The GCC was established in 1981 among Bahrain, Kuwait, Oman, Qatar, Saudi Arabia, and the UAE with the aim of fostering cultural, socioeconomic, industrial, military, tourism, environmental, and geopolitical cooperation while also promoting Islamic beliefs and common ideological objectives in the region. These six states constitute 2.5 million square kilometers of land mass, amounting to 82 percent of the total area of the Arabian Peninsula.[1] Over the past half century, these oil-rich Arab monarchies have undergone unprecedented technological and social change, while also surviving the 2011 Arab Spring uprisings without serious challenges to their political stability. These monarchies' strategic visions and economic growth, however, have relied exclusively and for too long on resource affluence and security. Thus, pursuing conflicting priorities has made GCC countries' relationship with climate change—in terms of sustainable development and living contexts—even more challenging.

Over the longer term, as experts warn, balancing such dramatic growth rates with environmental sustainability is likely to become extremely difficult—a reality that is emblematic of a larger problem of heavy dependency on fossil fuels. The variations in the structure of ecosystems in the region are less significant compared to catastrophic disasters such as hurricanes, tsunamis, and floods in other parts of the world. Yet, the economic impact in the GCC countries will be far more severe, as they largely depend on revenues from oil and gas exports. Understandably, but paradoxically, if the world shifts to other renewable sources of energy, the GCC countries will dramatically suffer from the loss of a major source of their revenue.[2]

The rentier systems that prevail in the GCC states severely handicap meaningful action on climate change. Revenues from the sale of oil and gas undergird the ruling bargain and social contract in GCC countries. When seen in this light, it becomes apparent that climate change poses an existential threat to the GCC countries.[3] Echoing a similar sentiment, Mari Luomi notes: "Not only is the Gulf Cooperation Council (GCC) states' environment under stress, but also is the continuation of their status quo of power and wealth."[4]

With economies largely dependent on the export of fossil fuels, and the persistence of an allocation system that protects the state-citizen rentier bargain, GCC member states have long avoided a robust commitment to climate change goals.[5] Given the link between climate change on the one hand and the persistent energy production and consumption on the other, environmentally sensitive projects, such as low-carbon energy and renewable technology, sustainable ecosystem, and alternative energy, have received greater attention than before. As the global economy gravitates slowly but steadily toward a low-carbon era, each monarchy in the region will have to take necessary mitigation and adaptation measures. These monarchies have begun to pursue environmentally sustainable plans with different latitude, speed, and intensity. The most innovative attempts at the so-called green diversification have emanated from Abu Dhabi and Dubai in the UAE, as well as from Qatar and Saudi Arabia.[6]

While GCC countries have publicly announced their plans for renewable energy and technologies that emit low levels of CO_2 emissions, or the so-called low-carbon emitting technologies such as hydrogen production, they have continued to rely heavily on the export of hydrocarbons. Coming to grips with climate change by the GCC countries could entail significant economic costs in the short term but would certainly have positive longer-term implications for the region and beyond. All GCC states plan to reduce emissions during this and next decades.[7] By 2030, however, Saudi Arabia will still set to account for 1 percent of global emissions, the eighteenth highest in the world, and the GCC region will account for 3.2 percent of global emissions, with only four countries—China, India, the United States, and Russia—contributing more.[8]

As largely rentier states, fossil fuels significantly contribute to GDP. Thus, policies to tackle climate change, such as carbon taxation and border carbon adjustments, will most likely hit GCC economies hard.[9] Oxford economists have estimated that the economy of the GCC could be nearly 20 percent smaller by 2050 if temperatures rise by 2 degrees Celsius.[10] The GCC countries, however, are capable of making the transition to solar energy. Green hydrogen also has potential to render that transition possible. Both solar and hydrogen technologies are now being developed in the UAE and Saudi Arabia by the Abu Dhabi National Oil Company and Saudi Aramco. This policy would also allow the region to maintain a competitive edge in sectors that are unlikely to totally surrender their dependence on oil.[11]

It is safe to argue that attitudes have shifted recently in the face of mounting international pressure. There is a growing focus on deploying the Circular Carbon Economy aimed at reducing, reusing, recycling, and removing GHG emissions from the ecosystem. This depends, among other things, on diversifying energy sources to renewables, nuclear energy, and hydrogen, enhancing energy efficiency, while also developing Carbon Capture and Storage and Direct Air Capture technologies.[12]

In this chapter, I attempt to answer several questions: How have GCC states responded to emerging climate and environmental changes? Can the GCC countries pursue low carbon development strategies given that their economic development plans have almost always been predicated upon their fossil fuel revenues? How committed are the GCC countries to planning for the post-oil era, given that the diversification of their economies away from oil will entail some short-term costs? Should the GCC member states be compensated for any climate change adaptation-related costs or for the revenue losses resulting from climate change mitigation?

This chapter offers an overview of the challenges that the GCC states are facing in the global low-carbon era and what could be done to help alleviate their transitional burdens. The chapter goes on to focus on several case studies of the UAE (Dubai and Abu Dhabi), Qatar, and Saudi Arabia. The ensuing section examines the post-petroleum region and its future scenarios. In the final and concluding sections, the issues of forced migration and climate displacement are explored with an eye toward explaining the possible initiatives GCC states could be taking.

The GCC in the Global Low-Carbon Era

Known for their water scarcity, arid lands, deserts, dust storms, and opulent fossil fuels, the Arab countries of the Persian Gulf region have come under tremendous stress to cope with new environmental realities in the face of climate change. The skyrocketing growth in energy consumption, along with a rise in GDP from oil and gas sales, has contributed to pollution in the region. At the same time, sandstorms and blowing dust have become recurring events throughout the year. These storms, which release millions of tons of dust into the atmosphere, tend to reduce soil fertility, damage crops, reduce solar radiation on the earth's surface, dismantle communications and mechanical systems, and cause lung disease and other public health issues. Global warming will make dust storms more intense in the region and beyond. The radiation effects of different types of dust storms will have dire climate change consequences.[13]

The region is atypical in its heavy reliance on thermal desalination and electricity generation from fossil fuel sources. For this reason, as some experts remind us, renewable energy (solar and wind) can be a viable alternative to fossil fuels in the region in the coming years and decades.[14] The region has one of the highest solar potential in the world owing to the large annual availability of solar radiation.[15] GCC countries also constitute good candidates for wind energy applications. In particular, the region's western mountains offer a plethora of wind energy resources.[16] Despite the sizeable solar and wind energy resources available in the GCC region, the use of such resources has been rather insignificant. The total electrical power generated from solar and wind energy, according to the 2019 International Renewable Energy Agency (IRENA), represented only 0.6 percent of total energy generation in the GCC.[17]

The Gulf Arab countries have often argued against using the per capita metric, suggesting that end users of fossil fuels, rather than producers, should be held responsible for hydrocarbon-related emissions. Qatar has noted its LNG exports have boosted global mitigation measures given the lower emission potency of natural gas compared to other fossil fuels.[18] On balance, energy efficiency has received great attention in policy discussions and measures of Gulf countries, even as it has not been fully discussed within a mitigation framework.[19]

According to the Paris Agreement (2015), NDCs include efforts by each country to reduce national emissions and adapt to the impacts of climate change. The Paris Agreement (Article 4, paragraph 2) requires each party to prepare, communicate, and maintain successive NDCs that it intends to fulfill. Member states shall pursue domestic mitigation measures, with a view toward making such contributions. Almost 80 percent of the initial NDCs submitted to parties in the Paris Agreement contain GHG reduction targets, and only a meager number contain "actions" only. Of the six GCC countries, according to one study, only Oman has established a quantitative emission reduction target.[20]

Given their high-income status, GCC states are on average better positioned to adapt to climate change and build resilience against the harsh impacts of climate change than their regional neighbors. Whereas in terms of per capita emissions, GCC countries rank among the top in the world, their combined renewable electricity capacity was noticeably low in global comparison, at 2.4 gigawatts in 2019. The UAE's solar energy volume, however, stood as the highest in the Middle East around that time.[21]

Governments in the region have set lofty sustainability targets, with the UAE Energy Strategy targeting 50 percent clean energy by 2050 (with 44 percent renewables, 38 percent gas, 12 percent clean coal, and 6 percent nuclear), while Saudi Arabia has announced its plan to generate over 27 gigawatts of clean power by 2023 and over 58 gigawatts by 2030.[22] At the international level, Saudi Arabia kept a major focus on its interests in oil sector, while the UAE has consistently moved forward toward becoming a regional leader in the sustainable energy transition.[23] On balance, the incremental approach of the GCC states to combat the impacts of climate change continues to supersede transformative strategies and actions.[24]

The UAE and Climate Change

Revenues from hydrocarbons have long undergirded the government's ability to sustain the so-called social contract, particularly in the UAE but in the GCC countries more generally—that is, lower taxes and subsidies have brought about economic growth and development in other sectors and industries in the face of global competition. Although the introduction of a corporate tax in the UAE has become part of efforts to raise government revenues in non-oil sectors, it is wrong to assume a fundamental shift of the country's economy at this juncture.[25]

In 2005, the UAE ratified the Kyoto Protocol to the UN Convention on Climate Change, becoming one of the first major oil-producing countries to do so. On October

7, 2021, the UAE announced a plan for NZEs by 2050 and would administer 600 billion dirhams ($163 billion) in investment in renewable energy.[26] This major policy pronouncement was made by Sheikh Mohammed bin Rashid Al Maktoum, vice president and prime minister of the UAE and ruler of Dubai: "We are committed to seize the opportunity to cement our leadership on climate change within our region and take this key economic opportunity to drive development, growth and new jobs as we pivot our economy and nation to net zero."[27]

The steady temperature rises caused by global warming have noticeably contributed to the UAE's water scarcity, drought, sea level rise, and dryness. The UAE's farmlands, which have long experienced intense aridity, rare precipitation, and high temperatures, have faced extended periods of dryness. In recent years, the UAE has reduced its emissions of per capita CO_2 due largely to deploying better technology and transition to more natural gas in power plants. To the extent that environmental sustainability is concerned, the UAE officials have attempted to diversify the government's revenue by moving away from oil. The UAE, for example, has led the other GCC countries in terms of providing an environment conducive to business and entrepreneurship. The UAE holds forty-five free zones, granting 100 percent foreign ownership.[28]

In January 2012, Sheikh Mohammed, the governor of Dubai, launched the Green Economy initiative under the slogan: "A green economy for sustainable development." In line with this policy, the UAE has made efforts to protect the environment by banning sea dumping, implementing policies and legislation, and making efforts toward sustainability. During the last few years, Dubai has taken many measures to contribute to the quality of the environment. It has built green buildings, enhanced the use of renewable energy, planted trees, and cut carbon emissions. In 2016, the UAE renamed the Ministry of Environment and Water as the Ministry of Climate Change and Environment, thus officially placing the management of climate change within the scope of this ministry.[29]

In recent years, the UAE and Saudi Arabia have been investing billions of dollars in solar farms around the world, and the Saudi Green Initiative aims to boost renewables to half of the country's energy mix, up from 0.3 percent in 2019, mostly using solar power. The skills these countries gain deploying large-scale solar projects at home and abroad can potentially make them global leaders in the implementation of solar energy.[30] The UAE's National Climate Adaptation Action plan is divided into three parts that are built on ensuring clear and targeted outcomes. These parts include (1) identifying climate trends and assessing their impacts, (2) identifying impacts and demanding urgent action, and (3) involving all sectors (private sector, research institutions, civil society and youth, and government entities) in implementation.[31]

The UAE has also been especially aggressive in aligning itself with global efforts to promote renewable energy. It hosts the UN-affiliated IRENA, and fifteen years ago launched Masdar, a platform for developing sustainable energy solutions based around what aspires to be a zero-carbon city. The strategy appears to be to not only align the UAE with current energy production patterns, but to get in early on future ones as well. One technology that could emerge as important in coming years is hydrogen. Commercializing hydrogen as a fuel source could draw on assets widely available in the Middle East—either natural gas for blue hydrogen or sunlight for green hydrogen—but the financial feasibility will depend on scientific advances that have not yet been made.[32]

It should be noted, however, that the UAE, which has set net-zero targets for 2050, has fallen short of its commitment to reduce oil production. While the government of the UAE has listed renewable and nuclear energy as important elements of reaching the net zero target, it has revealed no particular strategy and/or plan to move steadily toward achieving its targets for 2050. As of November 2021, the UAE has yet to submit a long-term strategy to the UNFCCC.[33]

Abu Dhabi's Energy Source Diversification

Attempts to address climate change in the UAE began in 2006 with the creation of Masdar City. Located in and owned by Abu Dhabi, Masdar (Arabic for "source") consists of a set of projects aimed at developing REA and carbon management. It is the first city in the world designed to be based on an ecological and sustainable foundation. It is also one of the world's most viable and renewable urban communities, structured on a low-carbon development, including residential neighborhood with restaurants, shops, and public green spaces. Local renewable sources provide the city with a solar power station, a wind power plant, geothermal energy, and the world's largest hydrogen power plant. Masdar City hosts the World Future Energy Summit every year.[34]

Germany has signed several hydrogen cooperation contracts with the UAE. The latter's ideal sunny weather conditions have made possible the production of cost-effective blue hydrogen from renewable energies. Germany has long sought, but especially in the aftermath of the Russian invasion of Ukraine, to import a significant share of the 3 million tons of "clean" hydrogen it aims to use by 2030. Thus, securing contracts with potential exporters such as the UAE has become a top priority.[35] In October 2022, UAE delivered its first shipment of ammonia derived from hydrogen to Hamburg, Germany. The latter plans to satisfy its growing demand for hydrogen largely through imports.[36]

The water management will be environmentally friendly by means of a solar-powered desalination plant and 80 percent wastewater recycling. This project has four main aims: (1) to diversify the economy, (2) maintain and expand Abu Dhabi's position in global energy technology, (3) transform the emirate into a developer and exporters of energy technology, and (4) contribute to sustainable development.[37] The most ambitious of all alternative energy initiative in the region was Abu Dhabi's announcement in 2008 to establish a civilian nuclear program. Both France and the United States assisted the UAE to promote its nuclear energy. Abu Dhabi stressed the peaceful nature of this project, while making a case for the need to develop additional sources of electricity to meet its increasing domestic demand.[38]

Several factors lay behind the UAE's nuclear program, some of which had to do with the regional geopolitics such as the rising security concerns regarding Iran's nuclear program and plans, as well as the long-standing dispute over three Islands near the Strait of Hormuz, namely: Abu Musa and the Greater and Lesser Tunbs.[39] There were, however, clear climate change-related factors accounting for seeking such a project. It is doubtful that international pressure to cut GHG emissions was in fact the central motif for launching such an adventure.

Qatar's Climate Change Governance

Qatar provides a prototypical example of an oil-rich rentier state: it is extremely wealthy, small in size, homogeneous in national population, and has the third highest natural gas reserves in the world. These characteristics have rendered the persistence of the ruling bargain and related patronage networks possible, with the autonomy of the state left intact. Qatar has a desert climate and often experiences long summers from May to September characterized by intense dry heat, with temperatures rising above 45 degrees Celsius. Winter temperatures are mild but may drop below 5 degrees Celsius. Rainfall is only experienced during the winter with the northern parts of the country receiving 30 percent more rainfall than the south. Qatar is impacted by Shamal winds, which cause sand and dust storms throughout the year.[40]

Qatar is the world's largest producer of LNG and aims to expand LNG production from 77 million tons in 2019 to 126 million tons per annum by 2027.[41] Qatari officials have insisted that their country's gas production helps curb climate change globally because it can help the world change from high-polluting industries reliant on oil and coal, to renewable energies.[42] A contrasting view provided by the IPCC states that Qatar has the highest anthropogenic CO_2 emissions per capita, even as it has contributed to reducing CO_2 by supplying the world with LNG.[43]

Since the Ministry of Environment was established in 2008, Qatar was expected to move closer to fulfilling its international environmental obligations. The persistent dominance of the energy and industrial sectors over environmental issues, however, undermined the authority of the newly established Ministry of Environment.[44] The architects of the Qatar National Vision 2030 admit that "even with Qatar's best efforts, it is impossible to entirely avoid harming the environment, given a development pattern that depends in its early stages on oil, gas, petrochemicals and heavy industries."[45]

In recent years, however, several initiatives have been taken to address urban water management, preserve biodiversity, and monitor air quality. Pollution in Qatar is a recurring issue. At the beginning of 2021, Doha was experiencing air quality that was classified as "Unhealthy for sensitive groups" with a US Air Quality Index reading of 160 with PM2.5 air pollution in Doha by August 1, 2024.[46]

In late 2021, Qatar launched a national climate change action plan aimed at achieving a 25 percent reduction in GHG emissions by 2030. The plan also envisioned reducing "carbon intensity" of its LNG facilities by 25 percent by the same year. Qatar's move follows other Gulf Arab states, including Saudi Arabia, which announced an NZE target by 2060 ahead of the COP26 climate change summit in Glasgow.[47]

Qatar's National Vision 2030

As a country that once hosted the UN Conference on Climate Change (COP18), Qatar dramatically increased its efforts to address climate change by creating the Ministry of Municipality and Climate Change and launching the Qatar National Environment and Climate Change Strategy (QNE) in 2021. In sync with Qatar National Vision 2030, Doha played an active role in the 26th session of the conference of the UNFCCC (COP26), in Glasgow, Scotland.

Qatar has also promoted sustainable socioeconomic development in an economy based largely on the export of LNG and related products. QNE has targeted five core environmental priority areas: GHG emissions, air quality, biodiversity, water and circular economy, and waste management and land use, putting in place a system capable of reducing global GHG emissions by 25 percent and increasing the number of biodiversity reserves in the near future.[48]

Qatar authorities also enacted a comprehensive plan to monitor the air quality around stadiums and training venues for the FIFA World Cup Qatar 2022, in cooperation with the Supreme Committee for Delivery and Legacy, in choosing the proper locations compatible with the highest international standards. Furthermore, Qatar implemented many environmental plans to confine the effects of climate change, including the "Plant a million trees" initiative prior to the FIFA World Cup Qatar 2022.[49] In recent years, Qatar has assigned air quality as one of its highest priorities and has adopted effective projects to complete it through the development of a unified national network for monitoring air quality. The country has forty air quality monitoring stations that are expected to reach over forty-five stations by the end of 2024.[50]

Consistent with its 2030 Agenda for Sustainable Development, under Sheikh Tamim's leadership, Qatar has pursued several National Development Strategies (NDS) to tackle socioeconomic (food and water security and human development) and environmental and climate changes (renewables and intergenerational justice), as well as shifting to green building projects such as the Qatar National Convention Center, Msheireb City and Lusail City as all their buildings are environment friendly.[51]

These strategies have targeted developing and implementing national plans for adaptation and mitigation of climate change impacts while elevating local, regional, and international strategic partnerships. These plans included NDS-1 (2011–16) and NDS-2 (2018–22), and the country's officials are currently determined to integrate this trend into NDS-3 (2023–7). Doha continues to focus on achieving sustainable results in all its development policy priorities by cooperating and forming partnership with the public and private sectors, CSOs, universities, and research centers.[52]

Saudi Arabia: Trapped in Fossil Fuel Dependency

The Kingdom of Saudi Arabia covers about four-fifths of the Arabian Peninsula. With a population of around 35 million, it constitutes an important land connection between Africa and southwest Asia. It is the largest petroleum exporter, with its long coastlines on the Persian Gulf and the Red Sea providing large shipping access through the Persian Gulf and the Suez Canal.[53] Saudi climate is described as a semiarid to arid desert environment. More than a third of the country is desert. With the exception of the Asir region, the country has experienced declining precipitation levels.[54] Average annual rainfall is approximately 9 centimeters (3.5 inches). Much of the Rub' al-Khali (Empty Quarter) continues to be "hyper-arid," with no rainfall for more than twelve consecutive months.[55]

The rising temperatures will most likely threaten water supplies, making large parts of the country and the GCC region unsafe and uninhabitable. The region remains heavily reliant on desalination to meet its water demand. Desalination, however, is particularly energy-intensive and also subsidized, leading to overuse and waste.[56] In 2019, nearly 60 percent of Saudi Arabia's water came from desalination, with nonrenewable groundwater (less than 40 percent) taking the bulk of the remaining share.[57]

Many experts note that heat-related deaths in Saudi Arabia have risen, with temperatures passing 50 degrees Celsius in some parts of the country during the summer. It is predicted that extreme heat will most likely make cities, including Jeddah, immensely difficult to live in within thirty years. Projections show that without serious plans and strategies to reduce emissions, the Arabian Peninsula will likely be too hot for human habitation within eighty years, as temperatures far exceed the physical limitations of human bodies.[58] Given this reality, the question arises: Can Saudi Arabia survive a global shift to green energy? The answer is a resounding "no." The fact remains that fossil fuels produced by Aramco accounted for approximately 4.8 percent of global emissions in 2018 and roughly 4.3 percent of total emissions since 1965.[59]

The Arabian Peninsula (and by extension Saudi Arabia) is one of the dustiest areas in the world. Three main sources of dust that can adversely affect the region during the year can be identified: (1) the Rub Al Khali Desert, which represents a large regional source of dust; (2) the deserts of Iraq, which constitute dust sources for northern and eastern Saudi Arabia; and (3) the Saharan Desert, which is the key distant source of dust for western Saudi Arabia.[60]

Due to its topography, drought, sandy topsoil, and insufficient fertile shield, Saudi Arabia is susceptible to dust and sandstorms, which are often caused by strong, steady winds on dry and uncompacted soil. Periods of extreme and extensive drought are likely to dramatically increase the likelihood of major dust storms.[61] In May 2022, sand and dust storms hit Riyadh and other parts of Saudi Arabia, resulting in the hospitalization of nearly 1,200 people suffering from breathing difficulties.[62] In the northern parts of the country, these storms combine with strong winds that overwhelm Kuwait. Some of these sand and dust storms are caused by the extremely dry *al-Bawareh* winds, which are stirred up by atmospheric pressure on the eastern shore of Mediterranean and moved rapidly onto the eastern Arabian Peninsula and the Rub al Khali Desert, as well as Iraq and Kuwait.[63]

Additionally, population growth and urbanization, economic development, consumption patterns, electricity generation, and increasing demand for water and energy have presented daunting environmental challenges for Saudi Arabia and the rest of the GCC countries. By 2017, Saudi's share of global emissions had unremittingly risen to about 1.62 percent of global CO_2 emissions.[64] Today, the Saudis burn more oil to generate electricity.[65] The Saudi government plans to generate 54 gigawatts of renewable energy by 2040.[66] Riyadh, according to *Arab News*, has highlighted its interest to reduce the emission of GHGs, revealing its commitment to achieve net carbon neutrality by 2060.[67]

Saudi Arabia updated its NDCs in October 2021, prior to COP26 in Glasgow, with the aim of publicizing its pledge to meet net-zero carbon emissions by 2060, reducing or avoiding GHG emissions by 278 million tons of CO_2 equivalent to a year by 2030 and invest $187 billion on climate project.[68] It is unclear, however, whether the target date of 2060 will cover CO_2 emissions only or all GHGs. Further, Saudi Arabia's NDC, based on the ultimate goal of the Paris Agreement's 1.5 degrees Celsius temperature limit, remains "critically insufficient."[69]

The Post-petroleum Middle East

One of the most frequently asked questions is: What is the proper strategy for a transition from an oil-based economy to a more reliable and sustainable economy in the region?[70] A great of deal of uncertainty surrounds the governance gap, as regards "the viability of these countries in the globalized economy if they can no longer rely on oil and gas exports, or if they run out."[71] Increasingly, GCC countries have come to realize that in order to reduce the environmental damage and climate impact from fossil fuel consumption, the move to low-carbon emission and alternative energy sources is inevitable.

Few within the region, however, are naive enough to think that making the transition to the post-oil era will be easy. Given that oil is a nonrenewable resource, the gradual depletion of petroleum reserves is also a concern for the region's oil-rich countries, especially if their governments are intent on building a sustainable economy in the future. The consequences of the failure to properly prepare for the gradual end to the oil era could be profound, even as oil and gas reserves continue to keep global interest in the region significant for the foreseeable future.[72] The transition toward a post-oil era, according to one expert, will be fundamentally different from the previous period of transformation that accompanied the entry into the oil era in the 1960s and 1970s. The key variable in the shift to a post-petroleum future is that the "socioeconomic and demographic constraints will limit regimes' capacity to co-opt political support and essentially buy off oppositional movements and groups."[73]

The Post-oil Future

Many experts in the US national security establishment continue to embrace the argument that Saudi Arabia's importance needs to be underscored in terms of its role in global oil economics, as well as providing a counterweight to Iran's expanding regional role.[74] As such, the ongoing US-Saudi relations would most likely override any other political considerations. The ruling elites of the GCC region have yet to fully come to terms with the reality that economic reforms and strict social control are incompatible and that economic and political security are mutually reinforcing. This may help explain why various governments' attempt to create wealth and jobs—factors essential to economic growth—that hinge upon the transparency of the regimes and the good governance that a lack of opacity represents.[75]

One lesson learned around the 2011 Arab Spring uprisings was that greater control over these societies via repressive policies has made matters worse and immensely complicated in ways that make a return to the old order possible only through violence and coercion. Throughout the region, expanding the private sector in the name of true entrepreneurship often runs counter to governmental policies designed to impose societal control.[76] Given the impending depletion of hydrocarbon resources, these countries' per capita income will dramatically decline, making them middle-income-level countries, resulting in the exit of many millions of foreign workers who have for so long sought temporary employment in these countries. This presents a major problem for the expansion of the private sector in the GCC countries because this sector has been dominated by migrants and expatriates.[77]

Another scenario necessitates introducing a value-added tax (VAT). While cutting spending may not be prudent given its anti-growth implications, imposing taxes—which at 5 percent would represent roughly 1.4 percent of the GDP of these economies—appears to offer the most politically acceptable long-term solution.[78] Because VAT is a tax on consumption—not profits—many governments deem it as a favorable means of raising public revenue without giving up much private-sector activity in the process. Introducing VAT, as experts concur, is less costly than imposing a corporate or personal income tax.[79]

From a strictly economic perspective, the call for reforming the public sector and increasing its productivity has become tempting. The problem, however, is that even non-oil sectors in many GCC countries, according to the IMF, are to some extent dependent on funding from oil revenues. Moreover, the private sector remains small in these countries. The energy industry is generally highly capital-intensive and thus less likely to become a sustainable source of productive jobs.[80] Economic diversification in this context requires a shift in focus, beginning with improving business climate, strengthening economic governance and transparency, and lessening regulatory barriers to competition. All these are essential for the private sector to grow.[81] The persistent overreliance of the region's oil exporters on oil rent makes these countries unlikely candidates to prosper in a decarbonizing world consistent with the Paris Agreement.

A looming question is: How soon will these oil-exporting countries implement economic diversification, given that such a strategy is the key to their sustainability in the post-carbon era? Consider, for example, Saudi Arabia's "Vision 2030." With nearly half of its population under the age of twenty-five and the size of its workforce expected to double by 2030, Saudi Arabia can no longer afford its passive and unproductive population. At the same time, if more Saudis are absorbed into the workforce and are expected to pay taxes, they may begin to seek expanded basic freedoms. That would have far-reaching implications for an absolute monarchy in which the old ruling bargain (cash-for-loyalty) will no longer work, and it should be dismantled in favor of an entirely new plan altogether.[82]

Interestingly enough, the younger generation of Saudis displays shifting attitudes toward adopting different energy consumption patterns aimed at ending wasteful behavior and fostering a commitment to energy efficiency. David Hobbs best summarizes such changing views: "They are marrying later and having few children, leading to lower consumption per household. In fact, they are generally more environmentally and ecologically conscious and are showing interest in climate change."[83]

From a macroeconomic standpoint, experts tend to question the outlook of developing an agricultural economy in the oil-rich Persian Gulf countries, arguing that water shortages and seasonal droughts have adversely affected the availability of cultivable land, making it impossible to expect significant growth of the agricultural economy in the region.[84] Neither could the prospects of developing the tourism industry in the oil-rich GCC countries be taken seriously, given the several successful competitors near the region.

The prospect for a knowledge-based economy in the region faces perhaps the biggest obstacle in the form of the existing traditional administrative infrastructure. The administrative systems in place in these countries cannot pave the way for a dynamic knowledge economy. The public bureaucratic structure has traditionally played a significant role in the economy and has incessantly consumed a large portion of income from the national GDP.[85]

That leaves the region with one credible alternative—that is utilizing the potential for trade development, facilitated by road development and the transportation industries. The attraction of this scenario has to do with rendering economic transactions, including supplying goods to consumers and travelers' transportation, far more affordable. It bears remembering that other development scenarios—including the use of renewable energy sources, agriculture, knowledge, and particularly tourism development—depend largely on road and transportation development of the region. Hence, the allocation of oil income to the improvement of economic infrastructures and the road development industry is widely regarded as one of the most effective ways to develop global trade in the region.[86]

Yet several factors account for why the predictions regarding the oil age can be uncertain. These include, among others, the Saudi-Iran cooperation over Yemen and Syria, US support for or its absence of involvement in the nuclear talks with Iran (also known as the JCPOA), but most importantly, the Russian invasion of Ukraine. The last factor has had paradoxical consequences. On the one hand, the Ukraine War has led to a broader consensus that the European dependency on Russian oil has made the former vulnerable and politically insecure. On the other hand, it has jacked up the prices of oil and gas. The prospect of Ukraine collapsing and falling under the Russian control cannot be entirely dismissed. Such an eventuality will have its own ominous and potentially destabilizing ramifications as well. Spikes in costs of staple food (grain) and fuel will entail both uncertain economic and political ramifications and positive economic consequences—the latter having to do with the rise in oil prices—that would benefit all GCC members.

Forced Migrants and Climate Displacement

It is estimated that the GCC member states, according to the World Bank, could potentially be the recipients of some of the 143 million climate migrants from Southeast Asia and sub-Saharan Africa, as well as Latin America by 2050.[87] Around the same time, some 80–100 million people in the MENA region, including the Persian Gulf region, will be exposed to a level of water stress. Some 400 million inhabitants of the Middle East will be at risk of exposure to extreme heat waves that could likely cause social and political tensions.[88] The costs to mitigate the impact of such climate migration are likely to dramatically increase as well.[89]

The GCC states lack a unified policy toward climate-displaced persons, asylum seekers, and refugees. Most notably, Qatar signed into law in September 2018 the GCC's first law outlining procedures and requirements for people to seek asylum in the country. While the law manifests Qatar's commitment to refugee rights and serves as a model for the rest of the GCC member states, it nevertheless falls short of Qatar's broader obligations, particularly with regard to its restrictions on freedom of movement and expression.[90]

The asylum law is a massive step forward in a region that has historically shut its doors to refugees. While the law underlines the notion that no one should be returned to a place where they face a well-founded fear of execution, torture, and inhuman and degrading treatment, it requires, contrary to international law, a connection to a particular ethnicity, religion, or a specific social group or political belief. It should be noted that Qatar has ratified the UN Convention Against Torture, which absolutely bans the forced return of any individual who would face a real risk of torture. The asylum law should be modified to ensure that return to torture and inhuman and degrading treatment is prohibited under all circumstances.[91]

In Saudi Arabia, by contrast, there is no asylum law. In 2022, 335 persons with refugee status and 13,483 asylum seekers were registered in Saudi Arabia. They mainly came from Syria (57.91 percent), Eritrea (10.75 percent), Iraq (9.55 percent), and Somalia (4.78 percent), and most of them were fleeing war and violence.[92] The Saudi law states that the "State shall grant political asylum if the public interest so requires."[93] There is no contingent asylum system for those fleeing persecution, as the country is not a party to the 1951 Refugee Convention.[94]

The Saudi government, however, has allowed refugees recognized by UNHCR to remain in the country only temporarily, pending a durable solution—that is, resettlement to a third country or voluntary repatriation. Additionally, the Saudi government has neither granted asylum nor accepted refugees for resettlement from third countries, due largely to its policy of denying refugee status to persons who are irregularly present in the country. As a result, these irregular asylum seekers face the danger of deportation.[95]

The government provides proven legal residents preferential access to education, health, public housing, and other social services. Refugees and asylum seekers, however, are generally unable to work legally except in the case of Syrians and Yemenis who hold a temporary visa and can access a visitor's card from the Ministry of Interior, granting them work permits. Rohingya refugees have to obtain residence permits to work in the country.[96] In 2020, 606 internal displacements were reported, mainly caused by natural disasters like floods and wildfires.[97]

The UAE lacks an asylum system but generally recognizes the principle of *non-refoulement*. The country endorsed the Global Compact for Safe, Orderly, and Regular Migration as well as the GCR in December 2018.[98] In the absence of an asylum system, it is difficult to estimate the number of refugees and assess their situations. The UAE faced more climate-related risks (relative to other countries) in 2018 than it did on average between 1999 and 2017. The UAE has faced many climate-induced pressures, including water scarcity, lack of natural freshwater resources, land degradation, temperature rise, extreme weather events, sandstorms, flash floods, and desertification.[99]

Some experts have even predicted the likely rise of sea levels and the concurrent shoreline retreat that is projected to happen in several decades in Abu Dhabi, threatening to submerge the city, as well as parts of Dubai. In 2017, there were 850 new disaster-induced displacements. In 2019, there were a further 220 incidents, of whom 20 were still displaced at the end of the year.[100] Storms and tornadoes have led to coastal flooding in the UAE, causing numerous evacuations. In May 2020, for example, four people died in Sharjah due to flash floods.[101]

On April 15, 2024, flood storms hit Dubai and parts of Oman. In Dubai, some areas recorded more than 250 millimeters of rain in less than twenty-four hours, exceeding all past records in daily rainfall. The UAE averages 140–200 millimeters of rainfall annually, while Dubai typically receives only 97 millimeters. The monthly average for April is only about 8 millimeters. As a result of these recent storms, four people in the UAE and some twenty people died in Oman. Dubai International Airport, the second busiest in the world, had to cancel hundreds of flights.[102]

Scientists believe that storms occurred after months of hotter-than-average sea surface temperatures partly brought on by El Niño. The higher ocean temperatures have added more moisture to the atmosphere, rendering torrential rain ever more likely. The future increases in rainfall intensity in the region, scientists concur, are linked to climate change. Furthermore, the way that cities have been constructed in the Gulf region, according to Roop Singh from the Red Cross Red Crescent Climate Centre, has also made the impact of the storm on humans more potent. Cities built without proper storm management, combined with very dry soils in the region that seek to absorb excess water, means that 85 percent of people in UAE live in flood-prone areas.[103]

Similarly, Oman is not a State party to the 1951 Refugee Convention and its 1967 Protocol. Oman is neither a State party to the 1954 Convention relating to the Status of Stateless Persons, nor to the 1961 Convention on the Reduction of Statelessness. The authorities apply visa restrictions to most countries, except for nationals of the GCC and Europe, the Americas, and Australia, who can obtain visas upon their arrival. All noncitizens in Oman fall under national immigration laws, also known as expatriate law, including refugees and asylum seekers who register with UNHCR. As in other GCC countries, obtaining sponsorship through work, investment, or family link is a prerequisite to securing legal residence in Oman.[104]

It should be noted, however, that Oman's Foreign Residency Law entails some provisions on political refugees' rights and obligations. Under Chapter 7, for example, political refugee status can be granted by the Sultan's royal decree. The political refugee will be afforded an ID describing their data and obligations. The political refugee is strictly banned from engaging in any political activity during their residency in Oman. Lastly, the country maintains discretionary power to withdraw the right of political asylum and deport them at any time.[105] According to the UNHCR, some 309 refugees have been placed under its mandate, of whom 296 are from Iraq, 8 from Syria, and 5 from Pakistan.[106] As of 2024, there were 254 asylum seekers (all from Iraq).[107] While Oman forbids refugees from working in the country, many tend to work illegally to send money back home. The government has provided temporary medical care and protection to Yemeni citizens who have been displaced due to the war. These forced migrants have demonstrated that they could not receive adequate care in Yemen and have thus been offered temporary stay for themselves and their immediate family in Oman during the treatment period.[108]

Looking Ahead

Will the GCC's rentier economies and ruling bargains, known for their subsidized oil and gas prices and authoritarian regimes, allow for a dynamic process of sustainable development and an environmentally friendly agenda? The contemporary politics of oil, Timothy Mitchell notes, has involved deploying violent means to secure arrangements for the production of oil and forms of representation that seem an equally essential aspect of the undemocratic politics of oil inside the GCC states and from the standpoint of US militarism to gain access to oil and even "as a project to bring democracy to the Middle East" under the George W. Bush administration.[109] What could and should give rise to further democratic opportunities largely hinges upon deploying the necessary and effective political tools to address the passing of the era of fossil fuel.[110]

Increasingly, experts argue that the CO_2 generated in its combustion has contributed to global warming on earth and in the ocean, threatening to cause disastrous climate change.[111] A broader consensus holds that without the reform of the subsidy regimes and transparent and participatory political system, achieving environmental sustainability is highly unlikely. The resource-rich GCC countries have in the past prioritized economic and developmental projects over social and environmental sustainability. Without adopting the necessary mitigation and adaptation measures, such an approach will prove to be an exercise in futility.

Relying on the oil-based economy is not a sustainable strategy over the longer term in the Arab Gulf states. The vagaries of oil markets, shifting geopolitical considerations, and a changing climate have made predictions regarding the future of the region notoriously unreliable. While it is still too early to predict with any certainty the future of the GCC region in the post-oil era, much remains unclear in the attempts and commitments of the region's ruling elites to diversify away from oil. Over-dependency on oil revenue, as marked by heavily subsidized economic models, appears to be a defining characteristic of the economies of these oil-rich countries.

Several environmental studies have shown alarming trends regarding the rapid growth of air pollution, a lack of access to water supply and sanitation services, land degradation, coastal zone degradation, and waste management. In the post-oil era, these studies suggest that the need to integrate the cost of environmental degradation into economic development decisions has never been more urgent.[112] Future tensions are likely to arise over water scarcity and the way in which the region's countries can secure the food needs of their population.

For now, the recognition of the post-oil era has not yet fully begun, even as the talk of investing in tourism, healthcare, and education continues to pervade the region. Other relevant ideas revolve around developing human capital, empowering women in the workforce, focusing on privatization, and investing in technologies, ideas, and innovation. To ensure multiple pathways to non-oil revenue generation, GCC governments must make large investments in the private sector.[113] But as long as GCC states demand full compliance and acquiescence from their citizens, without displaying at least a modicum of accountability and transparency, such reform-oriented plans remain difficult to achieve at best and inoperative at worst.[114]

None of these conditions are conducive to breaking away from the dependency on oil revenues. Given the centrality of oil to the prominence of the region's political economies as well as its ongoing role in facilitating politically unaccountable regimes to stay in power, the quality and structure of the post-oil era remain a key question—and perhaps more accurately a political conundrum.[115] As previously discussed, the oil-rich GCC countries are likely to apply reductions in water, fuel, electricity, and food subsidies, while also introducing a 5 percent VAT.

Some observers have recommended the creation of a fund in the GCC to address and mitigate the impact of climate-induced migration on affected populations.[116] UNHCR has recommended the adoption of sustainable climate laws and policies, strategies, and programs to predict potential natural disasters and tackle such events. Equally important is the recognition that a shift in attitudes among younger Arabs is not enough; rather, what is required is a shift in ruling elites' attitudes, values, and commitment to the ecological and environmental needs of their countries and the region.

That realization is significantly correlated with the planning and success of economic and structural reforms in the post-oil era. Increasingly, levying a substantial local carbon tax on organizations and companies that produce high rates of CO_2 emissions appears realistic and inevitable. Meanwhile, it is important to note two things: (1) many countries in Africa and Asia (especially China and India) remain heavily dependent on the consumption of oil, gas, and coal for the foreseeable future, and (2) using oil for industrial purposes will continue unabated, as most pharmaceuticals contain petrochemicals. Additionally, petroleum is also the key to transporting medicine so that patients can easily obtain it. In fact, this is the core area in which petroleum is used in the medical field.[117] Overtime, however, taking initiative to promote the development of decarbonization technologies is paramount in a region where most governments still rely heavily on petroleum.[118]

CHAPTER 11
MOROCCO

POLITICS, GREEN INVESTMENT, AND ADAPTATION PLANS

Located in North Africa and bordered by the Atlantic Ocean (to the west), the Mediterranean Sea (to the northeast), the Saharan desert (to the south), and Europe (to the north), Morocco enjoys a unique geographical position at the mouth of the Strait of Gibraltar and the entrance to the Mediterranean. The country lies between two main climate zones. The north is largely influenced by the Mediterranean and Atlantic climates, while facing a great deal of climate variability between the Atlas Mountains and the coast. The south, by contrast, is composed of both semiarid and arid regions that join the Sahara Desert.[1] Due to its unique location, Morocco has been a melting pot of different civilizations, including African, European, Middle Eastern, and Asian cultures. The country represents a hybrid socioeconomic milieu within which Arab, European, and African cultural traditions have blended with indigenous practices to make up the wide range of ideas, norms, and values of today's Morocco.[2]

Historians have described Morocco as a layered country, with each phase of its history influenced by cultural and political structural contexts. The Carthaginians and Romans brought the first cities, linking Morocco with the wider economic markets of the Mediterranean.[3] The coming of Islam perpetuated that process, while introducing Islam and the Arabic language as the predominant religio-cultural influences there. The great empires of the Almoravids and Almohads further deepened and smoothed the path for Marrakesh. Later, the rivalry with the Ottoman Turks settled the eastern frontiers of the country's borders. The dynasty that replaced the Sa'dis, known as the Alawis, has ruled the country ever since.[4] That dynasty has survived pressure by European powers, colonial occupation by the French and Spanish, and finally adjustment to the postcolonial world after independence in 1956.[5]

The country's ethnic and linguistic makeup represents a blend of Berber and Arab stocks. Arabic is the official and native language of about 65 percent of the population. The other 35 percent speak one of several Berber dialects. French is spoken extensively, as bilingualism and trilingualism are widespread.[6] Morocco's topography includes rich lowlands in the northwest and presents challenges to economically poor and sparsely populated mountains and plateaus in the east and south. In the country's interior the soil is poor, but the area around Khouribga (known as the phosphates plateau) is the locus of rich phosphate deposits, establishing Morocco as one of the world's largest producers of this mineral.[7]

Western and northern Morocco have mild winters and hot, dry summers. The climate is hot and dry, except for the humid Mediterranean coast and the mountains, where temperatures are much cooler. The average annual temperature in Morocco increased by 1.7 degrees Celsius between 1971 and 2017. That spike, however, varied markedly depending on geographical location and season. The central-northern and northern areas of the country saw the largest increase in temperature, over 2 degrees Celsius in cities like Ojuda, Taza, Errachidia, and Beni Mellal.[8]

The evidence suggests that the rise in temperature will continue in Morocco until the end of this century. Under a medium level of GHG emissions, the average annual temperature increase is predicted to reach from 1.5 degrees Celsius during the period from 1981 to 2018 to 1.8 degrees Celsius in 2050. Under a scenario of a high level of GHG emissions, the average annual temperature rise could reach between 2 degrees Celsius and 3 degrees Celsius. If such emissions are not mitigated, Morocco could experience more recurring and intense heat waves.[9] The ongoing increase in temperatures could demand the expansion of existing interconnected electricity networks, as well as the development of new national grid interconnectors—that is, new high-voltage cables that will connect the electricity systems of Morocco with those of Spain and Portugal.[10]

Morocco's rapid urbanization over the past half century, along with increasing resources devoted to agricultural production, has led to fast-growing rates of water usage, raising serious concerns in the most arid parts of the country.[11] Given this reality, climate impacts and their long-term consequences are likely to permanently change Morocco's ecological and socioeconomic realities. Thus, timely and long-term adaptation strategies to manage inexorable climatic shifts are needed if climate uncertainties are to be effectively addressed.

In this chapter, I first briefly review the migration history of Morocco, followed by an analysis exploring reasons for and against Morocco's political stability in the aftermath of the 2011 Arab Spring uprisings. In the next section, I examine emerging nonmilitary threats in the form of climatic disruptions. In this context, I particularly focus on climate change, precipitation and droughts, the energy sector, and food security. In the following section, I delve deeper into how the issue of gender equality is profoundly affected by climate change. The ensuing section explains the issue of green investment. The chapter concludes by looking at the future outlook, while reflecting upon various mitigation and adaptation strategies that might be considered to come to grips with and contain climate change in Morocco.

Migration History

Morocco presents an absorbing account of the migration history in North Africa. There was little migration from Morocco to Europe before the start of the twentieth century. In the twentieth century, the First World War and the Second World War were accompanied and followed by labor shortages, which led to the recruitment of Moroccan men in the French-occupied zone for the coal mines. The late 1950s and 1960s saw a great demand for Moroccan workers, not just in France but throughout Western Europe. As a result, thousands of unskilled Moroccan laborers migrated to France, Belgium, Germany, and the Netherlands through bilateral agreements signed after Morocco gained its independence in 1956. Following the 1972/3 freeze on importing labor migrants from Western Europe, a large wave of migration to Europe was initiated on the grounds of family reunification processes.[12] Since the 1980s, the Moroccan government has made it easier for remittances to flow into the country, without which poverty in Morocco would have been much higher.[13]

While generally known as a major emigration country, Morocco has also become a host and transit country for many who have migrated from neighboring countries in North and sub-Saharan Africa, and to some extent, from crisis-hit European countries. The increasing presence of immigrants from sub-Saharan Africa has also enhanced religious diversity and has, to some extent, reinvigorated Christian life in some cities of this chiefly Muslim country. Yet at the same time, the growing presence of immigrants confronts Moroccan society with a completely new set of social and legal challenges.[14]

Migrants and refugees from sub-Saharan Africa are often the target of violent racist attacks and discrimination in Morocco, especially in big cities and in improvised ad hoc camps close to Ceuta and Melilla. Some migrants have been arbitrarily deported via the Algerian border without safeguarding their right not to be sent back to places where their lives will be at grave risk, in a clear violation of the principle of *non-refoulement*.[15] In 2007, the UNHCR signed an *accord de siège* with the Moroccan government, bringing about limited improvements in the situation of refugees and asylum seekers. Since then, those holding UNHCR registration cards have been subject to less harassment and deportation. Their access to public health care and education has improved as well, although the Moroccan government continues to refuse to issue resident permits to refugees and asylum seekers. By mid-2013, the UNHCR office in Morocco had registered 874 refugees and 3,706 asylum seekers.[16]

Morocco continues to be an attractive destination due in large part to the fact that "it provides higher salaries, borders with Europe, and offers greater employment opportunities."[17] In 2014, 18,000 irregular migrants obtained residence permits to remain legally in the country. This number increased to 55,000 in 2018.[18] By 2020, the population of the eight Sahelian-Saharan countries had reached 166 million, compared to 132 million in 2010. Algeria, Libya, Morocco, and Tunisia had nearly 100 million inhabitants in total in 2020, compared with 66 million total for four Sahelian neighbors (Mali, Mauritania, Niger, and Chad). It is estimated that the population of these four Sahelian countries will reach 141 million by 2050, putting the Sahel's population at about 10 million more than that of the four countries north of the Sahara.[19]

As of August 31, 2023, Morocco has become both a transit and a host country for people seeking asylum, with a total of 19,133 refugees and asylum seekers (10,225 refugees and 8,908 asylum seekers) registered with UNHCR from across 78 localities. Many of these asylum seekers and refugees are from Syria, Guinea, Yemen, Cameroon, and the Ivory Coast.[20] UNHCR has worked closely with Moroccan authorities under the framework of the National Strategy for Immigration and Asylum. Also, in the context of the national emergency since the earthquake on September 8, 2023, UNHCR and several partners have actively worked to better protect the needs of asylum seekers and refugees.[21]

The migration issue in North Africa continues to be a contentious issue in the face of ongoing political and security instability in Libya, political tension and socioeconomic crisis in Tunisia, growing tensions between Algeria and Morocco, and the fragility of several countries in the Sahel and West Africa. Further compounding matters is the fact that decisions regarding migrants' transitory or final destinations will hinge on the various migration policies adopted by different governments.[22] Europe, however, is not the primary destination of African migrants. North Africa is indeed a prime destination for many. North African countries, such as Algeria, Libya, Morocco, and Tunisia, have been host to many thousands of migrants from Sahel and West African countries—some for short passage, some for long transit—by choice or necessity, especially to be financially able to continue their migration journey. They work as labor to continue emigration and some for long-term settlement.[23]

Two factors have led to an increase in the number of forced migrants in Morocco. First, Morocco has become a destination country for sub-Saharan migrants. Second, the country has served as a transit country for migrants traveling to Europe. Furthermore, due to increasing restrictions in Europe, many migrants have felt compelled to stay in Morocco. Between 2016 and 2019, the number of people registered with UNHCR as asylum seekers and refugees saw an increase of more than 64 percent. Of those registered, 34 percent are under the age of eighteen, more than half (55 percent) are from Syria, and most (75 percent) are Arabic speakers.[24]

Additionally, the EU has persuaded and even funded the Moroccan authorities to grant residency permits to migrants, making it possible for some to stay and anticipate a decent life in the country. Ghanaian musician Reuben Odoi, who founded Minority Globe, a music and theater production company, in 2009, has been able to support artistic expression by migrants and advocate for their cultural events. Put most simply, Odoi has become a staunch advocate of solidarity with migrants and refugees, paving the way for an inclusive identity in Morocco. How far the activists, such as Odoi, can go to accomplish this goal within the country's extant conservative, security-minded context remains to be seen.[25]

Within Morocco, refugees can find work in informal sectors, especially in locations close to urban cities such as Rabat and Casablanca. Women and child refugees living in Nador—a coastal city and provincial capital in the northeastern Rif region of Morocco—are more vulnerable to sexual and gender-based violence as well as human trafficking. Those who live in urban centers are often better protected by CSOs and NGOs such as Médecins Sans Frontières, which make it possible for refugees to enjoy better access to health services.[26]

Some media reports have been highly critical of the poor living conditions that many of the migrants endure in Morocco, which include provisional camps in Casablanca, with little to no basic sanitation, cold temperatures, and ubiquitous respiratory and infectious diseases—not to mention the perils of being exposed to traffickers. Apart from such difficulties, many migrants describe interactions with the Moroccan immigration officials as being hostile and inhumane, involving beatings and the use of excessive force.[27]

In contrast to other African countries, Morocco has had a relatively low number of IDPs, at least before the 2023 earthquake. In 2019, it was reported that around 200 people were displaced due to disasters. Conflict and disaster displacement were catastrophic in 2010, reaching over 15,000.[28] The September 8, 2023, earthquake in Morocco was the most devastating one in the country's history. While 380,000 people were gravely affected due to their proximity (less than 30 miles from the quake's epicenter), the overall number of impacted population was much greater. At least 500,000 people were displaced.[29]

Explaining Political Stability in Morocco?

Morocco has long been a politically stable monarchy led by King Mohamed VI, who came to power in 1999. Since 1996, the national legislature has been bicameral: the House of Representatives (*Majli al-Nuwab*) or the lower house is made up of 395 members who are elected directly for a five-year term. The upper house is called the House of Councillors (*Majli al-Mustacharin*) and has 120 members who are elected indirectly for a six-year term by two sets of electoral colleges.[30]

As a unitary semi-constitutional monarchy with an elected parliament, Morocco is considered a partly free and democratic country. Constitutional reforms in 2011 granted official status to Tamazight languages, which have been advanced in schools along with Amazigh culture. That said, Amazigh and other communities that do not conform to the dominant Arab culture tend to encounter major educational and economic drawbacks. Civil society groups that promote Amazigh rights have met with government backlash. Likewise, although gender equality has been formally sanctioned in the 2011 constitution, women still experience serious discrimination at the societal level and are underrepresented in the labor force.[31]

Meanwhile, LGBTIQA+ people experience harsh discrimination and occasional violence. Same-gender or same-sex sexual relations can be punished with up to three years in prison.[32] While the government has granted temporary residency permits to refugees and migrants in order to legalize and modulate their status and provide them with essential services, many registered refugees living in Morocco have yet to hold residency or work permits.[33]

At the political level, unlike the rest of the Arab world, Morocco saw a wave of reforms in the 1990s well prior to the eruption of region-wide (Arab Spring) uprisings. A set of liberalizing reforms—aimed at promoting human rights, allowing representation of Islamic groups in parliament and government, and fostering broader party participation and civil society activism—was initiated under the rule of King Hassan II. Since then, Bahey eldin Hassan, the former director of the Cairo Institute for Human Rights Studies, points out that Morocco has begun introducing reforms in a relatively more sustained way.[34]

This may explain why Morocco was one of the few states to remain undaunted and unaffected by the 2011 Arab Spring uprisings. When on February 20, 2011, thousands of Moroccans rallied in the capital city, Rabat, to demand political and economic reform, the country's king, Sidi Mohammed bin Hassan al-Alawi (also known as Mohammed VI), was able to relentlessly undermine them by building on the previously initiated reforms and adopting additional (albeit restricted) constitutional reforms. It is also important to recall the factors that boosted the efficacy of King Mohammed's appeasement strategy: the weaknesses of the February 20 protest movement, the unique role and legitimacy of a venerated monarch, the willingness of the ruling Justice and Development Party (PJD) to support the reform package, and regional events culminating in persistent protests that bent people's inclination toward the king's reform package.[35] These factors proved to be an effective way to ratchet down the growing political tensions that had engulfed a significant part of the region.

Several contrasting views about the efficacy of civil society and human rights organizations in Morocco demonstrate that part of the Moroccan regime's strategy of self-regeneration, adaptation, and cooptation has included an attempt to bring more NGOs into its political domain. Similarly, Driss Maghraoui argues that the government has found it prudent to contain the rising popularity of the Islamists by formal inclusion. The government's ability to coopt, integrate, and share parliamentary seats and even political power with moderate Islamists—notably the PJD and the socialists—has been part of a strategy to contain potential opposition to the regime.[36] This strategy has been and continues to be essential to how the monarchy functions in Morocco. That also means that the government has pursued these related strategies in order to involve civil society actors who are often outside the gamut of religious ideology and are determined to gain political autonomy for their activities.[37]

Although human rights issues have influenced the debate over the role of civil society in the public sphere, Maghraoui notes that the creation of the Advisory Council for Human Rights (CCDH) in 1990 under King Hassan II was part of a strategy to qualify human rights advocacy and to contain social and political activism. CCDH's financial dependence on the regime, for the most part, strictly deprived it of autonomy and legitimacy.[38] King Mohammed VI, for example, tends to use his office to prioritize the existence of civil society language and official discourse over full political expression or redress of societal grievances. Regardless of what strategies King Mohammed VI undertakes to engage civil society actors, such activities could potentially build sufficient legitimacy and credibility to become an effective venue for political participation.[39]

For the most part, political institutions in Morocco, such as parliament, are limited in their effectiveness. Saloua Zerhouni points out that parliament's role in legislation and government oversight is very limited and that it is more a space for "doing politics" than for "producing politics."[40] In a nutshell, parliament is a space for "managing the field of politics," and a forum for debates, while also resolving or containing conflicts within the society. Buttressing parliament, Zerhouni asserts, is neither a priority for the monarchical regime nor the political parties. The Islamists in parliament, for instance, have been forced to cooperate on the anti-terror law, despite its inconsistencies and contradictions in relation to human rights. The fact that Islamists have altered their position under pressure from the regime demonstrates the sort of limitations under which the opposition operates in parliament.[41]

For the Moroccan parliament to serve as an effective body, Zerhouni contends, it should transform itself from a channel for contestation and politics to a setting for policymaking and legislation. The change, Zerhouni insists, must come through constitutional reform, checks and balances, elite renewal, citizens' socioeconomic empowerment, and political democratization. The argument for constitutional reforms, however, remains impractically hopeful even to those who support the need for democratic change.[42]

Emerging Nonmilitary Threats

Nontraditional security issues pose serious challenges to the survival and well-being of peoples and states. A broad consensus holds that these challenges arise primarily from climate change, resource scarcity, infectious diseases, natural disasters, forced migration, food insecurity, human and drug trafficking, and transnational crime. These threats are generally transnational in nature and scope, evading unilateral remedies and requiring holistic political and socioeconomic responses.[43]

Yet, the ruling elites in the MENA region, notably in Egypt, Jordan, Lebanon, Morocco, and Tunisia, appear to have been and continue to be primarily concerned with demographic trends and priorities for economic growth rather than environmental degradation and changes. "The shift in gravity in Mediterranean water cooperation," Michael Mason writes regarding one such issue, "from UN-led hard-law to EU-driven soft-law processes has at the same time imported the slow, hesitant pace of European Neighborhood Policy in the region, favoring a lowest common denominator of (technical) collaboration and not challenging regional wielders of geopolitical power (e.g., Israel and Turkey), who are generating transboundary inequities in freshwater allocation."[44]

The major barrier to environmental effectiveness, however, stems from rapidly growing pollution sources—fueled in part by urbanization, industrialization, and population growth in southern Mediterranean countries. Political elites in these countries lack economic or political incentives, or even the capacity, to restrain growth.[45] For labor-abundant countries, such as Egypt, Iran, Iraq, Morocco, and Syria, both trade and foreign direct investment constitute a more realistic vision for the foreseeable future.[46] The oil-rich countries are likely to cut water and energy subsidies and introduce a new VAT.[47]

Climate-Related Disruptions

The 2014 IPCC report estimated that 25 percent of water shortages in the North African region in 2050 will be caused by climate change and 78 percent of water scarcity will be the result of socioeconomic issues.[48] In recent years, Morocco has grown increasingly more arid. During the period 1961–2017, although the northwest region of Morocco has received annual average precipitation of up to 1,200 millimeters (the highest), the southeast of the country has seen less than 50 millimeters per year on average. Due to declining precipitation in the southern part of Morocco, the Sahara Desert is expanding. Over the long term, it is predicted that the center of the country will have 30 percent less precipitation, while the Sahara region (southern Morocco) will see 5 percent precipitation increases.[49]

Apart from such changing precipitation rates, weather conditions have trended toward increasingly higher temperatures in Morocco. Since the 1980s, the country has seen a marked trend for more regular droughts and a visible shift toward warmer and drier conditions. Additionally, around the Sahara Desert, soaring temperatures and drought have led to growing arid sand and broader desertification. The result has been increasing food insecurity and social instability in certain parts of Morocco. In Al Hoceima, in the northern part of Morocco, for example, riots erupted in 2008 after a year of losses in food production, along with high global food prices and failed national policies. The country's increasing urbanization and robust population growth have made it necessary for the state to prevent price shocks and food insecurity.[50]

Other climate-related disruptions, such as salinization and soil erosion, also impact agriculture in far-reaching ways. It has been shown that in less than twenty years, long-irrigated soils can lose more than 50 percent of their productivity due to salinization, exacerbating the already dire impacts of climate change. Similarly, soil erosion is likely to complicate Moroccans' ability to adapt to climate change, as 75 percent of arable land in Morocco is already affected by erosion.[51] Decreasing precipitation has led to increasing concerns about droughts, which have increased in frequency, size, and duration. Climate projections point to more frequent and acute droughts in Morocco, especially in the central and southern regions. With some exceptions, almost all of Morocco is expected to become drier and its climate reshaped by an arid climate by the end of the twenty-first century.[52]

For too long, and especially since 1979, Morocco has been preoccupied with droughts causing enormous damage to crops and livestock.[53] At independence, some 80 percent of Moroccans lived in rural areas, largely at a subsistence level. With the mass departures from rural areas, precipitated by successive droughts, the number of urban dwellers had overtaken rural inhabitants by 1994, and that trend has been accelerating since then.[54] Given that nearly half of Morocco's population still depends on agriculture, the country's economic strength and sustainability is closely tied to the erratic rainfall.[55]

King Hassan II (1929–1999) sought the solution in irrigation. With the decline in annual precipitation, the reservoirs behind the dams were less than full. Furthermore, just 10 percent of farmers benefit from irrigation on the Atlantic littoral plains of the Gharb and Souss, which produce mainly export crops such as citrus fruits, frost-tolerant spring vegetables such as peas, broccoli, kale, cabbage, carrots, cilantro, and lettuce, as well as sugarcane. Most of the rural population lives in nonirrigated mountainous areas where they grow grains and raise and breed domestic sheep and other animals.[56]

In the past, during times of massive droughts, the pastoral nomadic population of Western Sahara would move out of the desert northward seeking pastures for their camels.[57] The successive droughts, the guerrilla war, and the defensive minefields, however, have significantly reduced camel herds.[58] Many observers have warned that the economic disparities between the north and the south could be manipulated to spread extremism. The country's King Mohammed VI has called for a nationwide mobilization to reduce "the plague of poverty."[59]

Climate change and the resultant frequent and intense droughts and floods have led to growing water scarcity for most of the population. The prolonged droughts and climatic disruptions have affected both rural and urban areas, presenting formidable challenges to the country's broader economy. It does bear stressing that approximately 21 percent of the land in Morocco is arable, with another 47 percent used for grasslands and pastures. About 91.3 percent of the country's electricity production comes from fossil fuels, of which 75 percent is coal. Only 8.7 percent of the country's electricity is generated by hydropower.[60]

Even though the country has been exposed to higher degrees of industrialization, the per capita rate of CO_2 emissions still remains relatively low. The per capita rate was 1 metric ton in 1990, rose to 1.2 metric tons in 2004, and to 1.5 metric tons in 2007.[61] The emissions per capita for Morocco reached 1.97 tons in 2021.[62] The increased use of liquid fuels includes the heavy use of oil, a quarter of which comes from Saudi Arabia and a fifth from the UAE, subsidized by some local production of oil and gas. The discovery of oil in Western Sahara is expected to generate additional wealth for the country, while also contributing to problems of increasing GHG emissions.[63]

Noteworthy is the fact that global warming and climate change are expected to result in longer droughts and diminishing crop production. The Moroccan government has consistently participated in the UNFCCC, signed in Rio de Janeiro in May 1992, and the ratification of the Vienna Convention in 1995. It has also accepted the Kyoto Protocol to the UNFCCC on January 25, 2002. The Moroccan government has also committed to the Clean Development Mechanism as part of its National Sustainable Development strategy.[64]

Storms have exacerbated climate-induced problems in Morocco. Recently, the number of dust and sandstorms has also been on the rise in the North Africa region. Environmental degradation and natural causes have also contributed to their increasing frequency.[65] Sea level rise is likely to become a serious and direct threat to Morocco's economy and infrastructure, given that the country's urban coastal zones are home to 70.2 percent of the population and 90 percent of industrial activities. Morocco is expected to experience a sea level rise of 0.4– 0.7 meters by 2100 depending on climate scenarios. It could cause permanent deluge of the lowest lying coastlines in Morocco, while accelerating the risks of coastal disintegration. Of the total length of Morocco's east Mediterranean coastline, nearly 42 percent would be at high risk by 2030.[66]

But perhaps most ominously, as scientists remind us, climate change has affected public health in Morocco in other ways, as a hotter climate has led to increased incidence, distribution, and severity of leishmaniasis, a skin disease spread by sand flies. Infectious diseases such as leishmaniasis, malaria, and schistosomiasis continue to present serious public health problems. Climate-induced changes are likely to lead to further increases in infectious diseases in the years ahead, even as the Moroccan government has offered effective domestic programs to combat parasitic diseases.[67]

Yet government officials in the past have given less serious consideration to the way in which climate change has impacted the spread of infectious diseases. As insect-borne diseases, the occurrence and spread of leishmaniasis are influenced by environmental changes, as well as by various socioeconomic and cultural factors—that is, the impact on the amount of human contact with the transmission. It is thus urgent that more strategic thinking goes into designing ways to control animals and insects such as mosquitoes, ticks, rodents, and fleas to reduce the risk of transmission of leishmaniasis that may be increased by climate change and other environmental factors.[68]

The Energy Sector

In the energy sector, less precipitation and more droughts have adversely affected hydropower generation, resulting in a dramatic decline in hydropower output from 3,631 gigawatt hours to 1,290 gigawatt hours between 2010 and 2020. The projected spike in aridity under a high-emissions scenario would have negative impacts on the country's hydropower capacity, which could further drop by over 30 percent between 2060 and 2099. To tackle the projected decrease in precipitation, Morocco has expanded its capacity for pumped-storage hydropower plants. The Office National de l'Electricité et de l'Eau Potable has, in recent years, launched projects for pumped-storage hydropower, including the construction of the Abdelmoumen pumped-storage facility (350 megawatts) and the announcement of plans for El Menzel II (300 megawatts) and Ifasha (300 megawatts) pumped-storage hydropower facilities.[69]

Accounting for around 75 percent of Morocco's electricity generation, coal power plants are also projected to face extremely arid climate in the coming decades. The projected surge in arid conditions under a high-emissions scenario would add mounting pressure to coal power plants that rely on water for cooling, which account for over 80 percent of the total installed capacity of coal-fired power plants in Morocco.[70]

The government's renewable energy targets constitute the key part of the country's decarbonization plans. In 2020, renewable energy amounted to 30 percent of electricity capacity. Morocco's electricity generation remains heavily dominated by coal and has not committed to a coal phase-out. Conversely, the government has unveiled plans to expand fossil gas infrastructure. In 2022, Algeria halted gas supply via the Maghreb-Europe pipeline, which accounted for all of Morocco's gas imports. To overcome this problem, Morocco has begun building LNG terminals to accommodate imports via Spain.[71]

Morocco has also sought offshore gas exploration, an approach that crucially undercuts the country's decarbonization plans and risks latching the country onto a high carbon emissions course. International support and local mobilization are essential for success if the Moroccan government is determined to make the best of its vast renewable energy potential and discard fossil gas and coal over the longer term.[72]

Food Security

Morocco plays an important role in global food security. It is one of the world's largest producers and exporters of phosphates and its derivatives, particularly fertilizers used in the production of agricultural goods for Global South countries. In 2021, Morocco produced 38 million tons of fertilizer—a production level second only to China. Morocco has the largest phosphate reserves, accounting for 70 percent of the world's reserves, making it a major country in the field of food security.[73]

Yet it bears stressing that access to long-term fresh water, food security, and sustainable development presents new challenges to Moroccan leaders. Hafed Al-Ghwell sheds light on the critical role that the agricultural sector plays in the Moroccan economy, arguing that the sector employs about 40 percent of the nation's workforce and accounts for nearly 14 percent of its GDP. Also, unrelenting reliance on water-intensive projects, along with a growing strain on water resources due to climatic shifts, has led to a hazardous situation not just for farmers but for the entire economic architecture of the country.[74] Furthermore, Al-Ghwell goes on to argue, "with dams nearly emptied and irrigation halted, farmers face the grim reality of jeopardized investments and crop failures."[75]

The aftermath of the 2023 earthquake, as evidenced by enormous damage to infrastructure, particularly roads and irrigation canals, has markedly disrupted access to water supplies. Reconstruction projects have diverted resources from potential investments in agricultural resilience, such as the adoption of drought-resistant crops and improved irrigation technology. The damage to agricultural activities, as well as the direct hit that tourism has taken, underlines the urgent need for a resilient and innovative economic strategy. Additionally, Al-Ghwell makes a compelling case for a transition to drought-resistant crops and the implementation of agricultural mechanisms that bolster soil moisture retention, capable of mitigating the effects of climate change. He argues that resistance to extreme heat and drought has made the argan tree a suitable plant to grow in areas of southeast Morocco. "Morocco's experiments with argan trees, which are well-adapted to arid climates," Al-Ghwell writes, "offer a promising avenue for sustainable agriculture."[76]

As noted in the preceding chapters, food insecurity in the broader Arab world was directly linked to the 2011 Arab Spring protests. Experts argue that macroeconomic indicators, such as unemployment, growth rate, GDP, and inflation, were not the only contributing factors to the poor Arab economies. The 2008 global financial crisis was accompanied by ensuing rounds of commodity price spikes, which energized the purportedly resigned Arab street into a formidable source of opposition against autocracies and their trade liberalization programs.[77]

The non-oil agricultural Arab states, such as Egypt and Morocco, have faced policies focused on increasing integration into global markets and the privatization of state enterprises at the expense of agricultural development policies, which have led to heavy dependency on food imports. For arid and semiarid Arab countries, several factors, including "a rapidly growing population, elimination of tariffs on agricultural imports, and lack of available water and fertile land," converted the region into the world's biggest food importer.[78] By 2011, Arab countries imported 58 percent of their cereal and 75 percent of their sugar consumption—nearly 61 percent of per capita calories—a fact that illustrated the extent of their vulnerability to the vagaries of world commodity markets.[79]

Today, food security and affordability are among the main concerns of the government in Morocco, as the country faces ongoing economic issues such as inflation and agricultural production. In 2022, five of the top twenty wheat-importing countries in the world (Egypt, Turkey, Algeria, Morocco, and Yemen) were located in the Middle East.[80] The number of malnourished people in Morocco, according to a recent UN report, amounted to 2.1 million in 2021, which was roughly 5.6 percent of the country.[81] Only 16.7 percent of Moroccans—that is 6.2 million people—were able to afford a healthy diet in 2020. The country's burgeoning dependence on core subsidized food imports poses an increasing threat to its development outlook and strategy.[82]

Looking more broadly, Morocco has designed the Country Strategic Opportunities Program 2022–7 that is supported by the International Fund for Agriculture Development (IFAD) to generate a sustainable boost in the incomes and food security of the rural population, particularly women and young people, and to bolster resilience to climate change in the country's mountainous and rural areas.[83] The broader goals, according to IFAD, are divided into three strategic areas:

> To promote equitable, inclusive development and build up the social and human capital of the rural population in vulnerable regions. Boost the competitiveness and diversify the production of solidarity farming and livestock systems on a sustainable and equitable basis in mountainous areas and strengthen their access, on an equitable footing, to markets and financial services. Promote the environmental sustainability of rural populations' livelihoods and their resilience to climate change.[84]

Gender Equality and Climate Action

A global consensus has emerged that the integration of women's rights and gender equality into the mitigation and adaptation to climate change is not only essential but enhances the efficacy of strategies dealing with environmental security. In 2001, during the Conference of Parties (COP7) meeting in Marrakesh, the parties to the UNFCCC agreed for the first time to incorporate in the text gender equality and women's rights and participation in their UN climate policies. Ever since, the Conference of Parties has adopted a gender-sensitive approach toward issues relating to climate change.[85]

It is important, however, to note that there was a growing recognition of the links between gender equality, women's rights, and climate change outside the

UNFCCC structure. In fact, the 1979 Convention on the Elimination of All Forms of Discrimination Against Women (CEDAW) had direct and profound implications for climate change policymaking. By empowering women in terms of participation in all levels of development planning, CEDAW has noticeably enhanced women's capacity to adapt to impacts of climate change, while participating in planning and implementation of climate change policies.[86] As a result, the three core principles of political and economic empowerment, participation, and accountability have been increasingly integrated into existing or planned climate change laws and policies at the national level.[87]

In the case of Morocco, women historically played an active role in the broader social and political resistance to end French colonial rule. Their participation depended on and extended gendered norms of women's roles in the family and community. Women operated under the French radar by taking advantage of traditional beliefs about their feminine identity—namely, that they lacked the ability to participate in political and military action. That proved wrong as Moroccan women occupied new roles within the resistance movement and new physical spaces outside the home, bringing new skills to bear in their relations with men.[88]

Most recently, it should be noted, the most visible tool for promoting gender equality has been gender-responsive financing and budgeting (GRB), with its aim being to guarantee equal opportunities and benefits for women and men. This requires an overall modification of taxation policies and the use of public funds to the extent they reflect the varying ways in which participation rates of women and men in the economy are measured. Gender ownership and transaction are another way to determine GRB. Some experts maintain that Morocco is an excellent example of GRB in practice. Since 2003, Morocco has "adopted a new finance law, legally institutionalizing gender equality through all budget processes in the country."[89]

Accordingly, Morocco produces an annual Gender Report, which provides information on the work done by different sectors. This report renders ministries accountable to the parliament and citizens. The 2004 Gender Report, for example, showed that women constituted 9 percent of the beneficiaries of the agricultural sector, albeit being responsible for 39 percent of the economic activity in rural areas. Thanks to this report, the government significantly enhanced the budget for women's productive activities in agriculture.[90] Subsequently, in 2013, the Ministry of Economy and Finance established a Center for Excellence on GRB, in partnership with UN-Women Morocco. As a result, women won 48 percent of the 70,000 jobs created in 2015 and represented 73 percent of the beneficiaries of training programs in 2016.[91]

Critics, however, have cast doubt on the credibility of such reports, arguing that the impacts of climate change are gender differentiated and that women's lives and livelihoods are at grave risk. In rural Morocco, Aya Kamil writes, grim climatic disruptions put in danger women's access to health services—especially in flood-prone areas such as Figuig, Ten Drara, and Bou Arfa—with negative consequences for their overall mental health.[92] Absent proper and timely attention to gender disparities, Kamil notes, Morocco's ability to adapt to climate change will remain woefully inadequate. The Moroccan Climate Change Policy, created in 2000, marked the country's contribution to sustainable development through strategic planning and deflates "the strategic importance of gender, in part by prioritizing climate adaptation and mitigation over climate risk prevention and reduction."[93]

For rural Moroccan women, reducing climate risk takes on far more significance than adaptation plans, as they need direct state interventions such as the distribution of drought-resistant crops and protection against deluge and violent storms. According to the most recent estimates, the government plans to spend $50 billion on mitigation programs for the effects of GHG emissions and another $35 billion on adaptation projects by 2030. In these estimates, as in the past, gender considerations are marginalized.[94] The fact remains that gender-sensitive climate adaptation and mitigation goals, as Kamil emphasizes, will demand a deeper and broader investment in social safety nets, access to green employment for women, and advancing climate risk-reduction programs. All this is manageable given that the country has the potential to lead the MENA region in the drive toward implementing gender-inclusive climate policies and practices.[95]

The Green Investment

Climate change in Morocco is likely to have a significant impact on tourism, and most importantly, such effects will be felt in the agricultural and fisheries sectors, which employ half of the population and account for 14 percent of GDP.[96] Given that 60 percent of the population lives on the coast and most of the country's industry operates there, sea level rise could pose a significant threat to major economic infrastructure. The government has pursued several national strategies to more effectively manage the water issue, enhance the use of nonconventional water resources, and modernize irrigation systems. It has also constructed several dams to reroute water from northern to arid southern regions. The government has successfully regenerated forests, including repopulating more than 130,000 hectares (320,000 acres).[97] By 2030, it is estimated that Morocco plans to generate 52 percent of its electricity from renewable energy—that is, 20 percent from solar energy, 20 percent from wind, and 12 percent from hydropower.[98]

To appreciate Morocco's commitment to green investment, it can also be noted that Morocco is the only Arab state that the Climate Action Tracker rated to meet the challenges of the warming target of 1.5 degrees.[99] In this context, Egypt, Saudi Arabia, and the UAE fall in the highly insufficient category, and Iran and Turkey are placed in the critically insufficient category.[100]

Mitigation and Adaptation Strategies

In the past few years, Morocco's efforts to mitigate climate change have evolved significantly, including enhancing the share of renewable energy use and building one of the largest solar parks in the world. Known as the Ouarzazate Solar Power Station, and alternatively called Noor Power Station, this station is a solar power complex with an auxiliary diesel fuel system near the city of Ouarzazate. In 2014, Morocco opened the largest wind farm in Africa, called Tarfaya wind farm.[101]

Subsequently, in 2018, the Moroccan government announced a stupendous solar power plant in the Sahara Desert with the aim of replacing imported oil and gas with renewable sources. This policy has rendered the kingdom less exposed to fossil fuel price fluctuations from its trading partners, while also eliminating many tons of emissions. Yet, the project is also water-intensive, in large part because solar energy production requires wet cooling. Moreover, implementing projects of this type often evokes resistance and pushback when the trade-offs and costs are not fully and properly disclosed to relevant patrons.[102]

Morocco is also the only country in North Africa that has vigorously pursued mitigation strategies by lifting all subsidies for diesel, gasoline, and heating oil, as well as by focusing on sustainable aquifer and ocean practices.[103] Regarding climate change adaptation strategies, the Moroccan federal state has focused particularly on agriculture and water management and control. It should be noted that the government has zeroed in on agriculture since the mid-twentieth century, promoting special crop mixes and harvesting strategies to ensure resilience in rain-fed agriculture. This attention intensified with the introduction of the Plan Maroc Vert across all sixteen regions of the country in 2008 that aimed at consolidating the integration of agriculture into national, regional, and global markets, reducing poverty while assisting small-scale farmers.[104] In tandem with these adaptation strategies, the government has also brought safe drinking water to rural communities and makeshift housing to urban and semiarid areas. Barley has been used instead of wheat in large part because barley consumes less water and grows faster, strengthening the farmers' fight against water deficit.[105]

It is generally argued that migration is a form of adaptation. Most environmentally related migration in Morocco, like anywhere else, will likely be limited to internal movements—that is, internal displacement. For most of these displaced individuals, urban areas will be the most likely destination. Given that 40 percent of the population is employed in agriculture and nearly 70 percent of the poor live in rural areas, environmental shocks and stress have detrimental effects on the livelihoods of many people who have limited mobility.[106]

It is noteworthy that the most immediate threat that climate change poses to Morocco's economy is in its tourism industry. The Moroccan beaches, long considered a major venue for tourism, have faced coastal erosion and other environmental changes. Unless effective measures are implemented, according to one study, the

Moroccan beaches will face significant retreat or loss in the context of climate change and sea level rise soon. Beaches that have already shrunk and experienced significant erosion due to human pressures (caused by dams, port landing stage, sand mining, and coastal urbanization) will be less resilient, failing to resist the additional pressures of climate change.[107]

Given that seaside tourism is a core contributor to the national economy and employment, the projected retreat of the beaches will certainly lead to significant economic losses. Moreover, narrowed beaches will fail to provide full protection as a buffer to the seafront infrastructures and population, directly exposing coastal residents to upsurge waves and storms. Absent proper climate change adaptation strategies, the loss of these coastal ecosystems and their valuable coastal protection will be unavoidable. Their protection and restoration thus should be included among the priorities of adaptation strategies and coastal management policies.[108]

The extent to which beach retreat transpires under different scenarios of sea level rise also shows the potential management strategies that will prevent the loss of the most vulnerable beaches. The geological structures of the Moroccan coast are very diverse, including sandy beaches and dunes, high cliffs, and flat rocky shores. Of the 3,411 kilometers of coastline, approximately 37 percent are low-lying coasts, of which only 2 percent are sandy beaches.[109]

The significant contribution of beach tourism to the country's national GDP demands urgent attention to protect and restore the beaches. In addition to the recreational offerings, sandy coasts safeguard coastal communities and tourist infrastructures from destructive deluge surges. The coastal primary sand dunes serve as protective barriers from submersion and erosion, while furnishing habitat for a variety of plants and animals.[110]

Underscoring the importance of adaptation and climate resilience, Morocco has developed several policy initiatives, notably the Moroccan Climate Change Policy, the 2030 National Strategy for Sustainable Development, the National Climate Plan 2020–30, and the updated NDC submitted in 2021.[111] The Moroccan Climate Change Policy published in 2014, which was deployed as a dynamic, inclusive, and adaptable tool for promoting a low-carbon climate-resilient future, has contributed to diminishing climate risks and vulnerability. Its adaptation elements are geared toward addressing the issues of water, agriculture, fisheries, health, forest and land degradation, biodiversity, and tourism.[112]

The 2030 National Strategy for Sustainable Development, which was designed in 2017, offers proposals to advance resilience-related objectives, such as "the development of a national plan for the prevention of and response to climate risks; the reinforcement of the scrutiny of these issues; and the promotion of innovation technologies."[113] The National Climate Plan 2020–30, launched in 2019, consists of various components, including "building resilience toward climate risks" as one of its core pillars: water constraints, agriculture, fisheries, health, biodiversity, and transport infrastructure.[114] One study demonstrated that local water demand in northern Morocco will be difficult to satisfy in the future, underlining the need to develop adaptation strategies.[115]

Conclusion

Moroccan leaders' attention has for some time been fixated on climate change, and the issue continues to be relevant if not central to the stability of King Mohammed's throne. Climate change has noticeably affected the living conditions of many Moroccans, particularly ongoing water scarcity, drought, and floods. The agriculture, fishery, and tourism sectors are the country's most important economic sectors, yet they are by far the most vulnerable sectors to climate change in the country. The new bitter fault line is no longer about economic growth but rather about climate change, as the latter poses a real challenge to the country's economic development and long-term sustainability practices.

These challenges are transforming the nature of debate between geopolitics and environmental security. An unrestricted reliance on fossil fuels will result in unreliable and flawed ways of seeking economic growth. It is vitally important that Moroccans build on new targets aimed at advancing their renewable energy sources to ensure that they are in sync with the 1.5 degrees Celsius limit of the Paris Agreement, commit to swift and steady coal phase-out from the power mix, and put a break on investing in coal and fossil gas.[116]

The gendered impact of climate change in Morocco cannot be underestimated. Just as women residing in Morocco's mountainous and desert areas are particularly vulnerable to climate change, so too are those who are employed in the agriculture and fisheries sectors. Those who live in rural areas are also visibly exposed to environmental shocks and stress. Given that two-fifths of the Moroccan population are employed in agriculture and less than three-fourths of the poor live in rural areas, extreme weather events will have a massive negative impact on many people's means of support.[117]

Evolving migration patterns have transformed Morocco from a country of origin of migration to a transit and destination country. Although holding environmental factors mainly accountable for forced migration is implausible, climate-induced disruptions have contributed to the movement of people and will do so in the future. The migration of these people, however, will be limited to internal displacement due to the cost and uncertainties of cross-border movement. Regarding mitigation and adaptation strategies, Morocco has achieved a considerable degree of success, even as newly emerging trends of internal displacement and forced migration point to multiple existing problems facing the Moroccan regime in this context.

CHAPTER 12
CONCLUSION

WEIGHING THE UNCERTAINTY

It is a familiar and repeated refrain in the conversation over climate change in both academic and policy circles that any rational geopolitical analysis should take climate variability into account, because to ignore it could have dire consequences for future political stability. If unchecked, extreme climatic disruption and chaos could affect human rights, freedom of movement, and regional as well as global economies. The risk to people in both rural areas and cities would be higher as climate change makes heat waves and hotter temperatures more common.

Global climate change has unleashed many dynamics that are likely to have direct and sweeping implications for human rights in the coming years and decades, not only affecting a great number of people but also causing species losses and a diminution in core ecosystems. Climate disasters, exacerbated by GHG emissions and the subsequent global warming, have led to climate migration around the world. According to the UNHCR, over 20 million people, on average, have been displaced each year by extreme weather events since 2008. Other estimates urge attention to the fact that over 200 million people could become climate refugees by 2050. Still other accounts estimate these figures could even go up to 1 billion people.[1]

The MENA region, Marwa Daoudy points out, has experienced more frequent droughts, heat waves, and other extreme weather events than the rest of the world. Flooding forced human migration and displacement in Egypt, Iran, and Tunisia in 2020, while wildfires have spread in Lebanon, Syria, and Turkey. Climate disruptions in the region, according to the IMF, have caused an annual average of $2 billion in direct physical losses and have affected over 7 million residents annually.[2]

Changing climate, as indicated by shifting temperature and precipitation patterns, has corresponded with a decrease in per capita income, largely because of damage to the agricultural sector.[3] Rainfall projections paint a distressing picture in the region. In Jordan, for example, rainfall is expected to decline by 30 percent and the likelihood of drought will triple by 2100.[4]

Ongoing drought will become a major concern, as countries in the region continue to struggle with depleted aquifers and groundwater shortages. Water scarcity will adversely affect marginalized groups and countries with the least capacity to adapt to dwindling resources in the region. This reduced water access results in more waterborne diseases and lower school attendance, resulting in decreasing educational opportunities for women and girls while also undermining their livelihoods, as they often are responsible for obtaining water for the rest of the family. The mismanagement of water resources will have serious consequences for marginalized groups in the region.[5]

As noted throughout this book, water scarcity and resultant food insecurity present a daunting challenge to the world today and in the future, especially in the face of population explosion in many parts of the Global South. More crucially, however, the war in Ukraine has led to a dramatic surge in the prices of food, fuel, and fertilizer, exacerbating an already global food insecurity. At the same time, climate change has adversely affected production levels, especially in the case of small farms and small-scale rain-fed land, which provide food for most people and households in poor and developing countries of the Global South.[6]

While most climate migration is projected to occur within a country's borders, cross-border migration also appears to be on the rise, forcing IDPs to find refuge beyond their countries of origin. Some analysts warn that drought-driven migration is likely to triple this century if international efforts fail to effectively deal with the growing climate crisis.[7] Climate migrants enjoy certain basic rights under existing international human rights law, but they lack many key protections afforded to traditional refugees. Governments are responsible for protecting internal climate migrants, but a parallel obligation to protect those crossing borders is noticeably lacking.[8]

The urgency of dealing with climate-induced vulnerabilities will most likely affect government policies in the MENA region, especially when it comes to providing water and food security—this in a region already experiencing violent conflicts and facing potential conflicts in the future. As previously mentioned, climate change will enhance risks of ethnic strife, civil war, and interstate conflict. Climate change poses clear and immediate risks to the regional countries' national security by intensifying "food and water shortages, pandemic diseases, and disputes over both refugees and resources."[9] The MENA region faces one of the worst heating and drying atmospheric conditions in the world, making it the globe's worst-affected region in this regard. Drought-related water and food scarcity requires long-term plans for managing such uncertainties.

Meanwhile, water has become increasingly securitized and weaponized in the region. Water scarcity and food security have been closely related to civil wars in Yemen and Syria—two countries whose severe water scarcity has prompted competing actors to deploy this precious resource as a tool in violent conflict to the detriment of millions of civilians. Amid these difficulties, forced migration and internal displacement have added a heavy burden on the economy of refugee-hosting countries. A recent UNHCR report on progress in indicators under the GCR showed that many refugees still lack access to decent work. At the global level, only 38 percent of refugees live in countries with unrestricted access to formal employment, including wage-earning jobs or self-employment.[10]

To be sure, across the world, we have seen a few success stories of the right to work and freedom of movement of migrants, with Uganda being one of them. Many sources in the West have described Uganda as "one of the best places to be a refugee."[11] While many refugee-hosting countries in Africa require refugees to live in camps and deny them access to labor markets, Uganda offers a self-reliance model that has three core parts: (1) its regulatory framework that allows refugees to work and choose their place of residence; (2) its assistance model allocates plots of land for refugees to cultivate within its villages and farms; (3) its model of refugees fosters integrated social service provision and market access.[12] These policies have made considerable difference to refugees' socioeconomic conditions in four areas: "greater mobility, lower transaction cost for economic activity, higher incomes, and more sustainable sources of employment."[13]

In this final chapter, I return to some of the major topics discussed in the book, including the outdated 1951 Refugee Convention, climate-human rights nexus, climate-related protest, environmental politics and green investment, and climate governance. I next address the issue of climate justice with a view toward exploring the climate change-human security link. I conclude by outlining policies and strategies that can expeditiously and effectively respond to evolving climate change in the MENA region.

The Archaic Refugee Convention

The global refugee system is not working today. It has become increasingly outdated as many refugees are now fleeing disorder, generalized violence, poverty, and environmental disasters. Despite several high-level conferences convened by the United Nations, as some experts point out, "there is still no clear strategy for the future of the global refugee system."[14] Yet the question persists: Why is there less support for "climate refugees" than those who have fled religious, political, and ethnic persecution?

Arguably, there needs to be more attention to the evolving definition of violence, as the interests of the vulnerable individual—faced with the threats of poverty, disorder, and climatic breakdowns—rather than the interests of states, must be included in the definition of refugees.[15] The concept of "climate refugees" has gained increasing currency in the early twenty-first century. While this category lies outside the 1951 Refugee Convention, lawyers have supported rationales for legal protections of people displaced by climate variability. The 2018 Sydney Declaration, for example, laid out principles of international protection for people displaced by rising sea levels that have threatened coastal communities. These threats have increased migration push factors, causing displacement and mass flight.[16]

To repeat a point made earlier and throughout this book, only a small minority of refugees live in camps. The camp model has failed to provide a sustainable solution to a large-scale influx of refugees. The UNHCR has increasingly provided refugees outside of camps with cash benefits rather than furnishing aid in the traditional manner.[17] Similarly, there has been growing attention on the increased opportunities for the refugees to work. It bears stressing that Articles 17–19 of the 1951 Refugee Convention provide and protect refugees' right to work and access to labor markets.[18] Signatory States shall accord to refugees lawfully staying in their territory the most favorable treatment accorded to nationals of a foreign country in the same circumstances, regarding the right to engage in wage-earning employment. In the late 1940s, the response of the UNHCR to the refugee situation was to keep them in the camps by providing food and shelter based on their needs.

Today, however, most refugees live outside these camps. As noted in the preceding chapters, Syrian refugees, according to the UNHCR, have sought asylum in more than 130 countries, but the vast majority of them live in neighboring countries, such as Turkey, Lebanon, Jordan, Iraq, and Egypt.[19] Nearly 92 percent of refugees who have fled to these neighboring countries live in rural and urban areas, with only 5 percent living in *refugee camps*. The more than 70 percent of Syrian refugees in these neighboring countries are living in poverty, with limited access to basic services, education, or job opportunities, and low prospects of ever returning home.[20]

In Jordan, approximately 85 percent of the refugees ignore the camps. Turkey hosts the largest number of Syrian refugees: 3.7 million, of whom, approximately 9 percent live in refugee camps. The remaining 91 percent are spread across Turkey, migrating from one place to another in search of economic opportunities. This reality renders the return of Syrians to their homeland even less likely.[21]

What refugees have sought in the past and wish to have today is the autonomy that comes from legal and gainful employment. Almost half of the Refugee Convention provisions relate to socioeconomic rights such as the right to work and freedom of movement. One of the most important components of autonomy is the right to earn a living. Some experts have argued that "Refugees—as refugees—need and should be entitled to expect three things: rescue, autonomy, and an eventual route out of limbo."[22] The refugee system has become a humanitarian system, and it lacks expertise in development or the tools to guide market-oriented solutions that can promote autonomy.[23]

Without rehashing one of this book's key arguments, it is a useful point to recall that stranded and displaced people and migrants, away from home due to changing environmental conditions, are likely to be exposed to abuse, exploitation, and human rights violations during their displacement. These displaced persons suffer from gaps in legal protection and human rights enforcement and often fall into a situation of "legal limbo" or a "legal grey zone."[24] The limits to their mobility, or total lack thereof, are generally attributed to their exclusion from the structures of power as well as their social disempowerment, which confine their access to the substantial means necessary to secure their livelihoods and mitigate their exposure to mistreatment related to work and income.[25]

Climate Change, Displacement, and Human Migration

Migration processes, behavior, and patterns, as Robert A. McLeman argues, are complex and difficult to specify given that circumstances on the ground may be different in any given situation. Migration decisions are shaped by wide-ranging forces, many operating beyond the control or influence of the migrants themselves.[26] In the context of vulnerability and adaptation to climatic disruptions, McLeman notes, migration is a possible outcome of processes through which vulnerable individuals and households adapt to climatic hazards.[27] In the case of extreme weather events, the poor are often at great risk of displacement, yet their limited financial capacity restricts their migration choices. The largest wave of migration is likely to be interregional, to destinations where gainful employment is available, and migration networks are already established.[28]

Climate migration has left many people homeless and at risk of food insecurity due to extreme weather events, as well as slower-moving climate challenges, such as sea level rise and growing water stress. Some experts have warned that "as climate change makes some parts of the earth uninhabitable, a climate migration crisis looms that international law is not prepared to address."[29] This form of migration is on the rise because the world has been unable to reduce GHG emissions and cease global average temperature rise, which leads to more climate disasters. Most climate migration is likely to occur within a country's borders, but cross-border migration will also rise.[30]

Also known as "environmental refugees," these climate refugees face greater political risks than traditional refugees who flee their homes due to conflict or political persecution. Unlike traditional refugees, climate refugees may be sent back to their ransacked homeland or forced into a refugee camp. As noted earlier, most climate refugees are internal migrants. Often, climate refugees are rural and coastal residents who have been forced to migrate to urban areas. These climate refugees must cope with multiple challenges. Their skills, primarily in herding and farming, are less in demand in urban areas. Rural farmers are often more self-sufficient than many urban dwellers; they are unfamiliar with depending on a corporation or other people for employment.[31]

The climate-related shock and stress in the context of the refugee crisis and forced migration have raised a key question: Whose responsibility—states, humanitarians, or international organizations—is it to tackle climate-based displacement? Some studies have placed the onus on states, arguing that among the key remedies available to policymakers are improvements in water-use efficiency and productivity in agriculture, better management and monitoring of groundwater resources, and multilateral agreements on managing and sharing the rivers that cross political borders.

The UN and Red Cross agencies face an equally massive humanitarian crisis under present circumstances. The UNHCR, constrained by current refugee law, appears helpless in the wake of these new crises. While the UNHCR promises to do its part to help those affected by natural disasters, it is not within its legal mandate to do so. Thus, the responses of relief agencies, NGOs, and donors, as well as suppliers of other forms of foreign aid, could prove to be vitally significant.

Large-scale migration from the MENA region to Europe and the rest of the world following the 2011 Arab Spring uprisings led to major disruptions in those areas. While the primary drivers of this forced migration in the region were economic upheaval, violence, and civil wars—especially in the cases of Syria, Yemen, Libya, Afghanistan, Iraq, and the Sahel countries—it is estimated that over the coming decades, climate change will increasingly contribute to transborder refugee flows.[32] In the MENA region, during 2021 alone, approximately 3 million people were forced to leave their homes due to natural disasters and climate-induced events. These displaced people epitomize the emerging notion of climate refugees.[33]

Climate-Related Uprisings and Protests

Climatic stresses, taken together with rapid urbanization and population growth, could become a source of climate-induced displacement, causing politically destabilizing and disruptive consequences. Policies aimed at improving the adaptive capacity of states to deal with climate change may mitigate the likelihood of civil unrest and forced migration in the future. The Syrian refugee crisis and the resultant forced displacement of several million people, caused partly by ongoing droughts (2006–11) and in large part due to the violent power struggle between sectarian groups, presented new challenges to the traditional description of a refugee according to the 1951 Refugee Convention.

Climate change disruptions in the MENA region will impact virtually everyone in the region, but they will have obvious deleterious effects on the most vulnerable—that is, the poor, women and children, minorities, and the indigenous people. While in some cases, protests have taken on nonviolent forms, in others, they have assumed the form of an uprising against the current governments; yet they have hardly ever assumed the form of an interstate war.

The 2011 Arab Spring uprisings bolstered scholarly and popular interest in developing a more multidimensional and deeper structural understanding of the dynamics of social protest in the region. The popular revolts gave way to complex civil wars in Libya, Syria, Iraq, and Yemen, as nonviolent and mass social mobilization and activism were overshadowed by such wars. Yet environmental mobilization did not plummet. Indeed, the attention to climate change, scarce water supplies, urbanization, and migration has significantly increased in most countries of the region, conferring legitimacy on social and institutional opportunities for activist strategies.[34]

Similarly, the Occupy Gezi movement in Istanbul's Taksim Gezi Park, which started on May 28, 2013, in protest against the implementation of an urban management project, reflected the discontent and frustration of ordinary citizens with the AK Party government. To be sure, as Nilüfer Göle remarks, just defending a few trees in Istanbul's Gezi Park was not the main pretext for political contestation. The plan to demolish this public park in order to construct a shopping mall caused a backlash against an oppressive plan, reflecting resistance to the government's extreme urban development approach. Within this movement, environmental sensitivity, criticism of capitalism, and the rejection of neoconservatism became interlinked.[35]

"For Istanbul citizens," Göle wrote, "the project of constructing a shopping mall in the middle of Gezi Park [was] nothing more than the confiscation of a public space by private capital."[36] Other observers, such as Coşkun Taştan, have disputed that idea emphatically, arguing that while the environmentalist discourse was cited to be at the very heart of the Gezi Park protests at first, it gradually dwindled away as the demonstrators expanded their demands and reach. While environmental concerns triggered the protests, they lost their original significance within the larger set of goals and demands.[37]

In Algeria, the Hirak protest movement (in Arabic: "the Revolution of Smiles") began in February 2019 after peaceful mass demonstrations engulfed the country, opposing then president Abdelaziz Bouteflika—a movement that eventually led to his ouster from office. These protests began in the south of the country against fracking in 2014 and 2015, paving the way for nationwide mobilization. While environmental concerns, including the use and pollution of groundwater, were at the center of these protests, they also contained an explicit political message. Demonstrators described fracking as a neocolonial project in which France, with its multinational corporations poised to exploit resources in the Sahara, was in cahoots with algerian's authoritarian regime to exploit natural resources to the detriment of the local population. These demonstrations strengthened the ties between northern and southern activists, giving further impetus to the emergence of more broad-based protests.[38]

Environmental Politics

All countries of the region have created environmental agencies or ministries to lead the way, albeit much weaker ones than other more traditional ministries such as the interior, industry, mining, and oil and gas. Turkey and Egypt have promoted state-sanctioned environmentalism. Local experts with donor assistance created the Egyptian Environmental Affairs Agency in 1982 and subsequently restructured it in 1994.[39] Later on, this agency was upgraded by the addition of a minister of state for environmental affairs at cabinet level in 1997. In Turkey, in 1991, a Ministry of Environment was created that had far-reaching jurisdiction.[40] In other parts of the region, ministries of water (the UAE and Yemen), sustainable development (Tunisia and Mauritania), housing or municipalities (Bahrain and Djibouti), or energy, mining, and water (Morocco) were established.[41]

There are approximately 313 NGOs supporting environmental causes including biodiversity, conservation, and protection in the MENA region.[42] Enraged citizens across the region, from Morocco to Iran, have taken to the streets to protest ecological damage, calling for policies in line with sustainable development and environmental protection. Climate activists constitute some of the most outspoken voices of opposition to ruling governments. "From Beirut's 'You Stink' movement to Iranian social media users who declared 'I am Lake Urmia' in an effort to revive a salt lake in northwest Iran," Will Todman remarks, "activists have highlighted the effects of environmental mismanagement."[43] But across the region, many others demand fundamental environmental adaptation. Academics, artists, and entrepreneurs implore their governments for innovative environmental policies and climate solutions.[44]

Jordanian groups have found it difficult to compete with well-connected agricultural families who are granted preferential access to water in Jordan. These families are part of the regime's patronage network and tend to resist efforts to distribute water in a more sustainable way. Likewise, environmentalists in Tunisia are acutely aware of the fact that it is immensely difficult for them to compete with the interests of the powerful electricity utility, the Tunisian Company of Electricity and Gas, which has previously thwarted government efforts to integrate renewables to protect its monopoly in the electricity sector.[45]

Some countries in the MENA region—Morocco, Qatar, Saudi Arabia, and the UAE—are among those that have invested in renewable energy, the so-called green investment, by promoting long-term mitigation and adaptation programs. These countries, which often impose significant restrictions on democratic freedoms and governance, have planned against the negative health and economic effects of air pollution while contributing to public safety and a healthy environment. They have done so even as they appear to be committed to different levels of curbing the adverse effects of climate change.[46]

In Morocco, climate change is likely to have a significant impact on tourism, and particularly such effects will be felt in the agricultural system and fisheries, which employ half of the population and account for 14 percent of GDP.[47] Given that 60 percent of the population resides on the coast, as does most of the country's industry, sea level rise poses a significant threat to Morocco's major economic plans. To appreciate Morocco's commitment to green investment, it is important to note that Morocco is the only Arab state that the Climate Action Tracker rated as meeting the challenges of the warming target of 1.5 degrees.[48]

Egypt, Saudi Arabia, and the UAE fall into the highly insufficient category, and Iran and Turkey are placed in the critically insufficient category.[49] Morocco is also the only country in North Africa that has vigorously pursued mitigation strategies by lifting all subsidies for diesel, gasoline, and heating oil, as well as by focusing on sustainable aquifer and ocean research, operations, and practices to support the UN Ocean objectives and goals. By 2030, this North African country plans to generate 52 percent of its electricity from renewable energy: 20 percent from solar energy, 20 percent wind, and 12 percent hydropower.[50]

Similarly, the member states of the GCC (Bahrain, Kuwait, Oman, Saudi Arabia, Qatar, and the UAE) have made some progress in implementing low-carbon energy investments and initiatives. Yet, there exists a substantial gap between the current projects and where the countries seek to be in a net-zero future (2050). To date, only Qatar, Oman, and the UAE, and to a lesser extent Saudi Arabia, have set forth a detail-oriented policy to meet their net-zero targets. These strategies include, among other things, the goal of achieving decarbonization levels by mid-century. These plans, however, lack a comprehensive strategy on how the decarbonization of each sector can be achieved from financial, technological, and institutional standpoints. The GCC pathway to net zero, one expert argues, without clever and innovative intervention, is likely to be unrealistic.[51]

The Qatar Foundation was established to promote a knowledge-based framework aimed at seeking possible routes to sustainable development. In recent years, more attention has been drawn to energy source diversification, economic diversification, energy efficiency, and technology and knowledge transfer in Qatar in order to survive in the global low-carbon era.[52] Qatar has played a constructive role in promoting its climate change program since its ratification of the Kyoto Protocol in 2005, and subsequently in 2012, hosting the 18th Conference of Parties (COP18).[53] Domestically, Qatar has a renewable energy plan of sourcing 20 percent of its total energy from renewable sources by 2024.[54]

The Saudi forward-thinking plan, known as "Vision 2030," seeks to reduce the kingdom's dependence on oil, diversify its economy, boost the role of the private sector, and develop public service sectors such as health, education, infrastructure, recreation, and tourism. Along these lines, Saudi Arabia plans to increase the share of non-oil sectors in national GDP from 16 percent in 2016 to 50 percent in 2030.[55] This new economic plan entails the development of NEOM, a technological vision focused on creating a global hub for energy and water, biotech, advanced technology, and food production. Additionally, to shift away from oil and diversify the economy and exports, Saudi officials argue that NEOM is designed to increase investment in renewable energy sources, with the aim of creating the world's largest renewable energy hub by 2032.[56]

To that end, Saudi Arabia has invested significant resources in promoting biodiversity of land, sea, and coastline. The kingdom's adaptation measures have been directed toward accelerating economic diversification. Furthermore, the so-called Saudi Green Initiative (2021) has sought to reduce carbon emissions, including by meeting 50 percent of the country's domestic energy needs from renewables by 2030. Achieving these targeted goals, however, will require implementing an elaborate road map with viable mitigation and adaptation programs.[57] Whether the Saudis, who have long relied on oil and hydrocarbons to build their institutions, will be able to dramatically pivot their economy away from such a dependency remains to be seen.

The most visible and noteworthy case in the region, however, is the UAE's climate change program. Since its independence in 1971, the UAE has been an active advocate of responsible and eco-conscious growth, moving consistently toward achieving sustainability. Over the past fifteen years, the country has pursued a strategy to diversify its economy by becoming less dependent on oil and gas, including massive investments in renewable energy. The city of Dubai has aimed to become the world's "lowest carbon footprint" city by 2050.[58]

The UAE has been at the vanguard of climate adaptation in the region by deploying the most effective measures, including expanding blue carbon ecosystems by planting 100 million mangrove seedlings by 2030, while also adopting new climate-friendly agricultural methods and devoting greater resources to bolstering sustainable production and consumption patterns.[59] In fact, the UAE became the first country in the region to sign the 2015 Paris Climate Accords as well as the first country in the region to commit to net-zero carbon emissions by 2050. The country launched a green hydrogen initiative and hosted COP28 in Abu Dhabi in the late fall of 2023.[60]

To promote its climate change strategy, the UAE has decarbonized its oil and gas sector, bolstered renewables, and transitioned its food and transportation systems. The government is forging public-private partnerships within the country while helping other countries in the region (e.g., Egypt) make this transition by providing access to finance and technology.[61] The Dubai Clean Energy Strategy 2050 aims to provide 25 percent of the city's total power output from clean energy by 2030 and 75 percent by 2050.[62]

Climate Governance

A growing body of literature supports the view that climate change and human rights have become inextricably intertwined.[63] International organizations, such as the UN Human Rights Council, the ECHR, and Inter-American Commission on Human Rights, have also noted that climate change poses a clear and immediate threat to the full enjoyment of human rights.[64] The Middle East and North Africa Civil Society Network for Displacement joins civil society actors including NGOs, academic institutions, faith-based organizations, artists, cultural influencers, and the private sector in their steadfast devotion to contribute to the protection of IDPs and host communities.

In this book, I have provided the reader with a broader and more detailed survey of how the issues of climate change, forced migration, and human rights interface with socioeconomic, political, and legal perspectives. The project to include refugees and IDPs in the broader protection of human rights and fundamental freedoms, however, has been frequently impeded by exclusionary inclinations of critics who have argued that not every human or socioeconomic problem merits being defined and resolved from a rights-based approach.[65]

It is important to note that compliance with human rights obligations demands promoting projects that offer potent remedies (mitigation and adaptation) to climate displacement. Effective climate change policies and actions must deliver climate justice (equity and fairness) for those who are unduly and disproportionately affected by environmental hazards and disruptions.[66] Throughout this book, I have argued that it is imperative to recognize, protect, and advance the rights of IDPs and refugees, the latter having crossed international borders due to either a fear of persecution, violent conflict, poverty, or climate-induced migration. In short, displaced people have faced multiple difficulties insofar as their right to seek and enjoy asylum is concerned.

Equally important is the need to recognize the rights of marginalized and at-risk populations and communities, including lower-income citizens, especially those who live in rural and coastal areas, women, and workers in the informal sectors. In Kuwait and Saudi Arabia, some demographic groups, such as migrant workers, sectarian minorities, and the so-called stateless Bidoon population, who lack access to many of the country's basic services, will most likely bear the brunt of the climate change impacts. Migrant workers in Saudi Arabia, with already low-quality living and working conditions, lacking legal protections, without the right to unionize, and with limited access to social services, will be more vulnerable to devastating climate change effects.[67]

Sectarian minorities, particularly Shiites, also face serious adaptive challenges. They are often excluded from significant decision-making processes. Those Shiites who are employed in agriculture or fishing, where climate breakdowns can significantly diminish their income and employment opportunities, are particularly vulnerable.[68] Similarly, the Bidoon population in Saudi Arabia, made up of nomadic tribes who have failed to register for Saudi nationality, are generally treated as stateless people. The Bidoon often rely on agriculture for their livelihoods. Droughts and desertification will adversely affect pastoralists among the Bidoon who typically raise livestock. Further, the anticipated rising sea levels and coastal flooding will have devastating effects on the Bidoon who live in coastal communities.[69]

In Jordan, given the chronic water scarcity issues, most of its water resources are transboundary. Water is broadly seen as a key security priority instead of merely a technical or sectoral challenge. Water shortages caused by ongoing droughts have contributed to social unrest in many parts of the country, as many believe that most groundwater resources located in the south of Jordan are utilized largely by residents in Amman and Zarqa.[70]

The most obvious signifier of climate change for Lebanon is its dwindling water supply. High temperatures and less snow on Lebanon's mountains have reduced groundwater, the country's key supply of water for consumption and agriculture. Water shortages have led to lower crop yields, significantly diminishing farmers' ability to maintain their livelihoods. Areas heavily dependent on irrigation systems in the Beqaa Valley, the principal agricultural region in Lebanon, have been impacted by an increase in groundwater stress levels and a depletion of water sources.[71] The inflow of more than 1 million Syrian refugees has adversely impacted climate variability. These refugees make up between 20 and 22 percent of Lebanon's total population, indicating a 20 percent increase in local water use. The rise in water consumption, along with the higher demand for other scarce resources and employment, has fueled further tensions between host communities and refugees.[72]

In Palestine, the Palestinian Authority maintains administrative jurisdiction over only a third of the West Bank. The Hamas authority in Gaza, by contrast, has more say over 2.2 million Palestinians there, but the enclave is under a complete Israeli-enforced blockade and has virtually no drinkable surface or groundwater. Absent an independent Palestinian state, Israeli occupation and hydro-hegemony pose a major obstacle to a comprehensive, just, and durable peace.[73] The lack of continuity in negotiations between the Israelis and the Palestinians for decades has allowed Israel to exploit transboundary waters without consideration for the needs of Palestinians in the besieged Gaza Strip or the West Bank, where the population has substantially increased since the Oslo II water allocations were decided in 1995. The interim agreement in the West Bank essentially conceded previous restrictions on Palestinian water exploitation, validating Israel's control of 80 percent of the groundwater in the West Bank.[74]

Following the Hamas attacks on Israel on October 7, 2023, in which some 1,200 Israelis were killed and more than hundreds taken hostage, Israel has killed more than 42,000 civilians (as of late October 2024) in Gaza, with two-thirds of them being children and women. Nearly 1.2 million Palestinians have been internally displaced because of the Israeli assault on northern Gaza, as they have taken shelter in the southern port city of Rafah.[75] A worldwide pushback against Israel, led by student protests on US campuses, has made divestment from Israel a core demand. As a result, divestment from companies that support Israeli military operations has increasingly gained prominence not only in the United States but also across the world. As the humanitarian crisis in Gaza deepens, more US university campuses become sites of continued protest against war crimes committed by Israel in Gaza.

Egypt's climate-induced water scarcity is likely to have negative consequences for the country's food security and jeopardize the economic livelihoods of farmers, who depend on rain-fed irrigation for agriculture and livestock grazing. Climate change will amplify existing economic inequality between rain-dependent peasants who rely on survival farming and better-resourced, large-scale commercial growers.[76] Given that many Egyptians rely on agriculture for food and employment, the government faces a real challenge in managing its resources while still alleviating rural and urban population vulnerabilities.[77]

Tunisia, according to the World Bank, "is considered to be one of the Mediterranean countries most exposed to climate change."[78] The country's biggest economic sectors—tourism and agriculture—are especially at risk due to climate change. Increased heat waves and decreased water supplies are likely to undermine Tunisia's tourist industry. Also, further plunges in precipitation could significantly undercut the livelihoods of the vast majority of Tunisians, who rely on the agricultural sector. Those Tunisians who live in the country's interior regions are particularly at risk, as they "earn a significant portion of their income from agriculture, ranging from 13.7 percent of the active population in Tataouine to 30.4 percent in Kasserine, compared to a national average of 16.5 percent."[79] Without exaggerating the impacts of climate change actions, it is important to note that an effective implementation could curb rising levels of unemployment and migration. Moreover, it could help keep in check political unrest and instability, as traditionally marginalized segments of the population will most likely suffer from climate change effects disproportionately.[80]

Climate Justice

Standing in the way of climate justice in the international ecological regime is the trade-off between justice and order, in which powerful states continue to marginalize equity and justice considerations in the name of preserving their superior geopolitical status within the global order.[81] Furthermore, some experts have cautioned that there are significant limits to a human rights approach, arguing that such an approach provides no effective form of accountability or redress framework in the case of climate change, where responsibility and damage are largely transnational.[82] Beyond rights is the necessity of solidarity—that is, the moral commitment of taking care of one's fellow humans. Seen in that light, solidarity is a collective responsibility not just to protect human rights but also to rectify *political and social instability caused by climate change.*[83]

Considering the uncertainties and inconsistencies in states' conduct within the existing treaty obligations, there remain many challenges to safeguarding the rights of forced migrants and refugees. One of the most daunting challenges facing the countries in the MENA region today is the refugees' right to engage in legal wage-earning employment. Refugees' right to work has been regularly recognized in international agreements (the 1951 Geneva Refugee Convention and the 2018 UN GCR), as many experts have repeatedly reminded us about the benefits of this right for refugees and their host countries alike. Yet most refugees today encounter significant legal and practical barriers to full economic inclusion in the labor markets of their host countries.

Syrian refugees in Turkey, Jordan, and Lebanon continue to face such barriers, as most of them work in the informal sector. Enforcing the proper policies is a necessary first step toward ensuring refugees' equitable access to fair and sustainable legal employment. Domestic laws should grant refugees the right to work and freedom of movement. The work permit should be viewed as an integral aspect of their status. Denying them work permits would push them toward engaging in the informal sector, a possibility fraught with exploitative workplace conditions, gender abuses, and many other labor-related violations.

Furthermore, this situation will most likely create more bureaucratic delays and confusion among employers and refugees alike. Protecting refugees' right to work in practice and supporting legal aid for refugees who experience unemployment can be reconciled with the national interests of host countries. I have also argued that external support—such as donations—can and should be linked to the refugees' right to work, mitigating opposition to labor market access for refugees. Hosting communities should stop treating refugees as a "burden" to their economy, but rather as an opportunity to benefit.

To argue that refugees have a right to work in a safe working environment is based not only on making strong ethical and emotional appeals but also on a logical appeal. In fact, with a timely investment and prudent policies, offering legal employment to refugees can yield significant economic gains. According to Article 23 of the UDHR: "Everyone has the right to work, to free choice of employment, to just and favorable conditions of work and to protection against unemployment."[84]

Finally, this book has proceeded from the assumption that migration is by far the most potent weapon in the struggle against global poverty.[85] I argue that just as states have the responsibility to protect their populations from genocide, war crimes, ethnic cleansing, and crimes against humanity, so too states have the primary responsibility to fight climate change. Nor can humanitarian agencies replace the role of states in offering political *solutions* to curb the *refugee* crisis. In fact, humanitarian aid and camps have never provided a long-term solution to refugees and displaced people and continue to fall short of doing so.

In this context, I have relied on the intersection of human rights laws and refugee laws. Refugees and asylum seekers are protected by international law, as long as they have fled their habitats to escape persecution, war, violence, poverty, or egregious human rights abuses. These rights include, among others, providing for the safety, basic health care, housing, and employment of refugees. Most notably, the right not to be returned to a country where a person would face persecution, punishment, and other irreparable harm (*non-refoulement*). In addition to the 1951 Convention and its 1967 Protocol relating to the status of refugees, refugee laws include other complementary laws such as international refugee law, international human rights law, and IHL.

Reconceptualizing the Climate Change–Human Security Nexus

As this study has demonstrated, several developments in the MENA region render people affected by environmental stressors explicitly prone to displacement: the lack of freshwater resources, the transboundary nature of this water, and slow but steady urbanization. These particularities have triggered mass displacement and mixed-migration flows within the region.[86] Climate change risks are being increasingly described as risks to the basic needs, human rights, and fundamental principles and values of individuals and communities.[87] This helps explain the climate change-human security nexus, as does the link between security and development. Broadening the definition of human security further to include freedom from fear, hunger, discrimination, racism, extrajudicial killing, natural disasters, and displacement has become a widely applied concept in international affairs.[88]

Human capital is universally deployed as a crucial tool to assist people in their attempts to adapt to climate-induced breakdowns. The economic losses from climate change are massive. In 2022 alone, according to the World Bank, recovery from natural disasters and climate-driven disruptions cost $300 billion globally.[89] The human cost of climate change has wide-ranging and enduring impacts. Climate breakdowns interfere with schooling and health services, affect people' income, and displace millions of people. Drought and desertification shrink crop yields, aggravate malnutrition, and push many families into poverty. Both across and within countries, the most disadvantaged and vulnerable are most exposed to climate risks.[90]

Some observers have argued that the accumulation of human capital should be viewed as a key driver of adaptation. Building up the human capital of the world's poor populations is likely to reduce the impacts of the climate crisis. Human capital expedites occupational change and enhances opportunities to migrate. The poor populations of the Global South face many constraints, as they are far from global labor markets and unaware of climate change disruptions. Under such circumstances, human investment is considered an investment in resilience—an asset that cannot be undercut by the weather.[91]

Other researchers and experts have underlined a sense of the global community given a cluster of environmental crises that threaten to derail people's livelihoods and personal security. We have, they argue, a better understanding of some—if not all—of these threats. One of the better understood challenges we face in our Anthropocene era is climate change. The challenges of human-induced climate change, however, are global, systemic, and invariably interconnected. While much focus has been placed on technological solutions, it is equally essential to cultivate "globally shared values of respect for all people, for the planet, and for truth."[92]

Still others, by throwing their support behind investment in human capital, have argued that the technical and economic dimensions of energy transition should be supplemented with educational and sociopolitical planning. Cultural elements can influence the adoption of technologies and practices, such as electric vehicles, digital efficiency measures, and recycling. Sustainable development education is particularly relevant in shaping how new qualifications and attitudes are required of today's and future citizens. A profound change in consumer behavior, awareness, and willingness to revamp consumption habits is key to transforming the sustainability culture across generations.[93]

The Boiling Frog Syndrome

The MENA region is like a "frog in a boiling pot" that is witnessing the gradual deterioration of its living conditions over time without much by way of foresight to invest its resources wisely and effectively manage a changing climate. It is important to note that the sources of a changing climate are as varied as the solutions needed to tackle it. No one strategy will provide a universal remedy for the unique human security challenges presented by climate change. But as countries across the MENA region grapple with this issue, deploying proper mitigation and adaptation strategies warrants special attention to keep the environmental degradation from spiraling out of control. Despite the recognition that the issues of climate change, forced migration, refugees, climate justice, and human rights have become increasingly interconnected, climate policies have yet to gain a considerably higher political priority. It is critical for countries in the MENA region to develop effective strategies to counter the detrimental effects of climate change. A comprehensive regional strategy focusing on counteracting climate change, however, is conspicuously missing.

The ongoing shortages of water and food supplies will most likely worsen due to climate change. The MENA region, as James L. Gelvin explains, is more vulnerable to climate change than most other regions for two reasons: (1) the region imports more food, particularly grains, than any other region in the world and (2) a large proportion of the region's population lives in coastal areas. Rising sea levels will have devastating effects on coastal populations, causing further salinization of freshwater sources. Populations that depend on coastal aquifers, such as Palestinians living in

Gaza, are particularly susceptible, due in large part to the fact that they depend on a single aquifer that they share with Israel.[94]

An intricate interplay of factors—including, but not limited to, foreign military interventions, regional wars, civil wars, uprisings, and climate change—has collectively resulted in unleashing forces conducive to a prolonged refugee crisis in the MENA region. As noted in several chapters of this book, most Syrian refugees from the thirteen-year conflict remain in the MENA region, some 3.7 million of whom live in Turkey, a country that continues to be the world's largest refugee-hosting country. Other Syrian refugees are scattered in Lebanon, Jordan, and Iraq.

Aside from the refugee problem in the region, the potential and real consequences of climate change throughout the MENA region represent a cause for concern. It is worth reiterating that this region provides a unique vantage point from which to consider the dimensions of climate-induced migration and how it connects the world's richest and poorest countries in an ongoing cycle of environmental degradation.[95] This all offers us as readers, activists, scholars, and concerned onlookers a portent for what a future of radical climate dynamism may include.

Some countries in the region have ensured that domestic laws grant refugees the right to work and freedom of movement, and that these rights are upheld in national laws or constitutions. As noted before, the right of refugees to work has been clearly articulated in many international legal instruments. As might be expected, refugees' right to work, as well as freedom of movement, tends to greatly contribute to the socioeconomic development of their host countries.[96]

Climate change and its consequences, including extreme weather events, will likely force people out of their homes and communities, creating climate migrants and refugees who lack legal status and yet deserve to be protected against pollution and other environmental hazards. Migration is a logical and legitimate means by which people adapt to climate-related, existential challenges.[97] There is a growing consensus among human rights scholars, activists, and legal experts that those displaced by climate change are, or at the very least should be, eligible for refugee status. Considering the increasing link between climate change and forced migration, it is thus essential to redefine and reconfigure traditional notions of "national security." Human security, by contrast, draws our attention to several dimensions related to feeling safe, secure, and stable, while enjoying the promised protections of fundamental rights and civil liberties.

A related challenge emerges in terms of seeking global governance, cooperation, and coordination to effectively tackle climate-induced migration. The time has come to amend the narrowly defined term of "refugee" under the 1951 Refugee Convention by adding a protocol directed toward conceptualizing climate migration and climate refugees, giving similar protections to those granted to traditional refugees. It is noteworthy that environmental refugees or climate refugees encounter far greater political risks than traditional refugees who have fled their homes because of political persecution. Unlike traditional refugees, who are typically supplied with legal, temporary residency, climate refugees may even be sent back to their ravaged homeland or kept in a refugee camp.

To recapitulate, the most formidable challenge facing these refugees and forced migrants across the region is that they lack the basic right to work. Numerous socioeconomic, legal, institutional, and practical barriers prevent their egalitarian access to job opportunities. Removing these barriers to equitable economic inclusion would afford displaced people greater agency and power, enabling them to more fully rebuild their lives and contribute to their host communities. Many observers have underlined the need to implement globally agreed norms—both in law and in practice.[98]

In this concluding section, I offer some explanations as to how best to cope with refugees and IDPs fleeing poverty, persecution, violence, and mass atrocity crimes. Absent the trio of durable solutions—voluntary repatriation, local integration, and resettlement—confining refugees solely to camps, as many observers have noted, leads to consequential and sustained loss of rights. "Among these," writes Carmen Gómez Martin, "are the inability to access decent housing and education and to exercise political and social rights, including the rights to work and to free movement."[99] Without such rights, Gómez Martin writes, "there is a risk of transforming the camps into cities without citizens."[100] This results, Gómez Martin maintains, "in more impoverished urban spaces with precarious and insecure pockets of disenfranchised populations."[101]

Charity and assistance, however, are not enough to deal with the refugee crisis in the MENA region. Granting refugees the right to work bolsters their financial autonomy, self-esteem, and development potential to contribute to the economic health of the society in which they reside. The denial of the right to work has had disastrous and tragic consequences for many refugees, leading to a long-term attrition of their skills and aspirations. Addressing the issues of equity and justice for climate refugees, and especially for those unfairly and disproportionately affected by climate breakdowns, has become pivotal to any level-headed approaches to handling the large-scale population displacements and forced migration projected for the coming years and decades. Grappling with the question of how best to respond to this challenge in the absence of durable solutions has kept both academics and practitioners searching for ways to improve the living conditions of refugees.

Resilient Ways Forward

Going forward, an overarching question is: How can countries in the MENA region improve their preparedness and resiliency to combat climate challenges? A great deal can be done to reduce these risks in climate-affected regions. To meet the climate challenge head-on, as some experts correctly insist, "requires a multisectoral merging of necessary, short-term mitigation and adaptation programming with long-term resilience building and sustainability frameworks, thus grounding climate change response firmly in humanitarian and development actors' mandates."[102]

Sustainability and resilience demand the creation of durable water supply systems, waste management, and water reuse to meet immediate and future needs. This cannot be done without cooperation from humanitarian, developmental, and government actors, who can devise new policies directed at increasing the resilience and livelihood opportunities for those who live in fragile contexts across the Middle East. Such a commitment has the potential to reduce further tensions/conflicts and forced migration in drought- and flood-prone areas.[103]

The need for adaptive and sustainable strategies that can meet the challenge of climate change has never been more urgent. Mitigating climate change can be tactically effective, but it does not amount to a long-term strategy of adaptation. Those countries that implement long-term reforms and prioritize resilient and low-carbon investments over increasing support for the fossil fuel industry appear to be far more poised and well-equipped to tackle climate breakdowns than those who lack such far-reaching plans. Thus, investing in bio-surveillance, healthcare delivery programs, economic development, and anticipatory adaptations, by both the private sector and governments, is the right place to start.[104]

The point to be noted here is that investing in digital technologies can help make carbon management more efficient, transparent, and effective. These technologies, according to the World Economic Forum, could reduce emissions by 20 percent by 2050 in the three sectors of energy, materials, and mobility. The adoption of digital technologies to produce low-carbon and sustainable energy would bring down emissions by 4–10 percent by 2030.[105]

Several countries in the MENA region face more formidable challenges posed by climate change. Noticeably, Yemen's vulnerability to climate change and its dire water scarcity are among the country's most serious and ongoing concerns. This scarcity has led to internal strife over water supplies, while also significantly affecting vulnerable populations. The countries of Egypt and Iraq, for example, are most prone to sea level rises. Egypt's Nile Delta and its coastal front on the Mediterranean are vulnerable to sea level rise and erosion due to climate change. Heavy rainfalls often result in deluges in Egypt, washing away property, claiming lives, and causing massive internal displacement.[106] Similarly, the Tigris and Euphrates delta in Iraq and Syria renders these countries extremely vulnerable to sea level rise. As a result of rising sea levels, according to one study, the saltwater along the Persian Gulf wedge will reach the Shatt al-Arab (the river formed by the convergence of the Tigris and Euphrates) in southern Iraq. At the same time, the decline in flows from the Euphrates will spur the sea's rise, leading to flooding and other problems both in Basra and in Iranian cities in the lower Euphrates basin.[107]

Similarly, climate change has had calamitous effects on Iran, forcing many to become internally displaced and some to migrate, while negatively affecting its largest refugee population made up mostly of Afghans. For the GCC states, climate change is likely to drive a retreat from oil, as their ruling elites have come to a sober realization: "The global economy will eventually become low carbon."[108] The challenges these countries face are many and varied. They include, among others, water and food insecurity, heat waves, sea level rise, dust and sandstorms, flash floods, and marine habitat destruction. In those countries that lack abundant fossil fuel resources, such as Jordan and Lebanon, their high dependence on imported energy resources renders their financial capacity to cope with climate disruptions even harder. In most cases, climate breakdowns are projected to negatively affect regional countries' GDP, fiscal balance, and public debt.[109]

In North Africa, climatic stressors, alongside socioeconomic crises and poor state responses, are likely to lead to protests and uprisings. The effects of insufficient water and food resources, when blended with economic disruption, inadequate urban planning, population growth, and demographic explosion, could spell trouble for existing political regimes. This may explain why countries in the MENA region continue to face dwindling resources and mismanagement, fueling greater public unrest. Should these governments choose to respond with tighter security measures, rather than sustainable reforms, they will face destabilizing political dynamics capable of unleashing new or renewed domestic or regional conflicts.[110]

One study has attributed the key obstacles to climate adaptation planning in regional countries such as Egypt, Morocco, and Tunisia to limited access to accurate climate risk data, inadequate institutional coordination and capacity, insufficient financial resources, and a dearth of holistic planning. It is vitally important that countries in the MENA region establish *national adaptation plans* to define medium- and long-term adaptation needs based on the most advanced climate science, in order to enable governments and societies to effectively meet climate challenges and develop proper strategies to address them.[111]

It has not gone unnoticed that many of those ruling elites in the MENA who are now underscoring the importance of climate change have previously sought to disregard such realities. The countries in the MENA region should define what it means to be "sustainable." Forestalling the unfolding climate change crisis—a task that involves a commitment to fighting for equality, inclusion, and social justice, in addition to establishing adaptation and resilience efforts—serves as a prudent reminder of our shared humanity. Doing so would not only help ruling elites to better cope with the consequences of extreme weather events and global warming today but would also effectively address their regional vulnerabilities, safeguard their people and resources, and promote their long-term stability and sustainability in the face of climate breakdowns.

Glossary

Adaptation The process of adjusting to the effects of climate change by attempting to moderate or avoid the harm done to people. Humans may also intervene to help adjustment for natural systems. Adaptation requires taking actions directed toward responding to current and future adverse climate change impacts.

Anthropocene epoch The current geological age. This era is seen as the period during which human activity has had the most significant influence on climate and the environment.

Anthropogenic climate change This theory hypothesizes that humans are causing most of the current changes to climate by burning fossil fuels such as coal, oil, and natural gas. Scientists have found compelling evidence regarding the impact of humans on the global environment, arguing that humans are responsible for virtually all of the global warming we have experienced since 1950.

Arab Spring uprisings The loosely related group of protests and, more accurately, sociopolitical movements in North Africa that started in late 2010 and expressed a desire for more transparent and democratic political systems, more open societies, and equitable economic systems. While these movements resulted in regime changes in several Arab countries, including Tunisia, Egypt, and Libya, they failed to upend other regimes in Syria and Bahrain.

Carbon dioxide (CO_2) A GHG emission from burning oil, coal, natural gas, and waste materials for energy use. CO_2 also enters the atmosphere from deforestation and from some industrial processes such as cement production. The emission of CO_2 is widely regarded as one of the main sources of air pollution, leading to serious concern due to its climatic effects as a GHG.

Cartagena Declaration on Refugees This Declaration on Refugees and the Protection of People Fleeing Armed Conflict and Other Situations of Violence in Latin America was drafted on November 22, 1984, in Central America by delegates from ten Latin American countries: Belize, Colombia, Costa Rica, El Salvador, Guatemala, Honduras, Mexico, Nicaragua, Panama, and Venezuela.

Clean Development Mechanisms (CDMs) Established by the Kyoto Protocol, CDMs are part of the UNFCCC. CDMs offer the public and private sector in high-income nations the opportunity to purchase carbon credits from offset projects in low- or middle-income nations. These mechanisms are intended to facilitate countries in reducing GHG emissions and to use a share of the proceeds from CDMs to assist developing states that are particularly vulnerable to the adverse impacts of climate change in defraying the costs of adaptation.

Climate change Climate change generally refers to long-term shifts in temperatures and weather patterns, sea level rise, and ocean acidification. These shifts may be natural, but climate scientists agree with the demonstrable evidence that they are due to human causes. Human activities are indeed causing global warming, primarily due to the burning of fossil fuels (including coal, oil, and gas) that produce heat-trapping gases.

Climate change policy Also known as the climate change initiative, this policy is designed specifically to tackle climate change and can be local, national, or international in scope. This policy helps governments and the private sector promote economic growth while addressing climate change.

Climate governance This term refers to the collaboration among civil society, policymakers, and private-sector actors to address constraints and opportunities in their countries for implementing more equitable climate actions. In general, international climate governance is organized around three pillars: mitigation, adaptation, and means of implementation.

Climate justice Refers to policies to address the disproportionate impact of climate change on certain vulnerable classes, groups, genders, communities, and states. Addressing and tackling this disproportionate environmental harm to humans assumes different forms and entails several components. Human rights scholars describe these interrelated elements as distributive justice, procedural justice, corrective justice, and social justice.

Climate-related migrants Also known as environmental migrants or refugees, climate-related migrants are those who are forced to relocate following droughts, floods, and extreme weather events. Climate-related migrants show different degrees and levels of adaptive capacity and adaptation insofar as facing cultural, socioeconomic, and political forces is concerned.

Common but Differentiated Responsibilities (CBDR) The UN Rio Declaration on Environment and Development (1992) outlined the basis for sharing responsibilities of states given their differing capabilities in addressing climate change, protecting the environment, and taking into account historical contributions to global environmental harm, such as pollution, when determining responsibility.

Conference of the Parties (COP) The central decision-making body of the UNFCCC. It includes representatives of all the countries that are signatories (or "parties") to the UNFCCC. It examines the effects of measures introduced by the parties to limit climate change and map out policy recommendations regarding containing the adverse effects of climate change. At least 197 countries have ratified the convention.

Copenhagen Climate Change Conference (COP15) World leaders agreed to a new political accord aimed at addressing climate change at the 2009 UN Climate Summit in Copenhagen, Denmark. The agreement called for explicit emission pledges by all major economies—including, for the first time, China and other major countries of the Global South.

Deforestation The act of perpetual removal of trees from a forest. This act involves the deliberate clearing of forested land for farming or livestock, or using the timber for fuel, construction, or manufacturing. Throughout history forests have been demolished and flattened to make space for agriculture and grazing of pastures, and to procure wood for fuel, manufacturing, and construction. As a result, many landscapes across the globe have undergone a fundamental change toward deforestation.

Desertification A process of ecological degradation driven by humans in which fertile land becomes arid and loses or diminishes its productivity. The result has been the transformation of drylands into desert on an unparalleled level globally, with grave repercussions.

Disaster risk reduction (DRR) This UN-based, global agreement is aimed at preventing new and reducing existing disaster risk and managing residual risk, with a view toward promoting resilience and sustainable development. The so-called Sendai Framework for Disaster Risk Reduction (2015–30) aims to achieve a significant reduction of disaster risk and loss of lives, livelihoods, and health of persons, businesses, communities, and countries over the next decade.

Distributive justice This approach to justice concerns the fair allocation of resources among diverse members of a community based on the notion that every person should have access to approximately the same level of material goods and services. The emphasis here is on equal socioeconomic outcomes and just distribution.

Durable solutions Refers to three routes for ending displacement—local integration, resettlement, and voluntary repatriation.

Environmental refugees People forced to leave or flee their homes due to environmental factors such as sea level rise, flood, tornado, cyclone, storm, or drought.

Euphrates and Tigris rivers These rivers originate in Turkey and flow through Syria and Iraq. While mountain snowfall in Turkey contributes a significant percentage of the Euphrates' water, neither Syria nor Iraq contributes substantially to the flow. Both countries depend on these rivers—Syria on the Euphrates and Iraq on both—but face water supply shortages as Turkey's demand for water increases upstream. Since the mid-1960s, Turkey has built several giant dams on the Euphrates, which have not only reduced downstream water flows but have also led to increased water consumption in Turkey itself. Both Syria and Iraq are thus greatly concerned about the future impact of the dams on their populations and have even labeled Turkey's claims of absolute territorial sovereignty over the two rivers violations of international law.

Food security Refers to having both physical and economic access to sufficient food to meet nutritional needs for a productive and healthy life. Affordability of food is also a central feature of stability in developing countries. Food insecurity, as many scholars note, was a key contributing factor to the 2011 Arab Spring uprisings in the MENA. Food security has become increasingly inseparable from many aspects of livelihood and human security, including health, economic growth, environment, and trade.

Food sovereignty Often synonymous with "food security," this term refers to the extent to which a country can provide sufficient food needed or demanded by its population. In some contexts, it is taken to mean food self-sufficiency. Different countries tend to adopt varying approaches toward maintaining their food sovereignty, some by seeking land purchases abroad, others by investing in agricultural and livestock sectors at home.

Forced migration An involuntary movement of people who fear violence, conflict, death, and destruction, usually based on religion and ethnicity. Disaster-driven causes, stemming from natural disasters, droughts, deluge, and famines, are also a cause of forced migration. The study of forced migration, according to some studies, has developed new perspectives and explanatory models for scholars to examine the livelihood opportunities of refugees and other forced migrants outside of camps.

Fossil fuel A fuel that contains carbon and hydrogen. Coal, oil, and natural gas, which are examples of fossil fuels, constitute by far the largest contributors to global climate change, accounting for the vast majority of global GHG emissions and CO_2 emissions. Increasingly, premature deaths from such air pollution are attributed to the burning of fossil fuels.

Fridays for Future This movement, founded in 2018, was led by Greta Thunberg, a Swedish environmental activist, who worked to address the problem of climate change.

Geoengineering This term refers to a set of emerging technologies that could manipulate or modify the environment and in part neutralize some of the impacts of climate change. While solar geoengineering is unlikely to reduce emissions or effectively deal with a changing climate, it could be valuable in supplementing these efforts, including the effort to counteract global warming.

Geopolitics The key factors, including geography, economics, demography, and security considerations, that influence the politics and especially the foreign policy of nation-states.

Gezi Park The site of demonstrations and civil unrest in Turkey (May 28, 2013), opposing the urban development plan for Istanbul's Taksim Gezi Park. The government forces violently put down these protests, sparking broad-based anger and frustration.

Global civil society Global, nongovernmental actors working to address social problems worldwide. The global public has increasingly shown a growing and tangible interest in addressing climate change and global warming. This interest has been fueled by activism arising in many places and by many nongovernmental organizations and social movement organizations.

Globalization The process of increasing economic, political, social, and cultural integration facilitated by the evolutionary process of technological advance and interconnectedness.

Global North Those socioeconomically and technically well-developed countries that are largely located in North America and Europe.

Global South Regions of the developing world, consisting of mid-income and low-income countries of Latin America, Asia, Middle East, Africa, and Oceania.

Global warming The prolonged and lasting increase in Earth's average temperature causing climate change. The observed warming of the planet, according to climate scientists, is due to human-caused GHG emissions.

Greenhouse Gas (GHG) emissions GHGs trap heat and make the planet warmer. Climate scientists believe that human activities are responsible for almost all of the increase in GHGs in the atmosphere over the last 150 years. The largest source of GHG emissions from human activities is generated from burning fossil fuels for electricity, heat, and transportation.

Gulf Cooperation Council (GCC) A council of cooperation formed in 1981 made up of the six countries of Bahrain, Kuwait, Oman, Qatar, Saudi Arabia, and the UAE.

Helmand River The 621-mile-long river that flows across Afghanistan and empties into lakes and wetlands in southeastern Iran. It is an essential source of water for both countries as it relates to agriculture, fishing, and human consumption for millions of people in Afghanistan and Iran's southeastern Sistan and Baluchistan provinces. The recent Talban regime has created several hydroelectric dams, diverting the water away from Iran, causing increasing tensions between the two countries.

Human security A general concept of security that includes the protection of individuals from poverty, violence, pandemics, ecological challenges, and humanitarian disasters. In addition to the provision of social well-being, this concept entails respect for dignity and the realization of human rights.

Hydropower The use of waterpower to produce electricity and to power machines.

Ikamet A residence permit obtained by a foreign citizen legally residing in Turkey. This document is temporary, although it can be extended frequently.

Inflation Reduction Act (IRA) This US federal law (2022) aims to curb inflation by reducing the federal government budget deficit, lowering prescription drug prices, and investing in domestic energy production while boosting clean energy. This legislation provides price reductions and discounts to low- and moderate-income households across the United States to electrify their homes.

Informal sector An economic sector in which the majority of workers are employed on a contact-intensive basis, such as domestic workers, market vendors, and taxi drivers. They are employed in insecure and exploitative jobs with no insurance coverage or paid leave. The workers' ability to join a labor union is nonexistent, and more often than not they lack any say in their wages, working hours, or certain basic rights.

Intergenerational Justice Refers to climate justice and fairness based on the idea that the pursuit of the interests of present generations should not undermine the needs of future generations.

Intergovernmental Panel on Climate Change (IPCC) An intergovernmental body of the United Nations focused on the advancement of scientific knowledge about climate change caused by human activities. The World Meteorological Organization and the UNEP established the IPCC in 1988.

Internally displaced persons (IDPs) Persons or groups of persons who have been forced to flee or to leave their homes or places of habitual residence, especially as a result of or in order to avoid the effects of armed conflict, violence, violations of human rights, or natural or human-made disasters. These persons are not considered "refugees" because they have not crossed an internationally recognized border.

Jordan River The Jordan River rises at the Syrian-Lebanese border, where the Hasbani River of Lebanon and the Banias River of Syria meet. From there the river flows south through northern Israel into the Sea of Galilee, eventually emptying into the Dead Sea. Geopolitically, the Jordan River is considered the border between the Israel and the West Bank. From its source north of the West Bank to its Dead Sea destination 124 miles away, the Jordan River covers 223 miles due to its winding course.

Kampala Convention The first ever legally binding regional instrument (2009) to provide a broad framework for response to internal displacement, whether prompted by armed conflict, generalized violence, natural disasters, or the disruptive effects of climate change.

Kimlik Issued throughout Turkey, all Turkish citizens carry these identity cards, which are biometric like passports. They can be used as a driver's license, bank card, and bus ticket. These cards are valid for only ten years and must be renewed afterward. Foreigners can be issued a *kimlik* for the purchase of real estate or other forms of investment in the Turkish economy.

Kyoto Protocol Adopted at the third session of the COP to the 1992 UNFCCC, held at Kyoto (Japan) from December 1 to 11, 1997, this protocol was the first international treaty to set legally binding targets to cut GHG emissions. The measures adopted in this treaty were directed at mitigating climate change and facilitating adaptation to climate change.

Liquified Natural Gas (LNG) Natural gas that has been cooled to a liquid state, about 260 degrees Fahrenheit, for shipping and storage. This process makes it possible to transport natural gas to places that pipelines do not reach and to use natural gas as a transportation fuel.

Loss and damage fund A fund created by the COP 28 based on an equity concern that requires addressing climate change impacts on countries that bear either no or little responsibility for climate harm. This fund makes it possible for countries, communities, and organizations to apply for money to address human displacement needs.

Low-lying areas Coastal lands are at, near, or below sea level. These areas face serious risks from climate change because of their relative elevation to current sea levels.

Migrant workers Persons who have been engaged in a remunerated activity in a country of which they are not a national. Migrant workers often have no intention to stay permanently in the country or region in which they work. A unique feature of the Arab countries of the Persian Gulf region has been their noticeable reliance on foreign labor, in large part because these small countries with their young population have limited domestic labor supplies.

Mitigation Taking actions and adopting policies aimed at avoiding and reducing emissions of heat-trapping GHGs into the atmosphere in order to prevent increases in global warming.

Nansen Initiative Initially launched by the governments of Switzerland and Norway in October 2012, this initiative is a state-led, bottom-up consultative process intended to forge consensus on addressing the needs of people displaced across international borders in the context of disasters and the adverse impacts of climate change.

Net zero The balance between the amount of GHG that has been produced and the amount that has been removed from the atmosphere. This balance, which is called net zero, will prevail when the amount of carbon added to the atmosphere does not exceed the amount removed. This target can be achieved through a combination of emission reduction and emission removal. Some scientists believe that we can stay below the dangerous threshold of 1.5 degrees Celsius, achieving NZEs even before 2040.

New York Declaration for Refugees and Migrants On September 19, 2016, the UN General Assembly unanimously adopted the New York Declaration for Refugees and Migrants. The declaration reaffirms the importance of the international refugee regime and contains a wide range of commitments by member states to strengthen and enhance mechanisms to protect people on the move. This declaration subsequently facilitated the adoption of two new global compacts in 2018: a GCR and a global compact for safe, orderly, and regular migration.

Nile River One of the longest rivers in the world, the Nile flows from south to north alongside eastern Africa. It begins in the rivers that flow into Lake Victoria (located in modern-day Uganda, Tanzania, and Kenya) and empties into the Mediterranean Sea more than 4,100 miles to the north. In addition to Egypt, the Nile runs through or along the border of nine other African countries, namely: Burundi, Tanzania, Rwanda, the Democratic Republic of the Congo, Kenya, Uganda, Sudan, Ethiopia, and South Sudan.

Non-refoulement A principle that asserts that refugees should not be returned to a country where they face serious threats to their life or freedom on the grounds of race, religion, nationality, ethnicity, political beliefs, or membership in a social group. This is considered customary international law.

Non-state actors (NSA) Individuals or organizations that have significant political influence but are not allied to any particular state. It is widely used to refer to nongovernmental actors.

Organization of African Unity (OAU) This organization was established on May 25, 1963, in Addis Ababa, Ethiopia, with thirty-two African states joining it. It has fifty-four members. Among its key objectives are the promotion of the unity and solidarity of African states, safeguarding the sovereignty and territorial integrity of member states.

Paris Climate Accords Adopted in 2015, this international treaty on climate change covers climate change mitigation, adaptation, and finance. Its all-encompassing goal is to hold the rise in the global average temperature to well below 2 degrees Celsius above preindustrial levels and foster efforts to limit the temperature increase to 1.5 degrees Celsius above preindustrial levels.

Populism This term refers to a political program or movement that claims to champion the common person, by contrast with elite, big businesses, and financial interests.

Procedural justice This form of justice refers to a fair process of resolving disputes and allocating resources. It is predicated upon the principle of due process, which calls for the equal administration of procedural and substantive law.

Refugees People who are forced to flee their own country and seek safety in another country. They are unable to return to their own country because of well-founded fear of persecution as a result of who they are (race and nationality), what they believe in or say (religions, political views, and social class), or because of armed conflict, violence, or serious public disorder.

Refugee regime The Convention Relating to the Status of Refugees was adopted in 1951 in Geneva. It is the key legal document governing international standards for refugee work and is administered by the UNHCR. The latter was established in 1950 to manage the millions of people displaced in the aftermath of the Second World War. Ratified by 145 states, the convention defines the term "refugee" and expounds on the rights of the displaced, as well as the legal obligations of nation-states to protect them.

Renewable energy Energy sources that are not depleted by use, including hydropower, solar energy, and wind power.

Rentier states Oil-rich nations of the Middle East in which the governments derive revenues from the export of oil and provide rent-subsidized welfare programs. In these states, citizens are neither taxed nor represented via elections, rendering these governments largely unaccountable to the public.

Responsibility to protect (R2P) A norm that the international community holds the moral obligation to protect populations from genocide, war crimes, crimes against humanity, and ethnic cleansing at the hands of their own state or others when the home state fails to provide security.

Restorative justice A type of justice primarily aimed at restoring the situation to that before the violation. Its primary objective is to determine what can be done to repair that harm while holding the person who caused it accountable for their actions. Accountability in this case involves not just accepting responsibility but also acting to repair the harm done.

Retributive justice A type of justice that involves a legal punishment that requires the offender to receive a punishment for a crime proportional or similar to its offense.

Riparian state A state located along a river or bordering a watercourse. A riparian state, according to international law, has the right to use water and an obligation to steer clear from interfering with the water rights of others.

Salinization Soil salinization is a global environmental problem that affects many aspects of our lives. The penetration of excessive salt in cultivable lands significantly diminishes crop productivity. Various salts account for salinity, including sodium, calcium, and magnesium combined with chloride, sulfate, and carbonate to form salt. Among other things, salinization leads to the loss of biodiversity in the long term.

Schengen Area This area, encompassing twenty-seven European countries (Austria, Belgium, Croatia, Czechia, Denmark, Estonia, Finland, France, Germany, Greece, Hungary, Iceland, Italy, Latvia, Liechtenstein, Lithuania, Luxembourg, Malta, the Netherlands, Norway, Poland, Portugal, Slovakia, Slovenia, Spain, Sweden, and Switzerland), allows more than 400 million people to travel freely between these member states without going through customs and border control.

Slow-onset disasters These events create crises through socioeconomic impacts of a disaster, albeit gradually and over time. They include sea level rise, drought, global warming, desertification, ocean acidification, glacial retreat, salinization, land and forest degradation, and biodiversity loss. Legal and social science scholars maintain that it is almost impossible to determine precisely at which point people's lives and livelihoods will be acutely threatened to the extent that they have little or no alternative but to move away from their homes.

Small island states These small island (developing) states are islands of the Caribbean Sea and the Atlantic, Indian, and Pacific Oceans. They are relatively remote, vulnerable to ecological challenges, such as climate change, and generally small in size. Due to sea level rise and extreme weather events, these states are disproportionately affected by climate change.

Soft law Represents agreements, norms, and declarations that are legally nonbinding. Soft law instruments are mostly found in the international sphere. UN General Assembly resolutions are typical. Such norms may or may not eventually become an obligatory law.

State sovereignty A principle for organizing political community that is based on the notion that sovereign states have predominant authority over their territory. In recent years, the issue of whether the existing state system and its assumed authority are absolute or relative has been intensely debated in domestic and international political and human rights circles.

Stateless Refers to individuals who do not belong to any state and thus have no passports or recognized rights of ordinary citizenship. Some people have become stateless in that their state of nationality has ceased to exist, or the territory on which they have lived has come under the control of another state. The Rohingya—a Muslim ethnic minority group who have fled persecution in Myanmar's Rakhine State, inciting a historic migration crisis—are the world's largest stateless population.

Sudden-onset disasters These kinds of calamities are marked by a hazard/disaster risk or devastating natural disaster that arises precipitously or unexpectedly. Sudden-onset disasters are likely to destroy livelihoods and leave the land unlivable. These disasters include, according to climate scientists, earthquake, volcanic eruption, tsunami, chemical blast, critical infrastructure failure, and transport accident.

Survival migration Describes the situation of people who have left their country of origin because of a failed state and the ensuing crisis of displacement, resulting in situations that pose an existential threat for which there is no domestic remedy.

Sustainable Development Goals (SDGs) Also known as the Global Goals, these goals were adopted by the United Nations in 2015 as a universal call to action to abolish poverty, hunger, AIDS, and discrimination against women and girls. They were intended to safeguard the planet and ensure that by 2030 all people enjoy peace and prosperity.

Threat multiplier Scholars and practitioners view climate change as a factor that exacerbates already dire conditions of inequality and injustice with the potential to cause political instability and economic decay over time, but not as the principal cause of forced displacement and mass migration.

Tragedy of the Commons A social and economic problem in which the individual consumes a resource at the expense of society. If each individual keeps acting in their own best interest, it can result in deleterious immoderation to the detriment of all. This phenomenon may result in a budget shortfall or total draining of a shared resource.

Transboundary strategy A strategy based on an agreement forged among countries that share similar and varied boundary concerns. A typical example is transboundary waters that are the aquifers, lakes, or river basins shared by two or more countries. Due to increasing water stress, mismanaged transboundary water supplies are likely to lead to social unrest and/or spark conflict.

Trans-local solidarities Shared notions of common threat emanating from the adverse effects of climate change and extreme weather events have led to the formation of solidarities along common ground to join in the struggle against climate injustice and human insecurity by forming groups, associations, and networks to do so.

United Nations Conference on Environment and Development (UNCED) This conference, also known as the "Earth Summit," was held in Rio de Janeiro, Brazil, June 3–14, 1992.

United Nations Environment Program (UNEP) The global authority for protecting the environment with programs, suggestions, and strategies that address and foster climate stability and a pollution-free future, as well as prevent biodiversity loss and promote sustainable development. The member states of the UNEP comprise 193 sovereign states.

United Nations Framework Convention on Climate Change (UNFCCC) The UNFCCC secretariat is a UN entity tasked with supporting the global response to the threat of climate change. Every year, member states of the UNFCCC meet to measure progress and negotiate multilateral responses to climate change. In 2024, there were 198 parties to the convention.

United Nations High Commissioner for Refugees (UNHCR) This organization is a UN agency mandated to protect and aid political refugees, forcibly displaced communities, and stateless people, and to assist in providing them with durable solutions—that is, voluntary repatriation, local integration, or resettlement to a third country.

Universal Declaration of Human Rights (UDHR) Adopted by the United Nations General Assembly in 1948, this declaration enshrines the rights and freedoms of all human beings. It contains thirty articles; the first twenty-one articles regard civil liberties and political freedoms, and nine other articles involve cultural, social, and economic rights.

Universal Periodic Review The process through which all UN member states review the human rights records of all other member states. The state under review is given an opportunity to report on human rights conditions within their own borders, including policies and actions that have been taken to address those concerns by other states.

Water scarcity Across the world, water scarcity has intensified as demand has surpassed water supply. A global issue on every continent, water scarcity affects poorer communities most harshly. As the world population surges and resource-based economies assume more saliency, many countries' water resources and infrastructure fail to meet growing demands.

Xenophobia The expression of anti-foreigner sentiments and violent attacks against foreigners, including refugees and migrants. This public attitude can be catalyzed by xenophobic politicians and political leaders merely for political gains during electoral cycles.

To prepare the terms defined in this glossary, I have relied on the following sources:

Rawan Arar and David Scott FitzGerald, *The Refugee System: A Sociological Approach*, London: Polity Press, 2023.

Sumudu Atapattu, *Human Rights Approaches to Climate Change: Challenges and Opportunities*, New York: Routledge, 2016.

Simon Behrman and Avidan Kent, eds., *Climate Refugees: Beyond the Legal Impasse?* New York: Routledge, 2018.

Alexander Betts and Paul Collier, *Refuge: Rethinking Refugee Policy in a Changing World*, New York: Oxford University Press, 2017.

John S. Dryzek, Richard B. Norgaard, and David Schlosberg, eds., *The Oxford Handbook of Climate Change and Society*, New York: Oxford University Press, 2011.

Sébastien Duyck, Sébastien Jodoin, and Alyssia Johl, eds., *Routledge Handbook of Human Rights and Climate Governance*, New York: Routledge, 2018.

Elena Fiddan-Qasmiyeh, Gil Loescher, Katy Long, and Nando Sigona, eds., *The Oxford Handbook of Refugee and Forced Migration Studies*, New York: Oxford University Press, 2014.

International Organization for Migration (IOM), "Environmental Migration Portal," available at https://environmentalmigration.iom.int/nansen-initiative.

Dimitra Manou, Andrew Baldwin, Dug Cubie, Anja Mihr, and Teresa Thorp, eds., *Climate Change, Migration, and Human Rights: Law and Policy Perspectives*, New York: Routledge, 2017,

Robert A. McLeman, *Climate and Human Migration: Past Experiences, Future Challenges*, New York: Cambridge University Press, 2014.

Joseph Romm, *Climate Change: What Everyone Needs to Know*, 2nd ed., New York: Oxford University Press, 2018.

Ashok Swain and Anders Jägerskog, *Emerging Security Threat in the Middle East: The Impact of Climate Change and Globalization*, Lanham, MD: Rowman & Littlefield, 2016.

VisitWorld.Today, and Wikipedia.com.

Notes

Chapter 1

1 *Concern Worldwide US*, "The 10 Largest Refugee Crises to Know in 2024," December 15, 2023, available at https://concernusa.org/news/largest-refugee-crises/. Accessed on February 8, 2024.

2 Ibid.

3 Ibid.

4 Dan Rabinowitz, *The Power of Deserts: Climate Change, The Middle East, and the Promise of a Post-Oil Era*, Stanford, CA: Stanford University Press, 2020, 119.

5 Joseph Romm, *Climate Change: What Everyone Needs to Know*, New York: Oxford University Press, 2018, pp. 9–12.

6 Ibid., p. 19.

7 "A World on Fire," Editorial Board, *The New York Times*, *Sunday Review*, January 2, 2022, pp. 2–3.

8 Stephen Humphreys, ed., *Human Rights and Climate Change*, New York: Cambridge University Press, 2009, pp. 7–8.

9 Somini Sengupta, "Colossal Stakes as Leaders Meet to Talk Climate," *The New York Times*, October 31, 2021, pp. 1 and 6.

10 Ibid.

11 Barry Buzan, Ole Waever, and Jaap de Wilde, *Security: A New Framework for Analysis*, Boulder, CO: Lynne Rienner Publishers, 1998, p. 199.

12 Rawan Arar and David Scott FitzGerald, *The Refugee System: A Sociological Approach*, London: Polity, 2023, p. 19.

13 Bill Frelick, "It is Time to Change the Definition of Refugee," *Human Rights Watch*, January 8, 2020, available at https://www.hrw.org/news/2020/01/28/it-time-change-definition-refugee. Accessed on January 2, 2024.

14 James L. Gelvin, *The New Middle East: What Everyone Needs to Know*, New York: Oxford University Press, 2018, p. 146.

15 ICRC, "Migrants, Refugees, Asylum Seekers," International Committee of the Red Cross, available at https://www.icrc.org/en/migrants. Accessed on January 2, 2024.

16 Gillian Brock, Justice for People on the Move: Migration in Challenging Times, New York: Cambridge University Press, 2020, p. 112.

17 Ibid., pp. 112–13.

18 Sumudu Atapattu, *Human Rights Approaches to Climate Change: Challenges and Opportunities*, New York: Routledge, 2016, p. 71; see also p. 155.

19 Ibid., p. 71.

20 Ibid., p. 71.

21 Siobhan McInerney-Lankford, Mac Darrow, and Lavanya Rajamani, *Human Rights and Climate Change: A Review of the International Legal Dimensions*, Washington, DC: The World Bank, 2011, p. 64.

22 Ibid., p. 158.

23 Atapattu, *Human Rights Approaches to Climate Change*, pp. 98–9.

24 Jason P. Evans, "21st Century Climate Change in the Middle East," *Climatic Change*, Vol. 92, 2009, pp. 418–32; see p. 418.

25 Ayokunle Olumuyiwa Omobowale, Olayinka Akanle, Olugbenga Samuel Falase, and Mofeyisara Oluwatoyin Omobowale, "Migration and Environmental Crises in Africa," in Cecilia Menjívar, Marie Ruiz, and Immanuel Ness, eds., *The Oxford Handbook of Migration Crises*, New York: Oxford University Press, 2019, pp. 315–29; see p. 324.

26 Celia McMichael, Carol Farbotko, and Karen E. McNamara, "Climate-Migration Responses in the Pacific Region," in Cecilia Menjívar, Marie Ruiz, and Immanuel Ness, eds., *The Oxford Handbook of Migration Crises*, New York: Oxford University Press, 2019, pp. 297–314; see p. 299.

27 McInerney-Lankford, Darrow, and Rajamani, *Human Rights and Climate Change*, p. 64.

28 Robert A. McLeman, *Climate and Human Migration: Past Experiences, Future Challenges*, New York: Cambridge University Press, 2014, p. 3.

29 Ibid.

30 Ibid., pp. 3–4.

31 Ibid., pp. 10–14, and p. 45.

32 *United States Environmental Protection Agency* (USEPA), "Causes of Climate Change," available at https://www.epa.gov/climatechange-science/causes-climate-change. Accessed on April 21, 2022.

33 Stefan Brönnimann, "Global Warming (1970-Present)," in Sam While, Christian Pfister, and Franz Mauelshagen, eds., *The Palgrave Handbook of Climate History*, New York: Palgrave/Macmillan, 2018, pp. 321–8; see especially p. 321.

34 Ibid., p. 322.

35 Ibid., pp. 323–4.

36 Ibid., pp. 325–7.

37 Zaigham M. Rizvi, "The Challenge of Affordable Housing for Low-Income City-Dwellers," *World Bank Blog*, September 8, 2016, available at https://blogs.worldbank.org/psd/challenge-affordable-housing-low-income-city-dwellers. Accessed on May 20, 2021.

38 Abrahm Lustgarten, "Refugees from the Earth," *The New York Times Magazine*, July 26, 2020, pp. 7–23, and 43.

39 Benoît Mayer, "Definitions and Concepts," in Robert McLeman and François Gemenne, eds., *Routledge Handbook of Environmental Displacement and Migration*, New York: Routledge, 2018, pp. 323–8; see especially p. 323.

40 Ibid., p. 324.

41 Frank Biermann and Ingrid Boas, "Preparing for a Warm World: Towards a Global Governance System to Protect Climate Refugees," *Global Environmental Politics*, Vol. 10, No. 1, February 2010, pp. 60–88; see especially p. 62.

42 Rawan Arar and David Scott FitzGerald, *The Refugee System: A Sociological Approach*, Cambridge: Polity Press, 2023, p. 38.

43 Ibid., p. 19.

44 Elizabeth Ferris, "Governance and Climate-Induced Mobility: International and Regional Frameworks," in Dimitra Manou, Andrew Baldwin, Dug Cubie, Anja Mihr, and Teresa Thorp, eds., *Climate Change, Migration, and Human Rights: Law and Policy Perspectives*, New York: Routledge, 2017, pp. 11–27; see especially pp. 14–15.

45 Ibid., p. 16.

46 Ibid., p. 23.

47 Cher Weixia Chen and Alison Dundes Renteln, *International Human Rights: A Survey*, New York: Cambridge University Press, 2023, p. 287.

48 Maria Trimarchi, "What Are Climate Refugees?," available at https://science.howstuffworks.com/environmental/green-science/climate-refugee.htm. Accessed on August 15, 2019.

49 Peter J. Hoffman and Thomas G. Weiss, *Humanitarianism, War, and Politics: Solferino to Syria and Beyond*, Lanham, MD: Rowman & Littlefield, 2018, p. 245.

50 Biermann and Boas, "Preparing for a Warm World," p. 83.

51 Mariëlle Snel, Nikolas Sorensen, and Reed Power, *Climate Change and Water Scarcity in the Middle East*, London: Taylor & Francis Human Geography, 2025, p. 5.

52 Romm, *Climate Change*, p. 133.

53 Ibid., p. 133.

54 Ibid., p. 134.

55 Ibid., p. 135.

56 Rene Zou, "The Climate Change in the Middle East," *Geopolitical Monitor*, July 24, 2017, available at https://www.geopoliticalmonitor.com/climate-change-in-the-middle-east/. Accessed on October 11, 2017.

57 Michael Hart, "Is Climate Change Driving Conflict in the Middle East?," *International Policy Digest*, October 29, 2016, available at https://intpolicydigest.org/2016/10/29/climate-change-driving-conflict-middle-east/. Accessed on October 11, 2017.

58 Ibid.

59 *Fars News*, "What is Going on in Khuzestan?" July 17, 2021, available at https://www.farsnews.ir/news/14000426000447. Accessed on September 3, 2021.

60 *Khabaronline*, "From Dehydration to Water War; Khuzestan Has Water But It Does Not Reach the People," July 20, 2021, available at https://www.khabaronline.ir/news/1537093. Accessed on September 3, 2021.

61 *ISNA*, Somayeh Rafiei, "Khuzestan Water Problem Has Much Deeper Roots than a Multi-year Management Period," July 20, 2021, available at https://www.isna.ir/news/1400042921463. Accessed on September 3, 2021.

62 Ibid.

63 Aya Batrawy, "Gulf Arab States, Squeezed by Climate Change, Still Tout Oil," *AP News*, October 26, 2021, available at https://apnews.com/article/climate-technology-business-environment-and-nature-middle-east-c9f2733f576c570f70dcc1ef631403c5. Accessed on November 1, 2021.

64 Ashok Swain and Anders Jägerskog, *Emerging Security Threats in the Middle East: The Impact of Climate Change and Globalization*, Lanham, MD: Rowman & Littlefield, 2016, pp. 30–1.

65 Ibid., pp. 157–8.

66 Sari Hanafi, "Forced Migration in the Middle East and North Africa," in Elena Fiddian-Qasmyeth, Gil Loescher, Katy Long, and Nando Sigona, eds., *The Oxford Handbook of Refugee and Forced Migration Studies*, New York: Oxford University Press, 2014, pp. 585–98; see especially p. 593.

67 Swain and Jägerskog, *Emerging Security Threats in the Middle East*, p. 159.

68 World Bank Group, *Beyond Scarcity: Water Security in the Middle East and North Africa*, Washington, DC: World Bank Group, 2018, p. 109.

69 Ibid., pp. 113–114.

70 John Wendle, "The Boiling Point of Climate Change and National Security," in John R. Wennersten and Denise Robbins, eds., *Rising Tides: Climate Refugees in the Twenty-First Century*, Bloomington, IN: Indiana University Press, 2017, pp. 157–88; see especially pp. 167–9.

71 Peter H. Gleick, "Water, Drought, Climate Change, and Conflict in Syria," *American Meteorological Society*, July 6, 2014, pp. 331–40; see 332.

72 Ibid., p. 336.

73 Ibid., p. 336.

74 Brad Plumer, "Drought Help Cause Syria's War: Will Climate Change Bring More Like It?," *The Washington Post*, September 10, 2013, available at https://www.washingtonpost.com/news/wonk/wp/2013/09/10/drought-helped-caused-syrias-war-will-climate-change-bring-more-like-it/?utm_term=.5f3c6ed6dd5c. Accessed on September 29, 2017. Brad Plumer interview Francesco Femia and Caitlin E. Werrell in this report.

75 Jeffrey D. Sachs, "Climate Change Refugees," *Scientific American*, June 1, 2007, available at https://www.scientificamerican.com/article/climate-change-refugees/. Accessed on May 30, 2022.

76 Plumer, "Drought Help Cause Syria's War."

77 Ozden Zeynep Oktav, "Turkey's Water Policy in the Euphrates-Tigris Basin," in Mohamed Behnassi and Katriona McGlade, eds., Environmental Change and Human Security in Africa and Middle East, Switzerland, Springer International Publishing, 2017, pp. 239–55; see p. 248.

78 Ibid.

79 Al-Monitor, "UN Releases Dire Warning About Water Scarcity in Middle East and North Africa," March 23, 2022, available at https://www.al-monitor.com/originals/2022/03/un-releases-dire-warning-about-water-scarcity-middle-east-and-north-africa#ixzz7OTIQZLpi. Accessed on March 24, 2022.

80 Ibid.

81 Ibid.

82 Ibid.

83 Ibid.

84 "Jordan River," *New World Encyclopedia*, available at http://www.newworldencyclopedia.org/entry/Jordan_River. Accessed on February 14, 2015.

85 Thomas Naff and Ruth C. Matson, *Water in the Middle East: Conflict or Cooperation?* Boulder, CO: Westview Press, 1984, pp. 30–3.

86 Ibid., p. 35.

87 Ibid., p. 47.

88 Ibid., p. 53.

89 Green Prophet, "Analyzing the Middle East Water Crisis: In Israel, Jordan, and Beyond," February 11, 2009, available at http://www.greenprophet.com/2009/02/water-israel-jordan/.

90 Dona J. Stewart, op. cit., p. 32.

91 Alex Stark, "Water Wars? The Role of Hegemony in the Jordan River, Nile River, and Columbia River Basin," *E-International Relations Students*, February 25, 2011, available at http://www.e-ir.info/2011/02/25/water-wars-the-role-of-hegemony-in-the-jordan-river-nile-river-and-columbia-river-basins/. Accessed on February 14, 2015.

92 Gidon Bromberg, Nader Khateeb, and Munqeth Mehyar, "From the Dead Sea to an Israel/Palestine Water Accord: 20 Years of Water Diplomacy in the Middle East," in UNESCO, ed., *Free Flow: Reaching Water Security Through Cooperation*, New York: Unesco Publishing, 2013, pp. 28–30; see especially p. 30.

93 Ruth Sherlock and Anas Baba, "The U.N. Says 6 Staff Were Killed in an Israeli Strike on a Gaza School," *NPR*, September 12, 2024, available at https://www.npr.org/2024/09/12/nx-s1-5109856/unrwa-israel-strike-school-gaza-hamas-war. Accessed on September 12, 2024.

94 Apoorva Mandavilli, "The Polio Vaccine Was Altered, and the Disease Surged," *The New York Times*, September 8, 2024, p. 6.

95 *The Iran Primer*, "Iran and Afghanistan Clash over Water Rights," *United States Institute of Peace*, May 30, 2023, available at https://iranprimer.usip.org/blog/2023/may/30/iran-and-afghanistan-clash-over-water-rights. Accessed on June 30, 2023.

96 Ibid.

97 Holy Dagres, "Iran and Afghanistan are Feuding over the Helmand River. The Water Wars Have No End in Sight," *Atlantic Council*, July 7, 2023, available at https://www.atlanticcouncil.org/blogs/iransource/iran-afghanistan-taliban-water-helmand/. Accessed on July 10, 2023.

98 For further information on this topic, see Mahmood Monshipouri and Reza Mohajer, "Climate Change and Forced Displacement in the Middle East: Emerging Threats to Security and Human Rights," *The Maghreb Review*, Vol. 49, No. 1, 2024, pp. 14–41.

99 Anja Mihr, "Climate Justice, Migration, and Human Rights," in Dimitra Manou, Andrew Baldwin, Dug Cubie, Anja Mihr, and Teresa Thorp, eds., *Climate Change, Migration, and Human Rights: Law and Policy Perspectives*, New York: Routledge, 2017, pp. 45–67; see especially p. 47.

100 Ibid.

101 Ibid.

102 Andrew Baldwin, "Conclusion: On the Politics of Climate Change, Migration, and Human Rights," in Dimitra Manou, Andrew Baldwin, Dug Cubie, Anja Mihr, and Teresa

Thorp, eds., *Climate Change, Migration, and Human Rights: Law and Policy Perspectives*, New York: Routledge, 2017, pp. 219–25; see especially p. 220.

103 Ibid., p. 224.

104 David Keith, "Should We Mess with the Atmosphere?," *The New York Times, Sunday Review*, October 3, 2021, p. SR 3.

105 Bradley C. Parks and J. Timmons Roberts, "Environmental and Ecological Justice," in Mechele M. Bestill, Kathryn Hoxhstetle, and Dimitris Stevis, eds., *Palgrave Advances in International Environmental Politics*, New York: Palgrave/Macmillan, 2006, pp. 329–60; see p. 329.

106 Ibid., p. 331.

107 Simon Evans, "Analysis: Which Countries are Historically Responsible for Climate Change?" *CarbonBrief*, October 5, 2021, available at https://www.carbonbrief.org/analysis-which-countries-are-historically-responsible-for-climate-change. Accessed on April 24, 2022.

108 Parks and Roberts, "Environmental and Ecological Justice," p. 345.

109 Ibid., p. 345.

110 *Humanitarian Action*, "The Climate Crisis is a Humanitarian Crisis," November 30, 2022, available at https://humanitarianaction.info/article/climate-crisis-humanitarian-crisis. Accessed on January 30, 2024.

111 Atapattu, *Human Rights Approaches to Climate Change*, p. 6.

112 Ibid., pp. 6–7.

113 Ibid., p. 2

114 Robin Bronen, "Climate-Induced Community Relocations: Institutional Challenges and Human Rights Protections," in Robert McLeman and François Gemenne, eds., *Routledge Handbook of Environmental Displacement and Migration*, New York: Routledge, 2018, pp. 338–402; see especially p. 398.

115 Ibid., p. 398.

116 Ibid., p. 400.

117 David P. Forsythe, *Advanced Introduction to the Politics of International Human Rights*, Cheltenham: Edward Elgar, 2021, p. 120.

118 Ibid., p. 121.

119 Richard P. Hiskes, *Suffer the Children: A Theoretical Foundation for the Human Rights of the Child*, New York: Oxford University Press, 2021.

120 Richard P. Hiskes, *The Human Right to a Green Future: Environmental Rights and Intergenerational Justice*, New York: Cambridge University Press, 2009, see pp. 48–68.

121 Atapattu, *Human Rights Approaches to Climate Change*, p. 294.

122 Ibid., p. 294.

123 Ibid., p. 297.

124 Brooke A. Ackerly, Mujibul Anam, Jonathan Gilligan, and Steven Goodbred, "Climate and Community: The Human Rights, Livelihood, and Migration Impacts of Climate Change," in Dimitra Manou, Andrew Baldwin, Dug Cubie, Anja Mihr, and Teresa Thorp, eds., *Climate Change, Migration, and Human Rights: Law and Policy Perspectives*, New York: Routledge, 2017, pp. 189–202; see especially pp. 198–200.

125 Ibid., p. 201.

126 Reports and Documents, *International Review of the Red Cross*, Vol 99: No 904, 2017, pp. 370–3; also available at <>cross/issue/B0B68D0AA2273D991857271739CFE710>>. Accessed on August 6, 2025.

127 Mohamed Behnassi, "Environmental Change and Human Conflict," in Sean Yom, ed., *The Societies of the Middle East and North Africa: Structures, Vulnerabilities, and Forces*, New York: Routledge, 2019, pp. 207–34; see p. 223.

128 Wendle, "Middle East," pp. 157–88; see especially pp. 180–1.

129 Swain and Jägerskog, *Emerging Security Threats in the Middle East*, p. 79.

130 Ibid.

131 David Wallace-Wells, "Energy 'Too Cheap to Meter,' Has Long Been a Sci-fi Dream. But as the Cost of Solar Power Plummets Toward Zero, It Just Might Become a Reality," *The New York Times Magazine*, September 1, 2024, pp. 18–19; see p. 18.

132 Mehran Kamrava, *Inside the Arab State*, New York: Oxford University Press, 2018, pp. 199–200.

133 James Hollifield, "Migration, Forced Displacement, and Human Development," *Wilson Center*, July 25, 2023, available at https://www.wilsoncenter.org/article/migration-forced-displacement-and-human-development. Accessed on January 30, 2024.

134 Ibid.

135 Swain and Jägerskog, *Emerging Security Threats in the Middle East*, p. 125.

Chapter 2

1 Zahra Babar, "Human Mobility in the Middle East," in Zahra Babar, ed., *Mobility and Forced Displacement in the Middle East*, New York: Oxford University Press, 2020, pp. 1–17; see especially p. 5.

2 Mohamed Behnassi, "Environmental Change and Human Conflict," in Sean Yom, ed., *The Societies of the Middle East and North Africa: Structures, Vulnerabilities, and Forces*, New York: Routledge, 2019, pp. 207–34; see p. 215.

3 Marc Lynch and Lauri Brand, "Refugees and Displacement in the Middle East," *Carnegie Endowment for International Peace*, March 29, 2017, available at https://carnegieendowment.org/2017/03/29/refugees-and-displacement-in-middle-east-pub-68479. Accessed on April 7, 2022.

4 Kate Fanning and Yossi Mekelberg, "The Coming Climate Migration Crisis in the Middle East and North Africa," *Newlines Institute.org*, December 7, 2021, available at https://newlinesinstitute.org/climate-migration/the-coming-climate-migration-crisis-in-the-middle-east-and-north-africa/. Accessed on April 11, 2022.

5 Zeynep Şahin Mencütek, "Forces Behind Lebanese Governance Patterns and Refugee Politics," in Zeynep Şahin Mencütek, ed., *Refugee Governance, State and Politics in the Middle East*, New York: Routledge, 2018, pp. 165–80; see especially pp. 176–7.

6 Zeynep Şahin Mencütek, "Lebanon's Responses to Syrian Mass Migration," in Zeynep Şahin Mencütek, ed., *Refugee Governance, State and Politics in the Middle East*, New York: Routledge, 2018, pp. 129–64; see especially p. 153.

7 Ibid.

8 Dan Rabinowitz, *The Power of Deserts: Climate Change, The Middle East, and the Promise of a Post-Oil Era*, Stanford, CA: Stanford University Press, 2020, p. 75.

9 George Zittis et al., "Business-as-usual Will Lead to Super- and Ultra-Extreme Heatwaves in the Middle East and North Africa," *npj/Climate and Atmospheric Science*, Vol. 4, No. 20, 2021, pp. 1–9.

10 Robert Kennedy, "Extreme Hotspot: What 60c Means for the Middle East," *Al-Jazeera*, November 8, 2021, available at https://www.aljazeera.com/news/2021/11/8/climate-hotspot-what-60c-means-for-the-middle-east. Accessed on March 30, 2022.

11 World Bank Group, *Beyond Scarcity* , p. 98.

12 Ibid., p. 99.

13 Ibid., p. 99.

14 Ibid., p. 102.

15 Douglas Broom, "How the Middle East is Suffering on the Front Lines of Climate Change," *World Economic Forum*, April 5, 2019, available at https://www.weforum.org/agenda/2019/04/middle-east-front-lines-climate-change-mena/. Accessed on April 2, 2022.

16 World Bank Group, *Beyond Scarcity* , p. 102.

17 The World Bank Group, "Adaptation to Climate Change in the Middle East and North Africa Region," 2013, available at http://web.worldbank.org/archive/website01418/WEB/0__C-152.HTM. Accessed on April 2, 2022.

18 Amin Dezfuli, "Rare Atmospheric River Caused Record Floods Across the Middle East," *Bulletin of the American Meteorological Society* (*BAMS*), Vol. 101, Issue 4, April 1, 2020, available at https://journals.ametsoc.org/view/journals/bams/101/4/bams-d-19-0247.1.xml. Accessed on April 27, 2022.

19 World Bank Group, *Beyond Scarcity* , p. 105.

20 Frederic Wehrey and Ninar Fawal, "Cascading Climate Effects in the Middle East and North Africa: Adapting Through Inclusive Governance," *Carnegie Endowment for International Peace*, February 24, 2022, available at https://carnegieendowment.org/2022/02/24/cascading-climate-effects-in-middle-east-and-north-africa-adapting-through-inclusive-governance-pub-86510. Accessed on April 2, 2022.

21 Ibid.

22 Ibid.

23 Ibid.

24 Ibid., p. 100.

25 Rachel Delia Benaim, "How the Middle East's Drought Cycle Will Probably Lead to Even More Refugees," *The Weather Channel*, June 5, 2019, available at https://weather.com/forecast-change/news/2019-06-04-how-the-middle-east-drought-and-flood-cycle-will-probably-lead-to. Accessed on April 1, 2022.

26 World Bank Group, *Beyond Scarcity*, p. 102.

27 Benaim, "How the Middle East's Drought Cycle Will Probably Lead to Even More Refugees."

28 Ibid.

29 Ibid.

30 Ibid.

31 Fanning and Mekelberg, "The Coming Climate Migration Crisis in the Middle East and North Africa."

32 James L. Gelvin, *The New Middle East: What Everyone Needs to Know*, New York: Oxford University Press, 2018, pp. 140–1.

33 Kennedy, "Extreme Hotspot."

34 Yara M. Asi, "Climate Change in the Arab World: An Existential Threat in an Unstable Region," Arab Center Washington, DC, March 2, 2022, available at https://arabcenterdc.org/resource/climate-change-in-the-arab-world-an-existential-threat-in-an-unstable-region/. Accessed on March 30, 2022.

35 Gelvin, *The New Middle East*, p. 148.

36 Ibid., p. 145.

37 Ibid., p. 145.

38 Ibid., p. 145.

39 Water Footprint Calculator, "What is Virtual Water?," December 20, 2019, available at https://www.watercalculator.org/footprint/what-is-virtual-water/. Accessed on April 3, 2022.

40 Ibid.

41 John Anthony Allan, *Virtual Water: Tackling the Threat to Our Planet's Most Precious Resource*, New York: I. B. Tauris, 2011, pp. 2–3.

42 Ibid., p. 48.

43 World Bank Group, *Beyond Scarcity*, p. 109.

44 Ibid., p. 118.

45 Ibid., p. 118.

46 Ibid., p. 120.

47 This section is based on Mahmood Monshipouri, *Middle East Politics: Changing Dynamics*, New York: Routledge, 2019, see chapter 9, pp. 218–44.

48 William R. Thompson and Leila Zakhirova, *Climate Change in the Middle East and North Africa: 15,000 Years of Crises, Setbacks, and Adaptation*, New York: Routledge, 2021. See Chapter 7, "Global Warming in the MENA Today and in the Coming Decades," pp. 137–61; see especially pp. 144–5.

49 Eckart Woerz, "Historic Food Regimes and the Middle East," in Zahra Babar and Suzi Mirgani, eds., *Food Security in the Middle East*, London: Hurst & Company, 2014, pp. 19–38; see pp. 36–7.

50 Rachel Bahn, Tariq Tell, and Rami Zurayk, "Social Life, From Rural to Urban," in Sean Yom, ed., *The Societies of the Middle East and North Africa: Structures, Vulnerabilities, and Forces*, New York: Routledge, 2019, pp. 33–78; see p. 56.

51 Ibid., p. 58.

52 Zahra Babar and Mehran Kamrava, "Food Security and Food Sovereignty in the Middle East," in Zahra Babar and Suzi Mirgani, eds., *Food Security in the Middle East*, London: Hurst & Company, 2014, pp. 1–18; see p. 2.

53 Ibid., p. 7.

54 Omar Saif, Toufic Mezher, and Hassan A. Arafat, "Water Security in the GCC Countries: Challenges and Opportunities," *Journal of Environmental Studies and Sciences*, Vol. 4, November 2014, pp. 329–46; see especially p. 330.

55 Matthew Machowski, "How the Gulf is Keeping Water Security Fears at Bay," *Rusi*, May 17, 2011, available at https://rusi.org/commentary/how-gulf-keeping-water-security-fears-bay. Accessed on September 26, 2017.

56 Babar and Kamrava, "Food Security and Food Sovereignty in the Middle East," p. 9.

57 Daniel Workman, "Wheat Imports by Country," *World's Top Exports*, available at https://www.worldstopexports.com/wheat-imports-by-country/. Accessed on April 3, 2022.

58 Sharif M. Taha, "Kingdom Imports 80% of Food Products," *Arab News*, April 20, 2014, available at http://www.arabnews.com/news/558271. Accessed on September 26, 2017.

59 Ibid.

60 Zahra Babar and Suzi Mirgani, eds., *Food Security in the Middle East*, New York: Oxford University Press, 2014.

61 May Ann Tetreault, Deborah L. Wheeler and Benjamin Shepherd, " Win-Win Versus Lose-Lose: Investments in Foreign Agriculture as a Food Security Strategy of the Arab States of the Persian Gulf," in Zahra Babar and Suzi Mirgani, eds., Food Security in the Middle East, London: Hurst, 2014, pp. 221-247; see especially pp, 226-7.

62 Elliot Spagat and Aya Batrawy, "Why Saudi Arabia Bought 14,000 acres of US Farm Land," *The Christian Science Monitor*, March 18, 2016, available at https://www.csmonitor.com/Environment/2016/0328/Why-Saudi-Arabia-bought-14-000-acres-of-US-farm-land. Accessed on September 27, 2017.

63 Mahmood Monshipouri, *Middle East Politics: Changing Dynamics*, New York: Routledge, 2019, p. 224.

64 Hussein A. Amery, *Arab Water Security: Threats and Opportunities in the Gulf States*, New York: Cambridge University Press, 2015, pp. 82–3.

65 Ibid., pp. 84–5.

66 Ibid., p. 86.

67 Ibid., p. 83.

68 Mary Ann Tetreault, Deborah L. Wheeler, and Benjamin Shepherd, "Win-Win versus Lose-Lose: Investments in Foreign Agriculture as a Food Security Strategy of the Arab States of the Persian Gulf," in Zahra Babar and Suzi Mirgani, eds., Food Security in the Middle East, London: Hurst Publishers, 2014, pp. 242–5.

69 Shreya Singh Kasana, "Is the Arab World Ready for a Post-Oil Future?," *The Daily Guardian*, January 15, 2022, available at https://thedailyguardian.com/%EF%BB%BFis-the-arab-world-ready-for-a-post-oil-future/. Accessed on April 2, 2022.

70 Salpie S. Djoundorian, "Middle East and North Africa: Civil Society and Environmental Activism in the Arab World," in Maria Grasso and Marco Giugni, eds., *The Routledge Handbook of Environmental Movements*, New York: Routledge, 2022, pp. 96–108; see p. 97.

71 Jeannie L. Sowers, "Environmental Activism in the Middle East: Prospects and Challenges," *Humanities Futures*, 2017, available at https://humanitiesfutures.org/papers

/environmental-activism-in-the-middle-east-prospects-and-challenges/. Accessed on April 24, 2022.

72 Aziza Moneer, "Environmental Activism in the Post-Arab Spring: It is not About A Mere Clean Environment," *10 Euromed Survey: IEMed*, December 2, 2020, pp. 66–73; see especially p. 66; available at https://www.iemed.org/wp-content/uploads/2020/12/02.Environmental-Activism-in-the-Post-Arab-Spring-It-is-not-about-a-Mere-Clean-Environment.pdf. Accessed on April 27, 2022.

73 Ibid., p. 67.

74 Dina Zayed and Jeanie Sowers, "The Campaign Against Coal in Egypt," *MERIP*, No. 271, Summer 2014, available at https://merip.org/2014/07/the-campaign-against-coal-in-egypt/. Accessed on April 27, 2022.

75 Moneer, "Environmental Activism in the Post-Arab Spring," p. 70.

76 Sowers, "Environmental Activism in the Middle East."

77 Ibid.

78 Behnassi, "Environmental Change and Human Conflict," pp. 225–6.

79 Ibid., p. 226.

80 Ibid., pp. 226–7.

81 Anne Barre, Irene Dankelman, Anke Stock, Eleanor Blomstrom, and Bridge Burns, "From Marrakesh to Marrakesh: The Rise of Gender Equality in the Global Climate Governance and Climate Action," in Sébastien Duyck, Sébastien Jodoin, and Alyssa Johl, eds., *Routledge Handbook of Human Rights and Climate Governance*, New York: Routledge, 2018, pp. 239–50; see especially pp. 240–1.

82 Ibid., pp. 242–3.

83 Deema AlSaffar Patterson, "Green Ghaya: The Youngest Environmental Activist Making Waves on Global Stage," *Al-Arabiya News*, January 4, 2024, available at https://english.alarabiya.net/features/2024/01/04/Green-Ghaya-The-youngest-environmental-activist-making-waves-on-global-stage. Accessed on May 18, 2024.

84 Ibid.

85 Sowers, "Environmental Activism in the Middle East."

86 Ibid.

Jamal Saghir, "The MENA Region Faces a Triple Threat," in Aisha Al-Sarihi, Jamal Saghir, Banafsheh Keynoush, Hafsa Halawa, Khaled Tarabieh, and Zena Agha, eds., *Key Environmental Challenges Facing the Middle East*, Middle East Institute, April 22, 2021, available at https://www.mei.edu/publications/key-environmental-challenges-facing-middle-east. Accessed on April 26, 2022.

87 Hafsa Halawa, "MENA's Environmental Challenges are Exacerbated by State Failures," in Aisha Al-Sarihi, Jamal Saghir, Banafsheh Keynoush, Hafsa Halawa, Khaled Tarabieh, and Zena Agha, eds., *Key Environmental Challenges Facing the Middle East*, Middle East Institute, April 22, 2021, available

at https://www.mei.edu/publications/key-environmental-challenges-facing-middle-east. Accessed on April 26, 2022.

88 Margaret Suter, "An Update on Yemen's Water Crisis and the Weaponization of Water," *Atlantic Council*, November 29, 2018, available at https://www.atlanticcouncil.org/blogs

/menasource/an-update-on-yemen-s-water-crisis-and-the-weaponization-of-water/. Accessed on April 7, 2022.

89 Behnassi, "Environmental Change and Human Conflict," p. 223.

90 Ashok Swain and Anders Jägerskog, *Emerging Security Threats in the Middle East: The Impact of Climate Change and Globalization*, Lanham, MD: Rowman & Littlefield, 2016, p. 162.

91 Mehran Kamrava, *Inside the Arab State*, New York: Oxford University Press, 2018, p. 199.

92 Saghir, "The MENA Region Faces a Triple Threat."

93 Swain and Jägerskog, *Emerging Security Threats in the Middle East* , p. 58.

94 Kennedy, "Extreme Hotspot."

95 Aziza Moneer, "Environmental Activism in the Post-Arab Spring: It is not About A Mere Clean Environment," Barcelona: IE Med., Euromed Survey of Experts and Actors: Civil Society and Social Movements in the Euro-Mediterranean Region, February 2020, pp. 66–71; see p. 66; available at https://www.iemed.org/wp-content/uploads/2020/12/Civil-Society-and-Social-Movements-in-the-EuroMediterranean-Region_.pdf#page=66. Accessed on May 17, 2024.

96 Ibid., p. 70.

Chapter 3

1 Alexander Betts, *The Wealth of Refugees: How Displaced People Can Build Economies*, New York: Oxford University Press, 2021, p. 228.

2 Center for the Study of Human Rights, *Twenty-Four Human Rights Documents*, New York: Columbia University, 1992, p. 11.

3 Joint Data Center on Forced Displacement, "Labor Market Access and Outcomes for Refugees," *JDC Quarterly Digest*, 7th Issue, January 2023, available at https://www.jointdatacenter.org/wp-content/uploads/2023/01/JDC-Quarterly-Digest-on-Forced-Displacement_January-2023.pdf. Accessed on February 1, 2024.

4 Ibid.

5 Alexander Betts and Paul Collier, *Refuge: Rethinking Refugee Policy in a Changing World*, New York: Oxford University Press, 2017, p. 10.

6 Thomas Ginn, Reva Resstack, Helen Dempster, Emily Arnold-Fernández, Sarah Miller, Martha Guerrero Ble, and Bahati Kanyamanza, "2022 Global Refugee Work Rights Report," *Center for Global Development*, July 28, 2022, available at https://www.cgdev.org/publication/2022-global-refugee-work-rights-report. Accessed on February 1, 2024.

7 Ibid.

8 *Refugees International*, "Right to Work," 2024, available at https://www.refugeesinternational.org/right-to-work/. Accessed on January 30, 2024.

9 Heba Gowayed, "Climate Change and Migration in the Middle East and North Africa," Arab Center, Washington DC, September 29, 2022, available at https://arabcenterdc.org/resource/climate-change-and-migration-in-the-middle-east-and-north-africa/. Accessed on July 10, 2023.

10 *UNHCR, 2021 Global Compact on Refugees: Indicator Report*, available at https://www.unhcr.org/global-compact-refugees-indicator-report/wp-content/uploads/sites/143/2021/11/2021_GCR-Indicator-Report_spread_web.pdf. Accessed on July 10, 2023.

11 Betts, *The Wealth of Refugees*, p. 361.

12 Ibid., p. 117.

13 Ibid.

14 Ibid., p. 140.

15 Roger Zetter and Héloïse Ruaudel, "Refugees' Right to Work and Access to Labor Markets—An Assessment," Part II: Country Cases, *KNOMAD: Global Knowledge Partnership on Migration and Development*, September 2016, pp. 1–235; p. iii, available at https://www.knomad.org/sites/default/files/2017-03/KNOMAD%20Study%201%20Part%20I-%20Assessing%20Refugees%27%20Rights%20to%20Work_final.pdf. Accessed on February 5, 2024.

16 Ibid.

17 Robert Guest, *Borderless Economics: Chinese Sea Turtles, Indian Fridges and The New Fruits of Global Capitalism*, New York: St. Martin's Griffin, 201; see p. 10.

18 Mia Prange, "Climate Change Is Fueling Migration. Do Climate Migrants Have Legal Protections?," *Council on Foreign Relations*, December 19, 2022, available at https://www.cfr.org/in-brief/climate-change-fueling-migration-do-climate-migrants-have-legal-protections. Accessed on July 10, 2023.

19 Ibid.

20 Robert A. McLeman, *Climate and Human Migration: Past Experiences, Future Challenges*, New York: Cambridge University Press, 2014, p. 230.

21 Masood Ahmad, "2022 Global Refugee Work Rights Report, *Refugees International*, July 28, 2022, available at https://www.refugeesinternational.org/reports-briefs/2022-global-refugee-work-rights-report/. July 10, 2023.

22 *National Geographic*, "Environmental Refugee," available at https://education.nationalgeographic.org/resource/environmental-refugee/. Accessed on July 10, 2023.

23 Benjamin Freedman, "Getting Ahead of the Middle East's Climate Refugee Conundrum," *Middle East Institute*, April 26, 2023, available at https://www.mei.edu/publications/getting-ahead-middle-easts-climate-refugee-conundrum. Accessed on July 10, 2023.

24 Ibid.

25 Center for the Study of Human Rights, *Twenty-Four Human Rights Documents*, pp. 6–9; see especially p. 8.

26 *The Refugee Convention, 1951*, available at https://www.unhcr.org/sites/default/files/legacy-pdf/4ca34be29.pdf. Accessed on January 30, 2024.

27 Ibid.

28 M. Murat Erdogan, "Syrian Refugees in Turkey: A Burden or Benefit," DCAF: A Center for Security, Development, and the Rule of Law, 2016, available at https://www.dcaf.ch/sites/default/files/publications/documents/DCAF-STRATIM-Paper-2.pdf.

29 Betts and Collier, *Refuge*, pp. 6–7.

30 Ibid., pp. 156–7.

31 Jane McAdam, "Human Rights and Forced Migration," in Elena Fiddaian-Qasmiyeh, Gil Loescher, Katy Long, and Nando Sigona, eds., *The Oxford Handbook of Refugee and Forced Migration Studies*, New York: Oxford University Press, 2014, pp. 203–14; see especially p. 203.

32 Ibid., p. 208.

33 Ibid., p. 203.

34 Ibid., p. 204.

35 Zetter and Ruaudel, "Refugees' Right to Work and Access to Labor Markets—An Assessment," p. 78.

36 Betts and Collier, *Refuge*, p. 77.

37 *Amnesty International*, "Middle East and North Africa 2022," available at https://www.amnesty.org/en/location/middle-east-and-north-africa/report-middle-east-and-north-africa/. Accessed on February 1, 2024.

38 Ibid.

39 *Norwegian Refugee Council* (NRC), "Guide to Employment Rights in Federal Iraq," Thomson Reuters Foundation, Trust Law, March 2023, available at https://www.nrc.no/globalassets/pdf/guidelines/guide-to-employment-rights/federal-iraq-employment-rights-guide---english.pdf. Accessed on February 1, 2024.

40 *Norwegian Refugee Council* (NRC), "Guide to Employment Rights in the Kurdish Region of Iraq," Thomson Reuters Foundation, Trust Law, March 2023, available at https://www.nrc.no/globalassets/pdf/guidelines/guide-to-employment-rights/kri-employment-rights-guide---english.pdf. Accessed on February 1, 2024.

41 Jared Malsin, "Why Some Arab States Refuse to Accept Syrian Refugees," *Time*, September 8, 2015, available at https://time.com/4025187/arab-states-syrian-refugees/. Accessed on February 6, 2024.

42 Ibid.

43 Zetter and Ruaudel, "Refugees' Right to Work and Access to Labor Markets—An Assessment," p. 78.

44 Ibid., p. 82.

45 Ibid., p. 83.

46 Ibid., p. 83.

47 *International Rescue Committee*, "Overview of Right to Work for Refugees Syria Crisis Response: Lebanon & Jordan," available at https://www.rescue.org/sites/default/files/document/987/policybrief2righttoworkforrefugees-syriacrisisresponsejanuary25.pdf. Accessed on July 12, 2023.

48 *UNHCR*, "Jordan Issues Record Number of Work Permits to Syrian Refugees," January 25, 2022, available at https://www.unhcr.org/news/news-releases/jordan-issues-record-number-work-permits-syrian-refugees. Accessed on July 10, 2023.

49 Ibid.

50 Ibid.

51 Betts and Collier, *Refuge*, p. 78.

52 Zetter and Ruaudel, "Refugees' Right to Work and Access to Labor Markets—An Assessment," p. 171.

53 Ibid.

54 Ibid.

55 Ibid.

56 Sarah Bidinger, "Syrian Refugees and the Right to Work: Developing Temporary Protection in Turkey," *Boston University International Law Journal*, Vol. 33, No. 223, 2015, pp. 224–49; see especially pp. 228–30.

57 Ibid., pp. 171–2.

58 M. Murat Erdoğan, "Thinking Outside the Camp: Syrian Refugees in Istanbul," *Migration Policy Institute (MPI)*, August 9, 2017, available at https://www.migrationpolicy.org/article/thinking-outside-camp-syrian-refugees-istanbul. Accessed on July 10, 2023.

59 Zetter and Ruaudel, "Refugees' Right to Work and Access to Labor Markets—An Assessment," p. 174.

60 Ibid.

61 Bidinger, "Syrian Refugees and the Right to Work," p. 235.

62 Ibid., pp. 234–5.

63 Jiwan Soz, "Syrian Working in Turkey: No Rights, No Legal Protections," *Majalla*, September 23, 2022, available at https://eng.majalla.com/node/257311/politicssyrians-working-turkey-no-rights-no-legal-protection. Accessed on November 23, 2022.

64 ECRE: European Council on Refugees and Exiles: https://ecre.org/the-reality-of-legal-employment-to-refugees-in-turkey-lack-of-access-and-discrimination/.

65 Zetter and Ruaudel, "Refugees' Right to Work and Access to Labor Markets—An Assessment," p. 119.

66 Ibid., p. 123.

67 Ibid., pp. 123–4.

68 Ibid., p. 124.

69 *Asylum Access*, "Refuge Work Rights Report: The Syrian Crisis and Refugee Access to Lawful Work in Greece, Jordan, Lebanon and Turkey," August 2017, pp. 1–24, see p. 19, available at https://asylumaccess.org/wp-content/uploads/2017/09/Middle-East-Refugee-Work-Rights-Syrian-Crisis.pdf. Accessed on February 5, 2024.

70 Ibid., p. 19.

71 Ibid.

72 Ibid., pp. 19–20.

73 Ibid.

74 Zetter and Ruaudel, "Refugees' Right to Work and Access to Labor Markets—An Assessment," p. 110.

75 *UNHCR: Iran*, "Refugees in Iran," available at https://www.unhcr.org/ir/refugees-in-iran/. Accessed on August 2, 2023.

76 Zetter and Ruaudel, "Refugees' Right to Work and Access to Labor Markets—An Assessment," p. 113.

77 Ibid., p. 113.

78 *Asylum Access*, "Refuge Work Rights Report," p. 16.

79 Ibid.

80 Elisa Massimino and Alexandra Schmitt, "A Rights-Centered Paradigm for Protecting the Forcibly Displaced," *The Center for American Progress (CAP)*, December 7, 2020, available at https://www.americanprogress.org/article/rights-centered-paradigm-protecting-forcibly-displaced/. Accessed on July 20, 2023.

81 Ibid.

82 Kirsten Schuettler, "Refugees' Right to Work: Necessary but Insufficient for Formal Employment of Refugees," *World Bank*, November 9, 2017, available at https://blogs.worldbank.org/peoplemove/refugees-right-work-necessary-insufficient-formal-employment-refugees. Accessed on August 2, 2023.

83 *Asylum Access*, "Refuge Work Rights Report."

84 Betts, *The Wealth of Refugees*.

85 Ibid., p. 372.

86 Ibid., p. 368.

Chapter 4

1 Environmental Justice Foundation (EJF), "Protecting Climate Refugees," available at https://ejfoundation.org/what-we-do/climate/protecting-climate-refugees?gclid=Cj0KCQiA64GRBhCZARIsAHOLriIq_NsOEn5v1yEUKuuiOvl7C8jWzgn4rR_zZ8_t6eM8aIu8a_Qo6MkaArEsEALw_wcB. Accessed on March 3, 2022.

2 Ibid.

3 Ibid.

4 Mark R. Amstutz, *International Ethics: Concepts, Theories, and Cases in Global Politics*, 5th ed., Lanham, MD: Rowman & Littlefield, 2018, p. 209.

5 *Internal Displacement Monitoring Center*, "Global Report on Internal Displacement 2021," available at https://www.internal-displacement.org/global-report/grid2021/. Accessed on March 18, 2022.

6 *Internal Displacement Monitoring Center*, "A Decade of Displacement in the Middle East and North Africa," available at https://www.internal-displacement.org/sites/default/files/publications/documents/IDMC_MenaReport_final.pdf. Accessed on March 18, 2022.

7 Ibid.

8 Ibid.

9 Ibid.

10 Jane McAdam and Marc Limon, "Human Rights, Climate Change, and Cross-Border Displacement: The Role of the International Human Rights Community in Contributing to Effective and Just Solutions," *Universal Rights Group: Policy Report*, 2015, pp. 1–28; see especially p. 15.

11 Ibid., p. 23.

12 Sean McAllister, "There Could be 1.2 billion Climate Refugees by 2050. Here's What You Need to Know," *Zurich.com*, January 14, 2024, available at https://www.zurich.com

/en/media/magazine/2022/there-could-be-1-2-billion-climate-refugees-by-2050-here-s-what-you-need-to-know. Accessed on February 9, 2024.

13 Issa Ibrahim Berchin, Isabella Blasi Valduga, Jéssica Garcia, and José Baltzar Salgueirinho Osório de Andrade Guerra, "Climate Change and Forced Migrations: An Effort Towards Recognizing Climate Refugees," *Geoforum*, Vol. 84, 2017, pp. 147–150.

14 Timothy Doyle and Sanjay Chaturvedi, "Climate Refugees and Security: Conceptualizations, Categories, and Contestations," in John S Dryzek, Richard B. Norgaard, and David Schlosberg, eds., *Oxford Handbook of Climate Change and Society*, New York: Oxford University Press, 2011, pp. 278–91; see p. 288.

15 *UNHCR: USA*, "Climate Change and Disaster Displacement," available at https://www.unhcr.org/en-us/climate-change-and-disasters.htm. Accessed on March 24, 2022.

16 Alex Randall, "Climate Refugees Definition: Can We Define a Climate Refugee?," *Climate and Migration Coalition*, available at https://climatemigration.org.uk/climate-refugees-definition/. Accessed on March 4, 2022.

17 Jane McAdam, *Climate Change, Forced Migration, and International Law*, New York: Oxford University Press, 2012.

18 "Climate Refugees," February 14, 2022, available at https://www.climate-refugees.org/why. Accessed on March 4, 2022.

19 Ibid.

20 Ibid.

21 Michelle Foster, *International Refugee Law and Socio-Economic Rights: Refuge From Deprivation*, Cambridge: Cambridge University Press, 2007.

22 Jan McAdam, "Review Essay: From Economic Refugees to Climate Refugees," *Melbourne Journal of International Law*, Vol. 10, No. 2, October 2009, pp. 579–95; see especially pp. 588–91.

23 K. F. Kuh, "Climate Change," in Dominick A. Dellasala and Michael I. Goldstein, eds., *Encyclopedia of the Anthropocene*, 2018, available at https://www.sciencedirect.com/topics/earth-and-planetary-sciences/united-nations-framework-convention-on-climate-change. Accessed on March 4, 2022.

24 Elizabeth Ferris, "Soft Law, Migration and Climate Change Governance," *Journal of Human Rights and the Environment*, Vol. 8, No. 1, March 2017, pp. 6–29; see especially p. 22.

25 Ibid., p. 23.

26 Ibid., p. 23.

27 Ibid., p. 24.

28 Ibid., p. 27.

29 Walter Kälin, "The Nansen Initiative: Building Consensus on Displacement in Disaster Contexts," *Forced Migration Review*, available at https://www.fmreview.org/climate. Accessed on March 4, 2022.

30 Ibid.

31 Ibid.

32 Ibid., p. 28.

33 Sumudu Atapattu, *Human Rights Approaches to Climate Change: Challenges and Opportunities*, New York: Routledge, 2016, p. 158.

34 *UNHCR: USA*, "Climate Change and Disaster Displacement," available at https://www.unhcr.org/en-us/climate-change-and-disasters.html. Accessed on March 24, 2022.

35 Dimitra B. Manou, "Climate Change and Human Rights," in Anja Mihr and Mark Gibney, eds., *The Sage Handbook of Human Rights*, Vol. 1, Los Angeles, CA: Sage Reference, 2014, pp. 238–52; see especially pp. 247–8.

36 Beatriz Felipe Pérez, "Beyond the Shortcomings of International Law: A Proposal for the Legal Protection of Climate Migrants," in Simon Behrman and Avidan Kent, eds., *Climate Refugees: Beyond the Legal Impasse*, New York: Routledge, 2018, pp. 214–29; see p. 221.

37 Atapattu, *Human Rights Approaches to Climate Change: Challenges and Opportunities*, p. 159.

38 Jane McAdam, "Swimming Against the Tide: Why a Climate Change Displacement Treaty is Not *the* Answer," *International Journal of Refugee Law*, Vol. 23 No. 1, January 10, 2011, pp. 2–27.

39 McAdam, *Climate Change, Forced Migration, and International Law.*

40 Ibid.

41 Jenny Poon, "Drawing Upon International Refugee Law: The Precautionary Approach to Protecting Climate Change-Displaced Persons," in Simon Behrman and Avidan Kent, eds., *Climate Refugees: Beyond the Legal Impasse*, New York: Routledge, 2018, pp. 157–71; see pp. 158–9.

42 Ibid., p. 159.

43 Ibid.

44 Ibid., p. 163.

45 Nadja Popovich and Brad Plumer, "Who Has the Most Historical Responsibility for Climate Change," *The New York Times*, November 12, 2021, available at https://www.nytimes.com/interactive/2021/11/12/climate/cop26-emissions-compensation.html. Accessed on March 28, 2022.

46 Joseph Romm, *Climate Change: What Everyone Needs to Know*, 2nd ed., New York: Oxford University Press,2018, p. 166.

47 Ibid., p. 133.

48 Ibid.

49 Ibid., pp. 136–7.

50 *Concern Worldwide US*, "Climate Refugees; The World's Forgotten Displacement Crisis," April 5, 2022, available at https://concernusa.org/news/climate-refugees-explained/. Accessed on March 27, 2024.

51 Heba Gowayed, "Climate Change and Migration in the Middle East and North Africa," Arab Center, Washington, DC, September 29, 2022, available at https://arabcenterdc.org/resource/climate-change-and-migration-in-the-middle-east-and-north-africa/. Accessed on March 27, 2024.

52 Ibid.

53 Ibid.

54 Banafsheh Keynoush, "Climate-Induced Migration in the GCC States: A Looming Challenge," *Middle East Institute*, January 13, 2023, available at https://www.mei.edu/publications/climate-induced-migration-gcc-states-looming-challenge. Accessed on March 27, 2024.

55 Ibid.

56 Jon Barnett, "Human Security," in John S. Dryzek, Richard B. Norgaard, and David Schlosberg, eds., *Oxford Handbook of Climate Change and Society*, New York: Oxford University Press, 2011, pp. 267–77; see p. 267.

57 Ibid. p. 274.

58 Ibid., p. 267.

59 Ashok Swain and Anders Jägerskog, *Emerging Security Threats in the Middle East: The Impact of Climate Change and Globalization*, Lanham, MD: Rowman & Littlefield, 2016, pp. 4–6.

60 Jane Arraf, "War and Climate Change Ravage Syria's Farms," *The New York Times*, February 20, 2022, p. 4.

61 Matthieu Aikins, "We've Never Been Smuggled Before," *The New York Times*, February 20, 2022, pp. 4–5 of the Sunday Review; see especially p. 5.

62 Ibid.

63 Alice Thomas, "Human Rights and Climate Discplacement and Migration," in Sébastein Duyck, Sébastien Jodoin, and Alyssa Johl, eds., *Routledge Handbook of Human Rights and Climate Governance*, New York: Routledge, 2018, pp. 110–27; see especially p. 115; see also Dug Cubie, "Human Rights, Environmental Displacement and Migration," in Robert McLeman and François Gemenne, eds., *Routledge Handbook of Environmental Displacement and Migration*, New York: Routledge, 2018, pp. 329–41; see especially p. 333.

64 Ibid., p. 116.

65 Ibid., p. 116.

66 Ibid., 117.

67 Sébastein Duyck, Sébastien Jodoin, and Alyssa Johl, "Integrating Human Rights in Global Climate Governance," in Sébastein Duyck, Sébastien Jodoin, and Alyssa Johl, eds., *Routledge Handbook of Human Rights and Climate Governance*, New York: Routledge, 2018, pp. 1–15; see especially p. 8.

68 Margaretha Wewerinke-Singh, "State Responsibility for Human Rights Violations Associated with Climate Change," in Sébastein Duyck, Sébastien Jodoin, and Alyssa Johl, eds., *Routledge Handbook of Human Rights and Climate Governance*, New York: Routledge, 2018, pp. 75–89; see especially p. 81.

69 Ibid., p. 84.

70 Margit Ammer, Manfred Nowak, Lisa Stadlmayr, and Gerhard Hafner, "Legal Status and Legal Treatment of Environmental Refugees," Environmental Research of the Federal Ministry of the Environment, Nature Conservation and Nuclear Safety, 54/2010; see pp. 1–14; see p. 3; available at file:///C:/Users/909186~1/AppData/Local/Temp/6671.pdf. Accessed on March 21, 2022.

71 Ibid., p. 5.

72 Jane McAdam, "Swimming Against the Tide: Why a Climate Change Displacement Treaty is Not *the* Answer," *International Journal of Refugee Law*, Vol. 23, No. 1, March 2011, pp. 2–27.

73 Rawan Arar and David Scott FitzGerald, *The Refugee System: A Sociological Approach*, Cambridge: Polity Press, 2023, p. 243.

74 Amstutz, *International Ethics: Concepts, Theories, and Cases in Global Politics* , p. 211.

75 Ibid., p. 212.

76 Ibid., p. 212.

77 Ibid., p. 212.

78 Bridget Lewis, "Human Rights as a Basis for Institutions for Future Generation," in Jan Linehan and Peter Lawrence, eds., *Giving Future Generations a Voice: Normative Frameworks, Institutions and Practice*, Cheltenham: Edward Elgar Publishing, 2021, pp. 62–84; see especially p. 70.

79 Sumudu Atapattu, "The Paris Agreement and Human Rights: Is Sustainable Development the 'New Human Rights'?," *Journal of Human Rights and Development*, Vol. 9, No. 1, March 2018, pp. 68–88; see especially p. 70.

80 Lewis, "Human Rights as a Basis for Institutions for Future Generation," p. 71.

81 Judith Blau, *The Paris Agreement: Climate Change, Solidarity, and Human Rights*, New York: Palgrave/Macmillan, 2017, pp. 111–12.

82 This statement is attributed to Elinor Ostrom, cited in Ibid., p. 71.

83 Ibid., pp. 72–82.

84 Vincente Silva Didier, "The Climate Emergency is a Human Rights Issue," *Health and Human Rights Journal*, December 2021, pp. 1–3; see especially p. 1.

85 Ibid.

86 Sumudu Atapattu, *Human Rights Approaches to Climate Change: Challenges and Opportunities*, New York: Routledge, 2016, p. 87.

87 Atapattu, "The Paris Agreement and Human Rights: Is Sustainable Development the 'New Human Rights?'"; see especially p. 84.

88 Francesco Francioni and Ottavio Quirico, "Untying the Gordian Knot: Towards the Human Rights to a Climatically Sustainable Environment," in Ottavio Quirico and Mouloud Boumghar, eds., *Climate Change and Human Rights: An International and Comparative Law Perspective*, New York: Routledge, 2016, pp. 133–56.

89 Silbhán McInerney Lankford, Mac Denow, and Lavanya Rajamani, *Human Rights and Climate Change: A Review of the International Legal Dimensions*, Washington, DC: World Bank, 2011, p. 10.

90 Ibid.

91 Ibid.

92 Tracey Skillington, *Climate Justice and Human Rights*, New York: Palgrave/Macmillan, 2017, p. 157.

93 Ibid., p. 158.

94 Ibid., pp. 158–9.

95 Ibid., p. 159.

96 Ibid., p. 160.

97 Ibid., p. 16.

98 Ibid., pp. 16–17.

99 Ibid., p. 17.

100 This argument is attributed to Marc Limon, in Doreen Stabinsky, "Climate Change and Human Rights," in Sébastein Duyck, Sébastien Jodoin, and Alyssa Johl, eds., *Routledge Handbook of Human Rights and Climate Governance*, New York: Routledge, 2018, pp. 280–91; see p. 287.

101 Ibid., p. 289.

102 Atapattu, "The Paris Agreement and Human Rights: Is Sustainable Development the 'New Human Rights?'"; see especially pp. 87–8.

103 Annalisa Savaresi, "Climate Change and Human Rights: Fragmentation, Interplay, and Institutional Linkages," in Sébastein Duyck, Sébastien Jodoin, and Alyssa Johl, eds., *Routledge Handbook of Human Rights and Climate Governance*, New York: Routledge, 2018, pp. 31–42; see especially pp. 37–8.

104 Helen Dempster and Kayly Ober, "New Zealand's 'Climate Refugees' Visas: Lessons for the Rest of the World," *Devpolicyblog*, January 31, 2020, available at https://devpolicy.org/new-zealands-climate-refugee-visas-lessons-for-the-rest-of-the-world-20200131/. Accessed on April 4, 2022.

105 Sean McAllister, "There Could be 1.2 billion Climate Refugees by 2050."

106 *Environmental Justice Foundation*, "Climate refugees urgently need international legal protection: New report," October 15, 2021, available at https://ejfoundation.org/news-media/climate-refugees-urgently-need-international-legal-protection-new-report. Accessed on March 22, 2022.

107 Anxhela Mile, "Protecting Climate Migrants: A Gap in International Asylum Law," *Earth Refuge*, January 7, 2021, available at https://earthrefuge.org/protecting-climate-migrants-a-gap-in-international-asylum-law/. Accessed on March 17, 2022.

108 Nicholas Micinski, "5 Steps Biden Can Take to Protect Climate Migrants," *WBUR*, January 21, 2021, available at https://www.wbur.org/cognoscenti/2021/01/21/immigration-climate-change-refugees-nicholas-micinski. Accessed on March 24, 2022.

109 Elizabeth G. Ferris and Katharine M. Donato, *Refugees, Migration and Global Governance: Negotiating the Global Compacts*, New York: Routledge, 2020.

110 Thomas, "Human Rights and Climate Discplacement and Migration," p. 122.

Chapter 5

1 Colbert C. Held and John Thomas Cummings, *Middle East Patterns: Places, Peoples, and Politics*, 6th ed., Boulder, CO: Westview Press, 2014, p. 603.

2 Abdullah Gokhan Yilmaz and Monzur Alam Imteaz, "Climate Change and Water Resources in Turkey: A Review," *International Journal of Water*, Vol. 8, No. 3, January 2014, pp. 299–313; see especially p. 303.

3 *Climate Change Knowledge Portal*, "Turkey," available at https://climateknowledgeportal.worldbank.org/country/turkey. Accessed on June 27, 2022.

4 *World Atlas*, "What Type of Climate Does Turkey Have?," available at https://www.worldatlas.com/articles/what-type-of-climate-does-turkey-have.html. Accessed on June 28, 2022.

5 Tomer Barak and Hay Eytan Cohen Yanarocak, "Confronting Climate Change, Turkey Needs 'Green,' Leadership Now More Than Ever," *Middle East Institute*, January 25, 2022, available at https://www.mei.edu/publications/confronting-climate-change-turkey-needs-green-leadership-now-more-ever. Accessed on June 28, 2022.

6 Ibid.

7 Umar Farooq, "Is Turkey Really Serious About Climate Change?," *Al Jazeera*, October 5, 2021, available at https://www.aljazeera.com/features/2021/10/5/turkeys-climate-sceptical-country-is-serious-on-climate-change. Accessed on June 28, 2022.

8 *The World Bank*, "Key Highlights: Country Climate and Development Report for Türkiye," June 13, 2022, available at https://www.worldbank.org/en/country/turkey/brief/key-highlights-country-climate-and-development-report-for-turkiye. Accessed June 28, 2022.

9 *Climate Action Tracker*, October 22, 2021, available at https://climateactiontracker.org/countries/turkey/. Accessed on June 28, 2022.

10 Umar Farooq, "Is Turkey Really Serious About Climate Change?" *Al Jazeera*, October 5, 2021, available at https://www.aljazeera.com/features/2021/10/5/turkeys-climate-sceptical-country-is-serious-on-climate-change. Accessed on June 28, 2022.

11 Ibid.

12 Held and Cummings, *Middle East Patterns*, p. 618.

13 Akgün Ilhan, "Istanbul's Water Crisis," *Green European Journal*, November 8, 2021, available at https://www.greeneuropeanjournal.eu/istanbuls-water-crisis/. Accessed on June 27, 2022.

14 Ibid.

15 Ibid.

16 Ibid.

17 Mattieu Théry, "Water Crisis," available at http://www.matthieuthery.com/energy/water-crisis/. Accessed on February 15, 2015.

18 Clement M. Henry, "Political Economy of the Maghreb," in David S. Sorenson, ed., Interpreting the Middle East: Essential Themes, Boulder, CO: Westview Press, 2010, pp. 185-212; see p. 201

19 Held and Cummings, *Middle East Patterns*, p. 618.

20 Ryan Wilson, "Water Shortage Crisis Escalating in the Tigris-Euphrates Basin," *Future Directions International*, August 28, 2012, available at http://www.futuredirections.org.au/files/FDI_Strategic_Analysis_Paper_-_28_August_2012%28129.pdf. Accessed on February 15, 2015.

21 Deborah Amos, "Middle East Water Crisis Brings Misery, Uncertainty," *NPR*, January 7, 2010.

22 Ufuk Utku Turunçoğlu, Murat Türkeş, Deniz Bozkurt, Barış Önol, Ömer Lütfi Şen, and Hasan Nüzhet Dalfes, "Climate," in Selim Kapur, Erhan Akça, and Hikmet Günal, eds., *The Soils of Turkey*, Cham: Springer International Publishing, 2018, pp. 25–44; see especially p. 42.

23 Ayşegül Kibaroğlue, "Türkiye's Water Security: Energy, Agriculture and Transboundary Issues," in Muhittin Ataman and Gloria Shkurti Özdemir, eds., *Different Dimensions of Environmental Security in Türkiye and Beyond*, Ankara: SETA, 2023, pp. 179–97; see pp. 184–5.

24 Yilmaz and Imteaz, "Climate Change and Water Resources in Turkey: A Review," p. 309.

25 Ibid., p. 309.

26 Ibid., p. 310.

27 Kibaroğlue, "Türkiye's Water Security," pp. 186–7.

28 Begum Toprak, "Turkish Farmers Are Squeezed by Government Neglect," *Syndication Bureau*, February 23, 2020, available at https://syndicationbureau.com/en/turkish-farmers-are-squeezed-by-government-neglect/. Accessed on June 27, 2022.

29 Ibid.

30 Ibid.

31 *The World Bank*, "Employment in Agriculture (% of total employment) (Modeled ILO estimate)—Turkey," January 29, 2021, available at https://data.worldbank.org/indicator/SL.AGR.EMPL.ZS?locations=TR. Accessed on June 30, 2022.

32 *Statista*, "Distribution of Employment in Turkey from 2014 to 2023, by Economic Sector," available at https://www.statista.com/statistics/1386449/turkey-share-of-employment-by-economic-sector/. Accessed on August 17, 2024.

33 Kayam and Muslum Beyazgül, "New Approaches for Effective Drought Risk Assessment," in Saeid Eslamian and Faezeh Eslamian, eds., *Handbook of Drought and Water Scarcity: Principles of Drought and Water Scarcity*, New York: CRC Press, 2017, pp. 345–59; see p. 346.

34 Ibid., p. 347.

35 Ibid., p. 349.

36 Ibid., p. 350.

37 Ibid., p. 350.

38 *Daily Sabah*, "Drought Fears Deepen in Turkey's Second Largest Lake Tuz," October 28, 2021, available at https://www.dailysabah.com/turkey/drought-fears-deepen-in-turkeys-second-largest-lake-tuz/news. Accessed on June 30, 2022.

39 Ibid.

40 Ken Sofer, "The Slow Motion Crisis: The Destabilizing Effects of Climate Change in Turkey and Iraq Through 2050," Policy Paper, *New America*, May 16, 2017, pp. 1–14; see p. 6, available at https://www.newamerica.org/resource-security/policy-papers/relationship-between-climate-change-and-security/. Accessed on June 30, 2022.

41 Ibid., pp. 7–10.

42 Ibid., p. 9.

43 Karim Elgendy and Kadri Tastan, "Turkey's Climate Opportunities and Challenges," *Chatham House*, May 9, 2022, available at https://www.chathamhouse.org/2022/05/turkeys-climate-opportunities-and-challenges. Accessed on July 12, 2022.

44 Ibid.

45 Ibid.

46 *Daily Sabah*, "Turkey Drafting Long-Term Strategy to Combat Climate Change," May 15, 2022, available at https://www.dailysabah.com/turkey/turkey-drafting-long-term-strategy-to-combat-climate-change/news. Accessed on June 29, 2022.

47 Ibid.

48 Jeannie Sowers, "Environmental Activism in the Middle East and North Africa," in Harry Verhoeven, ed., *Environmental Politics in the Middle East*, New York: Oxford University Press, 2018, pp. 27–52; see p. 41.

49 Ibid., p. 27.

50 Om Marathe, "Explained: What is Kanal Istanbul, and Why is Erdogan Keen on Seeing the Project Through," *The Indian Express*, August 23, 2021, available at https://indianexpress.com/article/explained/explained-what-is-kanal-istanbul-and-why-is-erdogan-so-keen-on-seeing-the-project-through-7465902/. Accessed on July 13, 2022.

51 Ibid.

52 Dorian Jones, "Controversy Growing in Turkey over Erdogan's Massive Canal Project," *Voice of America*, January 15, 2020, available at https://www.voanews.com/a/europe_controversy-growing-turkey-over-erdogans-massive-canal-project/6182662.html. Accessed on July 13, 2022.

53 Ibid.

54 Hamid Firat Buyuk, "Istanbul Mayor Vows to Resist Erdogan's Disastrous Canal Project," *Balkan Insight*, June 28, 2021, available at https:/balkaninsight.com/2021/06/28/istanbul-mayor-vows-to-resist-erdogans-disastrous-canal-project/. Accessed on July 14, 2022.

55 Izzet Ari, "Türkiye's Climate Policy: Aligned with International Climate Politics," in Muhittin Ataman, and Gloria Shkurti Özdemir, eds., *Different Dimensions of Environmental Security in Türkiye and Beyond*, SETA, 2023, pp. 129–151; see p. 151.

56 Ibid.

57 Mehmet Emin Birpinar and Çiğdem Tuğac, "Climate Security and Migration: An Evaluation for the World and Turkey," *Insight Turkey*, Vol. 24, No. 1, Winter 2022, pp. 105–33; see especially p. 125.

58 Ibid., p. 125

59 Ibid., p. 126.

60 Ibid., p. 127.

61 *Mercy Corps*, "The Facts: What You Need to Know About the Syria Crisis," March 12, 2021, available at https://www.mercycorps.org/blog/facts-syria-crisis. Accessed on May 22, 2024.

62 Rebecca Kilberg, "Turkey's Evolving Migration Identity," *Migration Policy Institute (MPI)*, July 24, 2014, available at https://www.migrationpolicy.org/article/turkeys-evolving-migration-identity. Accessed on July 3, 2022.

63 *The World Bank*, "10 Years On, Turkey Continues Its Support for an Ever-Growing Number of Syrian Refugees," June 22, 2021, available at https://www.worldbank.org/en/news/feature/2021/06/22/10-years-on-turkey-continues-its-support-for-an-ever-growing-number-of-syrian-refugees. Accessed on July 1, 2022.

64 This section is based on Mahmood Monshipouri, Burcu Ellis, and Cassidy Renee Yip, "Managing the Refugee Crisis in the Era of the COVID-19 Pandemic," *Insight Turkey*, Vol. 22, No. 4, Fall 2020, pp. 179–200.

65 UNHCR, "UNHCR Turkey – Fact Sheet September 2021 [EN/TR]," *ReliefWeb*, September 17, 2021, available at https://reliefweb.int/report/turkey/unhcr-turkey-fact-sheet-september-2021-entr. Accessed on July 7, 2022.

66 Kemal Kirisci and M. Murat Erdogan. "Order from Chaos, Turkey and COVID-19: Don't Forget Refugees." *Brookings, The Brookings Institution*, April 20, 2020, available at www.brookings.edu/blog/order-from-chaos/2020/04/20/turkey-and-covid-19-dont-forget-refugees/. Accessed on May 1, 2020.

67 Dominic Evans, "Turkey's Plan to Settle Refugees in Northeast Syria Alarms Allies," *Reuters,* October 8, 2019, available at https://www.reuters.com/article/us-syria-security-turkey-refugees-graphi/turkeys-plan-to-settle-refugees-in-northeast-syria-alarms-allies-idUSKBN1WN28J. Accessed on February 21, 2020.

68 Ömer Yilmaz, "Turkey's Buffer Zone in Northern Syria and Its Impact on Irregular Migration," in Muhittin Ataman and Ibrahim Efe, eds., *On Climate Migration: Exploring Cases from Turkey and Beyond*, Ankara: SETA Publications, 2023, pp. 101–22; see especially p. 117.

69 Ibid.

70 Saraçoğlue and Bèlanger, op. cit., p. 282.

71 *Turkey iResidence*, "Turkey Residence Permit Application to Live and Work, or Study, for Business," available at https://residencepermitturkey.com/. Accessed on May 22, 2024.

72 Jinan Bastaki, "Temporary Protection Regimes and Refugees: What Works Comparing the Kuwaiti, Bosnian, and Syrian Refugees Protection Regimes," *Refuge: Canadian Journal of Refugees*, Vol. 34, No. 2, 2018, pp. 73–84; see especially p. 77.

73 Ibid.

74 *Danish Refugee Council*, "Syrian Refugees' Perception of the (Formal) Labor Market in Southeast Turkey," August 2021, pp. 1–14, available at file:///C:/Users/909186~1/AppData/Local/Temp/Livelihoods%20Study_Aug2021.pdf. Accessed on July 5, 2022.

75 Ibid., p. 1

76 Saraçoğlue and Bèlanger, op. cit., p. 282.

77 Ibid., p. 282.

78 Izza Leghtas, "Insecure Refugees: Deportations and Lack of Legal Work for Refugees in Turkey," *Refugees International*, September 2019, pp. 1–20; see p. 8, available at https://static1.squarespace.com/static/506c8ea1e4b01d9450dd53f5/t/5d82e33643e3fb135e2a1c6d/1568858939191/Turkey+Report+in+English+-+Livelihoods+-+September+2019+-+final.pdf. Accessed on July 1, 2022.

79 Ibid., pp. 78–9.

80 Ibid., pp. 282–3.

81 Ibid., p. 283.

82 Ibid., pp. 8–9.

83 Ibid., p. 14.

84 Ibid., p. 14.

85 Ibid., p. 15.

86 Nefise Ela Gökalp Aras, "A Game Changer in EU-Turkey Relations: The Opportunities and Pitfalls of Migration Policy," *The International Spectator*, Vol. 54, No. 4, 2019, pp. 47–61; see especially p. 54.

87 Ibid., p. 55.

88 Ibid., p. 58.

89 Saraçoğlue and Bèlanger, op. cit., p. 288.

90 *International Rescue Committee*, "What is the EU-Turkey Deal?," March 18, 2022, available at https://eu.rescue.org/article/what-eu-turkey-deal. Accessed on July 1, 2022.

91 I interviewed a few experts at a think tank or policy institute, called SETA, in Ankara, on August 16, 2024, who were highly critical of the EU's decision in this context.

92 This view was articulated by Professor Ali Zafer Sagiroglü, the director of the Migration Policies Center at Ankara Yildirim Beyazit University, in an interview with me on August 16, 2024, on the campus of the Middle East Technical University, Ankara.

93 Leghtas, "Insecure Refugees," p. 17.

94 Ibid., p. 17.

95 Sarah Kenyon Lischer, "The Global Refugee Crisis: Regional Destabilization & Humanitarian Protection," *Daedalus: The Journal of the American Academy of Arts & Sciences*, Vol. 2, 2017, pp. 85–97; see especially p. 92–5.

96 Quoted in Kyilah Terry, "The EU-Turkey Deal, Five Years On: A Frayed and Controversial But Enduring Blueprint," *Migration Policy Institute*, April 8, 2021, available at https://www.migrationpolicy.org/article/eu-turkey-deal-five-years-on. Accessed on July 3, 2022.

97 Ibid.

98 This section is also based partly on Monshipouri, Ellis, and Yip, "Managing the Refugee Crisis in the Era of the COVID-19 Pandemic."

99 Reuters, "Turkey Confirms First Coronavirus Case, Wins WHO Praise for Vigilance," *Physicians Weekly*, March 11, 2020, available at https://www.physiciansweekly.com/turkey-confirms-first-coronavirus/. Accessed on May 1, 2020.

100 Carlotta Gall, "Turkey Orders All Citizens to Wear Masks as Infections Rise," *The New York Times*, April 7, 2020, available at https://www.nytimes.com/2020/04/07/world/europe/turkey-virus-erdogan-masks.html. Accessed on July 2, 2022.

101 John Hopkins University and Medicine: Coronavirus Resource Center, "COVID-19 Dashboard by the Center for Systems Science and Engineering (CSSE) at John Hopkins University (JHU)," *John Hopkins University*, May 1, 2020, available at https://coronavirus.jhu.edu/map.html. Accessed on May 1, 2020.

102 Daily Sabah, "Turkey Completes Over 1 Million COVID-19 Tests; Number of Patients in Critical Care Declines," *Daily Sabah*, April 30, 2020, available at https://www.dailysabah.com/turkey/turkey-completes-over-1-million-covid-19-tests-number-of-patients-in-critical-care-declines/news. Accessed on May 2, 2020.

103 Tuvan Gumrukcu, "Turkey's First Coronavirus Case Shuts Schools, Impact Sports," *Reuters,* March 12, 2020, available at https://www.reuters.com/article/us-health-coronavirus-turkey/turkeys-first-coronavirus-case-shuts-schools-impacts-sports-idUSKBN20Z2RH. Accessed on May 2, 2020.

104 Reuters, "Turkey Confirms First Coronavirus Case, Wins WHO Praise for Vigilance."

105 Sasha Toperich, "Turkey Emerges as Key Player in Global COVID-19 Fight," *The Hill*, April 12, 2020, available at https://thehill.com/opinion/international/492628-turkey-emerges-as-key-player-in-global-covid-19-fight. Accessed on May 1, 2020.

106 Daily Sabah with agencies, "Stimulus Package Shields Public from Pandemic as Turkey Charts Own Path to Protect Economy", *Turkuvaz Haberlesme ve yayincilik*, April 12, 2020, available at https://www.dailysabah.com/business/economy/stimulus-package-shields-public-from-pandemic-as-turkey-charts-own-path-to-protect-economy. Accessed on May 2, 2020.

107 Ibid.

108 Kirisci and Erdogan, "Order from Chaos, Turkey and COVID-19."

109 Ibid.

110 Mahmut Bozarslan, "Turkey's Syrian Refugees Among Hardest Hit Amid Coronavirus Pandemic," *Al-Monitor*, April 27, 2020, available at https://www.al-monitor.com/pulse/originals/2020/04/turkey-syria-refugees-most-affected-by-coronavirus-jobless.html. Accessed on May 2, 2020.

111 Ibid.

112 Ibid.

113 Gokhan Ergocun, "Turkey Distributes Free Face Masks to Citizens," *Anadolu Agency Turkey*, April 5, 2020, available at https://www.aa.com.tr/en/turkey/turkey-distributes-free-face-masks-to-citizens/1793420. Accessed on May 2, 2020.

114 Ibid.

115 Monshipouri, Ellis, and Yip, "Managing the Refugee Crisis in the Era of the COVID-19 Pandemic," p. 187.

116 *Daily Sabah*, "School Closures in Turkey Extended until May 31 amid COVID-19 Pandemic," *Daily Sabah*, April 29, 2020, available at https://www.dailysabah.com/turkey/school-closures-in-turkey-extended-until-may-31-amid-covid-19-pandemic/news. Accessed on May 2, 2020.

117 Anadolu Agency, "Remote Courses Keep Education Open in Turkey While Schools Shut," *Daily Sabah*, March 22, 2020, available at https://www.dailysabah.com/turkey/education/remote-courses-keep-education-open-in-turkey-while-schools-shut. Accessed on May 2, 2020.

118 Kirisci and Erdogan, "Order from Chaos, Turkey and COVID-19."

119 *Daily Sabah*, "Turkey Drafting Long-Term Strategy to Combat Climate Change."

120 Brian Michael Jenkins, "Consequences of the War in Ukraine: Two Areas of Contention—Turkey and the Balkans," *The Rand Blog*, March 6, 2023, available at https://www.rand.org/pubs/commentary/2023/03/consequences-of-the-war-in-ukraine-turkey-and-the-balkans.html. Accessed on February 13, 2024.

121 Iliya Kusa, "Turkey's Goals in the Russian-Ukraine War," *The Wilson Center*, June 13, 2022, available at https://www.wilsoncenter.org/blog-post/turkeys-goals-russia-ukraine-war. Accessed on July 12, 2022.

122 Elgendy and Tastan, "Turkey's Climate Opportunities and Challenges."

123 Ayşegül Kibaroğlu, "Türkiye's Water Security Policy: Energy, Agriculture and Transboundary Issues," in Muhittin Ataman and Globria Shkurti Özdemir, eds., *Different Dimensions of Environmental Security in Türkiye and Beyond*, Ankara: SETA, 2023, pp. 179–97; see p. 197.

124 Ibid., p. 196.

125 Izzet Ari, "Turkey's Climate Policy: Aligned with International Climate Politics," in Muhittin Ataman and Gloria Shkurt Özdemir, eds., *Different Dimensions of Environmental Security in Türkiye and Beyond*, Ankara: SETA, 2023, pp. 129–51; see p. 151

126 Mehmet Emin Birpinar and Cigdem Tugac, "Climate Security and Migration: An Evaluation for the World and Turkey," in Muhittin Ataman and Ibrahim Efe, eds., *On Climate Migration: Exploring Cases from Turkey and Beyond*, Ankara: SETA Publications, 2023, pp. 141–70; see especially p. 167.

127 Ibid., p. 167

128 Ibid., p. 168.

129 Ibid., p. 169.

130 Kemal Kirişci, "As EU-Turkey Migration Agreement Reaches the Five-Year Mark, Add a Job Creation Element," *Brookings*, March 17, 2021, available at https://www.brookings.edu/blog/order-from-chaos/2021/03/17/as-eu-turkey-migration-agreement-reaches-the-five-year-mark-add-a-job-creation-element/. Accessed on July 11, 2022.

131 Ibid.

Chapter 6

1 *Climate Change Knowledge Portal*, "Egypt," available at https://climateknowledgeportal.worldbank.org/country/egypt. Accessed on July 5, 2022.

2 *Climate Change Adaptation: UNDP*, "Egypt," available at https://www.adaptation-undp.org/explore/northern-africa/egypt. Accessed on July 5, 2022.

3 *Climate Center*, "Egypt: Country Overview," 2021, pp. 1–11; see p. 1; available at https://www.climatecentre.org/wp-content/uploads/RCCC-ICRC-Country-profiles-Egypt.pdf. Accessed on July 13, 2022.

4 *World Bank Group*, "Climate Risk Country Profile: Egypt," 2021, pp. 1–23; see p. 6; available at https://climateknowledgeportal.worldbank.org/sites/default/files/2021-04/15723-WB_Egypt%20Country%20Profile-WEB-2_0.pdf. Accessed on July 6, 2022.

5 Ibid., pp. 6–9.

6 *Climate Change Adaptation: UNDP*, "Egypt."

7 Justin Corfield, "Egypt," in S. George Philander, ed., *Encyclopedia of Global Warming & Climate Change*, 2nd ed., Los Angeles, CA: Sage Reference, 2012, p. 495.

8 Nils Gilman, Doug Randall, and Peter Schwartz, "Climate Change and 'Security,'" in John S. Druzek, Richard B. Norgaard, and David Schlosberg, eds., *Oxford Handbook of Climate Change and Society*, New York: Oxford University Press, 2011, pp. 251–66; see p. 256.

9 Ibid., p. 261.

10 Soha M. Mostafa, Osama Wahed, Walaa Y. El-Nashar, Samia M. El-Marsafawy, Martina Zeleňáková, and Hany E. Abd-Elhamid, "Potential Climate Change Impact on Water Resources in Egypt," *Water*, Vol. 13, No. 1715, June 21, 2021, pp. 1–14.

11 Ross Michael Pink, *Climate Change Crisis Solutions and Adaptation for a Planet in Peril*, New York: Palgrave/Macmillan, 2018, see the chapter on "The Middle East: Egypt, Israel, Jordan," pp. 185–216; see especially p. 189.

12 Ibid., pp. 190–1.

13 *Climate Change Adaptation: UNDP*, "Egypt."

14 *Climate Change Knowledge Portal*, "Egypt."

15 Soha M. Mostafa, Osama Wahed, Walaa Y. El-Nashar, Samia M. El-Marsafawy, Martina Zeleˇnáková, and Hany F. Abd-Elhamid, "Potential Climate Change on Water Resources in Egypt," *Water*, Vol. 13, June 21, 2021, pp. 1–15; see p. 13, available at file:///C:/Users/909186933/Downloads/water–13-01715-v3.pdf. Accessed on May 23, 2024.

16 Sarah Schaefer, "Climate Change: The Nile River," *Middle East Policy Council*, available at https://mepc.org/commentary/climate-change-nile-river. Accessed on February 13, 2024.

17 UN: High-Level Political Forum on Sustainable Development, *Egypt*, 2018, available at https://hlpf.un.org/countries/egypt/voluntary-national-review-2018. Accessed on February 13, 2024.

18 El-Sayed Ewis Omran and Abdelazim M. Negm, "Introduction to Climate Change Impacts on Agriculture and Food Security in Egypt," in El-Sayed Ewis and Abdelazim M. Negm, eds., *Climate Change Impacts on Agriculture and Food Security in Egypt: Land and Water Resources: Smart Farming, Livestock, Fishery, and Aquaculture*, Cham: Springer, 2019, pp. 3–19; see especially p. 9.

19 *World Bank Group*, "Climate Risk Country Profile: Egypt," pp. 1–23; see p. 3.

20 Ibid., p. 3.

21 Hassen Hussein, "Egypt and Ethiopia Spar Over the Nile," *Al Jazeera America*, February 6, 2014, available at http://america.aljazeera.com/opinions/2014/2/egypt-disputes-ethiopiarenaissancedam.html. Accessed on February 14, 2015.

22 Mwangi S. Kimenyi and John Mukum Mbaku, "The Limits of the New 'Nile Agreement,'" *Brookings*, April 28, 2015, available at https://www.brookings.edu/blog/africa-in-focus/2015/04/28/the-limits-of-the-new-nile-agreement/. Accessed on July 6, 2022.

23 Ibid.

24 M. A. Mahmoud, "The Impact of Climate Change on the Agricultural Sector in Egypt," in Abdelazim M. Negm, ed., *Conventional Water Resources and Agriculture in Egypt*, Cham: Springer International Publishing, 2019, pp. 213–28; see p. 214.

25 Ibid., p. 222.

26 Samiha A. H. Ouda and Abd El-Hafeez Zohry, eds., "Introduction," in Samiha A. H. Ouda and Abd El-Hafeez Zohry, eds., *Management of Climate Induced Drought and Water Scarcity in Egypt: Unconventional Solutions*, Cham: Springer, 2016, pp. 1–4.

27 Samiha A. H. Ouda, Tahany Noreldin, and Alia Amer, "Rain Fed Areas in Egypt: Obstacles and Opportunities," in Samiha A. H. Ouda and Abd El-Hafeez Zohry, eds., *Management of Climate Induced Drought and Water Scarcity in Egypt: Unconventional Solutions*, Cham: Springer, 2016, pp. 27–46; see p. 27.

28 For a detailed and illuminating volume on this subject, see Abdelazim M. Negm, ed., *Conventional Water Resources and Agriculture in Egypt*, Cham: Springer International Publishing, 2019.

29 Schaefer, "Climate Change."

30 Raoudha Mougou, Ayman Abou-Hadid, Ana Iglesias, Mahmoud Medany, Amel Nafti, Riadh Chetali, Mohsen Mansour, and Helmy Eid, "Adapting Dryland and Irrigated Cereal Farming to Climate Change in Tunisia and Egypt," in Neil Leary, James Adejuwon, Vicente Barros, Ian Burton, Jyoti Kulkarni, and Rodel Lasco, eds., *Climate Change and Adaptation*, London: Earthscan, 2009, pp. 181–95; see p. 187.

31 Ibid.

32 Mosaad Khedr, "Challenges and Issues in Water, Climate Change, and Food Security in Egypt," in Abdelazim M. Negm, ed., *Conventional Water Resources and Agriculture in Egypt*, Cham: Springer International Publishing, 2019, pp. 229–44; see p. 240.

33 Heba Elbasiouny and Fathy Elbehiry, "Rice Production in Egypt: The Challenges of Climate Change and Water Deficiency," in El-Sayed Ewis and Abdelazim M. Negm, eds., *Climate Change Impacts on Agriculture and Food Security in Egypt: Land and Water Resources: Smart Farming, Livestock, Fishery, and Aquaculture*, Cham: Springer, 2019, pp. 295–319; see p. 309.

34 Ibid., p. 309.

35 Schaefer, "Climate Change."

36 Ibid.

37 Khedr, "Challenges and Issues in Water, Climate Change, and Food Security in Egypt," p. 241.

38 Nader Noureldeen Mohamed, "Egyptian Food Insecurity under Water Shortage and Its Socioeconomic Impacts," in Abdelazim M. Negm, ed., *Conventional Water Resources and Agriculture in Egypt*, Cham: Springer International Publishing, 2019, pp. 245–74; see p. 245.

39 Daniel Workman, "Wheat Imports by Country," available at https://www.worldstopexports.com/wheat-imports-by-country/. Accessed on July 7, 2022.

40 Mohamed, "Egyptian Food Insecurity under Water Shortage and Its Socioeconomic Impacts," p. 249.

41 *World Bank Group*, "Climate Risk Country Profile: Egypt," pp. 1–23; see pp. 13–14.

42 *Climate Center*, "Egypt: Country Overview," 2021, pp. 1–11; see especially p. 2; available at https://www.climatecentre.org/wp-content/uploads/RCCC-ICRC-Country-profiles-Egypt.pdf. Accessed on July 13, 2022.

43 International Organization of Migration (IOM): *Migration Research Series*, "Migration and Climate Change," No. 31, 2008, Geneva, pp. 1–60; see especially p. 18.

44 Nobuhiko Daito, Dahlia Lotayef, and Arturo Ardila-Gomez, "Reducing Air Pollution in Greater Cairo Involves Switching from Private Vehicles to Improved Public Transport," *World Bank Blogs*, September 20, 2021, available at https://blogs.worldbank.org/

arabvoices/reducing-air-pollution-greater-cairo-involves-switching-private-vehicles-improved-public. Accessed on July 5, 2022.

45 Mona Rabia, "The Impact of Climate Change on International Tourism: Evidence from Egypt," *International Journal of Energy Economics and Policy*, Vol. 13, No. 2, March 2023, pp. 379–90.

46 AKM Ahsan Ullah, "Climate Change and Climate Refugee in Egypt: An Overview from Policy Perspectives," *TMC Academic Journal*, Vol. 7, No. 1, 2012, pp. 56–70; see p. 61.

47 *Climate Change Knowledge Portal*, "Egypt," available at https://climateknowledgeportal.worldbank.org/country/egypt. Accessed on July 5, 2022.

48 Ayman F. Abou Hadid, "Climate Change and Food Sovereignty in the Nile Valley and Red Sea," in Mannava V. K. Sivakumar, Rattan Lal Ramasamy Selvaraju, and Ibrahim Hamdan, eds., *Climate Change and Food Security in West Asia and North Africa*," Cham: Springer 2013, pp. 193–206; see especially p. 199.

49 Marina Wes, "Egypt: Acting Against Climate Change for a Healthier, More Prosperous Future," *The World Bank*, April 19, 2022, available at https://www.worldbank.org/en/news/opinion/2022/04/19/-egypt-acting-against-climate-change-for-a-healthier-more-prosperous-future. Accessed on July 11, 2022.

50 George Mikhail, "Egypt Announces Plan to Address Climate Change Following Heatwave," *Al-Monitor*, August 23, 2021, available at https://www.al-monitor.com/originals/2021/08/egypt-announces-plan-address-climate-change-following-heat-wave#ixzz7YmpnZWOl. Accessed on July 11, 2022.

51 Ashok Swain and Anders Jägerskog, *Emerging Security Threats in the Middle East: The Impact of Climate Change and Globalization*, Lanham, MD: Rowman & Littlefield, 2016, p. 113.

52 Ibid.

53 Jeannie Sowers, "Resources and Revenues: The Political Economy of Climate Initiatives in Egypt," in David Held, Charles Roger, and Eva-Maria Nag, eds., *Climate Governance in the Developing World*, Malden, MA: Polity Press, 2013, pp. 199–217; see p. 214.

54 Robert Springborg, "The Challenge of Squaring the Energy-Environment-Growth Triangle," in Robert E. Looney, ed., *Handbook of Transitions in Energy and Climate Security*, New York: Routledge, 2017, pp. 272–83; see pp. 272–3.

55 Ibid., pp. 272–3.

56 Ibid., p. 273.

57 Ibid., p. 280.

58 Mahmoud Adel Hassaan, Mohamed Abdel Karim Aly Abdrabo, Hadeer Ahmed Hussein, Azza Abdallah Abdelhamid Ghanem, and Hany Abdel-Latif, "Potential Impacts of Climate Change on Renewable Energy in Egypt," *Environmental Monitoring and Assessment*, 2024, pp. 196–268, available at https://link.springer.com/content/pdf/10.1007/s10661-024-12428-1. Accessed on May 23, 2024.

59 Jacob J. Torba, "Climate Change, Environmental Migration and Conflict: The Case of Egypt and Sudan," *Calhoun.NPS*, Thesis Submission, December 2020, pp. 1–73; see p. 61, available at https://calhoun.nps.edu/bitstream/handle/10945/66735/20Dec_Torba_Jacob.pdf?sequence=1&isAllowed=y. Accessed on July 23, 2022.

60 Ibid., p. 62.

61 *Information and Decision Support Center*, Khaled Kazem Abo Douh, "Climate Change and Environmental Refugees in Egypt," March 24, 2021, available at https://www.idsc.gov.eg/DocumentLibrary/View/4573. Accessed on July 8, 2022.

62 Wagdy Sawahel, "Climate Change in Egypt 'to Force Millions to Migrate,'" *SciDevNet*, June 27, 2005, available at https://www.scidev.net/global/news/climate-change-in-egypt-to-force-millions-to-migr/https://www.scidev.net/global/news/climate-change-in-egypt-to-force-millions-to-migr/. Accessed on July 14, 2022.

63 Enterprise, "What Is Air Pollution Costing Egypt? At least EGP 47 bn a Year, the World Bank Estimates," April 20, 2021, available at https://enterprise.press/stories/2021/04/20/what-is-air-pollution-costing-egypt-at-least-egp-47-bn-a-year-the-world-bank-estimates-38534/. Accessed on July 21, 2022.

64 K. Warner, M. Hamza, A. Oliver-Smith, F. Renaud, and A. Julca, "Climate Change, Environmental Degradation and Migration," *Natural Hazards*, Vol. 55, No. 3, December 2010, pp. 689–715; see especially pp. 702–3.

65 Ibid., p. 704.

66 *Middle East Monitor*, "Egypt: 5 Million Could be Displaced by Climate Change," June 21, 2018, available at https://www.middleeastmonitor.com/20180621-egypt-5-million-could-be-displaced-by-climate-change/. Accessed on July 8, 2022.

67 Warner, Hamza, Oliver-Smith, Renaud, and Julca, "Climate Change, Environmental Degradation and Migration," pp. 689–715; see especially p. 708.

68 Ibid., p. 709.

69 Julia Kloos and Niklas Baumert, "Preventive Resettlement in Anticipation of Sea Level Rise: A Choice Experiment from Alexandria, Egypt," *Natural Hazards*, Vol. 76, 2015, pp. 99–121; see p. 100.

70 Ibid., p. 101.

71 Yehya Serag, "Waves of Displacement in Egypt Impacts of the Internal and External Displacement Waves on the Egyptian Cities," Conference Paper, November 2016, available at file:///C:/Users/909186~1/AppData/Local/Temp/WavesofDisplacementinEgypt-1.pdf. Accessed on July 6, 2022.

72 Ibid.

73 Ibid.

74 *UNHCR*, "Refugee Context in Egypt," available at https://www.unhcr.org/eg/about-us/refugee-context-in-egypt. Accessed on July 8, 2022.

75 Ibid.

76 Ayman Zohry, "Egypt's International Migration After the Revolution: Is There Any Change?," *Dossier*, No. 87, Autumn 2013, pp. 47–54; see especially p. 53, available at file:///C:/Users/909186~1/AppData/Local/Temp/COME_087_0047.pdf. Accessed on July 16, 2022.

77 Brian Dabbs, "Refugees Stuck in Limbo at Desolate Egypt-Libya Border," *The Atlantic*, May 17, 2011, available at https://www.theatlantic.com/international/archive/2011/05/refugees-stuck-in-limbo-at-desolate-egypt-libya-border/238993/. Accessed on July 16, 2022.

78 Hend Hafez, "The Arab Spring and Migration in Egypt, One Year On: Impacts, Perceptions, and Attitudes," *Cairo Studies on Migration and Refugees*, The Center for

Migration and Refugee Studies, American University at Cairo, Paper #5, October 2012, available at https://documents.aucegypt.edu/Docs/GAPP/Hend-Hafez_Paper.pdf. Accessed on July 16, 2022.

79 *UNHCR*, "Refugee Context in Egypt," available at https://www.unhcr.org/eg/about-us/refugee-context-in-egypt. Accessed on July 8, 2022.

80 Karen Jacobsen, "Livelihoods and Forced Migration," in Elena Fiddian-Qasmiyeh, Gil Loescher, Katy Long, and Nando Sigona, eds., *The Oxford Handbook of Refugee and Forced Migration Studies*, New York: Oxford University Press, 2014, pp. 99–11; see pp. 106–7.

81 Jeannie L. Sowers, "Environmental Activism in the Middle East and North Africa," in Harry Verhoeven, ed., *Environmental Politics in the Middle East*, New York: Oxford University Press, 2018, pp. 27–52; see pp. 32–33.

82 Jeannie L. Sowers, *Environmental Politics in Egypt: Activists, Experts, and the State*, New York: Routledge, 2013, p. 1.

83 Ibid., p. 1.

84 Sowers, "Environmental Activism in the Middle East and North Africa," p. 45.

85 Ibid., p. 46.

86 Ibid.

87 Dina Zayed, "On the Margins: Civil Society Activism and Climate Change in Egypt," *Carnegie Endowment for International Peace*, May 2, 2024, available at https://carnegieendowment.org/research/2024/05/on-the-margins-civil-society-activism-and-climate-change-in-egypt?lang=en. Accessed on May 23, 2024.

88 Samar Samir, "Egypt Launches National Strategy for Climate Change 2050," *Egypt Today*, May 19, 2022, available at https://www.egypttoday.com/Article/1/116031/Egypt-launches-National-Strategy-for-Climate-Change-2050. Accessed on July 14, 2022.

89 Ibid.

90 Ibid.

91 Gerasimos Tsourapas, "Egypt: Migration and Diaspora Politics in an Emerging Transit Country," *Migration Policy Institute*, August 8, 2018, available at https://www.migrationpolicy.org/article/egypt-migration-and-diaspora-politics-emerging-transit-country. Accessed on July 15, 2022.

92 Shardul Agrawala, Annett Moehner, Mohamed El Raey, Declan Conway, Maarten van Aalst, Marca Hagenstad, and Joel Smith, "Development and Climate Change in Egypt: Focus on Coastal Resources and the Nile," *OECD*, 2004, COM/ENV/EPOC/DCD/DAC(2004)1/FINAL, pp. 1–68; see p. 48.

93 Ibid., p. 49.

94 Cited in Wagdy Sawahel, "Climate Change in Egypt 'to Force Millions to Migrate,'" *SciDevNet*, June 27, 2005, available at https://www.scidev.net/global/news/climate-change-in-egypt-to-force-millions-to-migr/https://www.scidev.net/global/news/climate-change-in-egypt-to-force-millions-to-migr/. Accessed on July 14, 2022.

Chapter 7

1 Britannica, "Syria," available at https://www.britannica.com/place/Syria. Accessed on May 23, 2024.

2 Sam White, John Brooke, and Christian Pfister, "Climate, Weather, Agriculture, and Food," in Sam White, Christian Pfister, and Franz Mauelshagen, eds., *The Palgrave Handbook of Climate History*, New York: Palgrave/Macmillan, 2018, pp. 331–53; see especially p. 336.

3 Dagomar Degroot, "Climate Change and Conflict," in Sam White, Christian Pfister, and Franz Mauelshagen, eds., *The Palgrave Handbook of Climate History*, New York: Palgrave/Macmillan, 2018, pp. 367–85; see especially pp. 367–368.

4 Ibid.

5 Ibid.

6 Ibid., p. 380.

7 White, Brooke, and Pfister, "Climate, Weather, Agriculture, and Food," p. 336.

8 For an illuminating explanation on this subject, see Peter H. Gleik, "Water, Drought, Climate Change, and Conflict in Syria," *Weather, Climate, and Society*, Vol. 6, No. 3, July 2014, pp. 331–40; see p. 333.

9 *Deutsche Welle* (DW), "How Climate Change Paved the Way to War in Syria," February 26, 2021, available at https://www.dw.com/en/how-climate-change-paved-the-way-to-war-in-syria/a-56711650. Accessed on June 14, 2021.

10 Gleick, "Water, Drought, Climate Change, and Conflict in Syria," p. 334.

11 Rami Zurayk and Anne Gough, "Bread and Olive Oil: The Agrarian Roots of the Arab Uprisings," in Fawaz A. Gerges, ed., *The New Middle East: Protest and Revolution in the Arab World*, New York: Cambridge University Press, 2014, pp. 107–31; see especially p. 112.

12 Jane Arraf, "War and Climate Change Ravage Syria's Farms," *The New York Times*, February 20, 2022, p. 4.

13 Francesco Femia and Caitlin Werrell, "Syria: Climate Change, Drought and Social Unrest," *The Center for Climate and Security*, February 2012, available at https://climateandsecurity.org/2012/02/syria-climate-change-drought-and-social-unrest/. Accessed on June 13, 2022.

14 Ibid.

15 Ibid.

16 Ibid.

17 Ashok Swain and Anders Jägerskog, *Emerging Security Threats in the Middle East: The Impact of Climate Change and Globalization*, Lanham, MD: Rowman & Littlefield, 2016, p. 143.

18 Sari Hanafi, "Forced Migration in the Middle East and North Africa," in Elena Fiddian-Qasmiyeth, Gil Loescher, Katy Long, and Nando Sigona, eds., *The Oxford Handbook of Refugee and Forced Migration Studies*, New York: Oxford University Press, 2014, pp. 585–98; see especially p. 592.

19 Swain and Jägerskog, *Emerging Security Threats in the Middle East*, p. 143.

20 Marc Lynch, *The New Arab Wars: Uprisings and Anarchy in the Middle East*, New York: Public Affairs, 2016, p. 209.

21 *Mercy Corps: Syria*, available at https://www.mercycorps.org/where-we-work/syria. Accessed on June 13, 2022.

22 *World Vision*, "Syrian Refugee Crisis: Facts, FAQs and How to Help," available at https://www.worldvision.org/refugees-news-stories/syrian-refugee-crisis-facts. Accessed on June 14, 2022.

23 Ibid.

24 Catherine Panter-Brick, "Inclusive Partnerships: Building Resilience Humanitarianism with Syrian Refugee Youth in Jordan," in Marcia C. Inhorn and Lucia Volk, eds., *Un-Settling Middle Eastern Refugees: Regimes of Exclusion and Inclusion in the Middle East, Europe, and North America*, New York: Berghahn Books, 2021, pp. 261–80; see especially p. 265.

25 Ibid., p. 266.

26 Ibid., p. 266.

27 *World Vision*, "Syrian Refugee Crisis."

28 Ibid.

29 *Statista*, "Arrivals of Syrian Refugees in the U.S. FY 2011–2021," August 9, 2021, available at https://www.statista.com/statistics/742553/syrian-refugee-arrivals-us/. Accessed on June 19, 2022.

30 Somini Sengupta and Hwaida Saad, "Syria's Young Saw the Worst, But Dangers Lurk Beyond 1. ISIS," *The New York Times*, August 1, 2017, pp. A1 and A5.

31 Lynch, *The New Arab Wars*, p. 210.

32 Mahmood Monshipouri, "Europe's Borders, Refugees, and the Islamic State," *Sustainable Security*, September 30, 2016, available at https://sustainablesecurity.org/2016/09/30/europes-borders-refugees-and-the-islamic-state/.

33 *Middle East Political and Economic Institute*, "UAE Calls on the UN to Assist Syria in Fighting Against COVID-19," available at https://mepei.com/uae-calls-on-the-un-to-assist-syria-in-fighting-against-covid-19/. Accessed on April 3, 2021.

34 http://www.dailymail.co.uk/news/article-3752718/ISIS-uses-Greek-refugee-camp-distribute-fake-passports-terrorists-hide-migrants-attack-west.html.

35 Ibid.

36 Betts Alexander and Paul Collier, available at https://www.foreignaffairs.com/articles/middle-east/2016-04-28/jordans-refugee-experiment.

38 Mahmood Monshipouri, "Paris Attacks: Echoes of Terrors to Come," *Tehran Times*, November 21, 2015, available at

http://www.tehrantimes.com/index_View.asp?code=251047. Accessed on November 21, 2015.

39 Neil MacFarquhar, "Kerry Says Brussels Attacks Renew Urgency for US-Russia Efforts in Mideast," *The New York Times*, March 22, 2016, available at http://www.nytimes.com/2016/03/25/world/europe/kerry-russia-talks.html?ribbon-ad-idx=7&rref=world/europe&module=ArrowsNav&contentCollection=Europe&action=swipe&region=FixedLeft&pgtype=article. Accessed on March 24, 2016.

40 Ian Bremmer, "The Migrant Issue is Dividing Europe," *Time*, November 23, 2015, p. 16.

41 Steven Simon and Daniel Benjamin, "Could Paris Happen Here," *The New York Times*, November 16, 2015, p. A21.

42 Beth Macy, "The Refugees of Roanoke," *The New York Times, Sunday Review*, November 22, 2015, p. SR3.

43 Ian Bremmer, "These 5 Facts Explain the Worrying Rise of Europe's Far-Right," *Time*, October 15, 2015, available at http://time.com/4075396/far-right-politics-rise-europe/. Accessed on March 25, 2016.

44 Ibid.

45 Mabel Berezlin, "Xenophobia and the New Nationalism," in Gerard Delanty and Krishan Kumar, eds., *The Sage Handbook of Nations and Nationalism*, Los Angeles, CA: Sage, pp. 273–84; see especially p. 274.

46 The Conversation, "Globalism, Refugee Crisis is Fueling Xenophobia," September 21, 2015, available at http://theconversation.com/globalism-refugee-crisis-is-fueling-xenophobia-47313. Accessed on March 25, 2016.

47 Mike DeBonis, "House Passes Bill to Tighten Flow of Syrian Refugees over Obama's Objections," *The Washington Post*, November 19, 2015, available at https://www.washingtonpost.com/news/powerpost/wp/2015/11/19/house-moves-to-tighten-flow-of-syrian-refugees-over-obamas-objections/. Accessed on November 24, 2015.

48 Frank Bruni, "How ISIS Defeats Us," *The New York Times, Sunday Review*, November 22, 2015, SR3.

49 Allen S. Keller, "A Letter to the Editor, *The New York Times*, November 24, 2015, p. A 24.

50 Madeleine Albright, "Refugees are Not the Enemy," *Time*, November 30–December 7, 2015, p. 61.

51 Ibid.

52 Mark Gibney, "Refugees," in David P. Forsythe, ed., *Encyclopedia of Human Rights*, Vol. 4, New York: Oxford University Press, 2009, pp. 315–22; see especially p. 315.

53 Ibid., p. 321.

54 Azubike Onuora-Oguno, "Migration, Refugees, Asylum and Uprooted Peoples' Rights," in Anja Mihr and Mark Gibney, eds., *The Sage Handbook of Human Rights*, Vol. 1, Los Angeles, CA: Sage, 2014, pp. 253–66; see especially p. 260.

55 Ibid., p. 261.

56 Bruni, "How ISIS Defeats Us," SR3.

57 Ibid., SR3.

58 Michael D. Shear and Peter Baker, "Obama Says Strategy to Fight ISIS Will Succeed," *The New York Times*, November 17, 2015, p. A11.

59 Ibid., pp. 50–1.

60 Mohsen Milani, "Why Tehran Won't Abandon Assad (ism)," *The Washington Quarterly*, Vol. 36, No. 4, 2013, pp. 79–93; see especially p. 87.

61 Michael Jansen, "Da'esh Campaign Sparks Russian Fears," *Gulf Today: Panorama*, September 25–October 1, 2015, pp. 40–1; see especially p. 41.

62 Alexis Xydias and Sara Miller Llana, "Chaos Grows over Asylum Seekers," *The Christian Science Monitor*, September 14, 2015, p. 10.

63 James Kanter and Tim Arango, "EU Must Do More to Contain Syria Conflict, Turkish Leader Asserts," *The New York Times*, October 6, 2015, p. A6.

64 Ibid.

65 Zack Weissmueller, "What Americans Really Think about ISIS," *Reason.com*, October 16, available at https://reason.com/reasontv/2014/10/16/what-americans-really-think-about-isis-r. Accessed on October 10, 2014.

66 Giselle Lopez, "Responsibility to Protect at a Crossroads: The Crisis in Libya." *Humanity in Action*, available at http://www.humanityinaction.org/knowledgebase/580-responsibility-to-protect-at-a-crossroads-the-crisis-in-libya. Accessed on October 3, 2015.

67 Ibid.

68 John J. Mearsheimer, "America Unhinged," *The National Interest*, January/February 2014, pp. 9–30; see especially pp. 13–19.

69 Richard Falk, *Humanitarian Intervention and Legitimacy Wars: Seeking Peace and Justice in the 21st Century*, New York: Routledge, 2015, p. 140.

70 Ibid., p. 141.

71 Mohammad Ayood, *Will the Middle East Implode*? London: Polity Press, 2014, p. 11.

72 Robert G. Rabil, *The Syrian Refugee Crisis in Lebanon: The Double Tragedy of Refugees and Impacted Host Communities*, Lanham, MD: Lexington Books, 2016. See especially chapter 4.

73 Aleksandar Keševljevic and Rok Spruk, "Estimating the Effects of the Syrian Civil War," *Empirical Economics*, Vol. 66, 2024, pp. 671–703; see p. 676.

74 Jan Selby and Mike Hulme, "Is Climate Change Really to Blame for Syria's Civil War," *The Guardian*, November 29, 2015, available at https://www.theguardian.com/commentisfree/2015/nov/29/climate-change-syria-civil-war-prince-charles. Accessed on February 14, 2024.

75 Madhuri Karak, "Climate Change and Syria's Civil War," *JSTOR Daily*, September 17, 2019, available at https://daily.jstor.org/climate-change-and-syrias-civil-war/. Accessed on February 14, 2024.

76 Rabil, *The Syrian Refugee Crisis in Lebanon*, pp. 101–11.

77 Sumudu Anopama Atapattu, "A New Category of Refugees: 'Climate Refugees' and a Gaping Hole in International Law," in Simon Behrman and Avidan Kent, eds., *Climate Refugees: Beyond the Legal Impasse?* New York: Routledge, 2018, pp. 34–51; see especially p. 36.

78 Quoted in Rabil, *The Syrian Refugee Crisis in Lebanon*, p. 110.

Chapter 8

1 Weather and Climate, "Climate and Average Weather in Yemen," available at https://weather-and-climate.com/average-monthly-Rainfall-Temperature-Sunshine-in-Yeme. Accessed on July 26, 2022.

2 Sarah Boslaugh, "Yemen," in S. George Philander, ed., *Encyclopedia of Global Warming & Climate Change*, 2nd ed., Thousand Oaks, CA: Sage Reference, 2012, pp. 1503–4; see p. 1503.

3 Melani Cammett, Ishac Diwan, Alan Richards, and John Waterbury, *A Political Economy of the Middle East*, 4th ed., Boulder, CO: Westview Press, 2015, p. 220.

4 Ibid.

5 Colbert C. Held and John Thomas Cummings, *Middle East Pattern: Places, Peoples, and Politics*, 6th ed., Boulder, CO: Westview Press, 2014, p. 551.

6 Sarah G. Phillips, "Yemen," in Ellen Lust, ed., *The Middle East*, 14th ed., Thousand Oaks, CA: Sage, 2017, pp. 895–916; see p. 903.

7 *Trading Economics*, "Yemen Unemployment Rate," available at https://tradingeconomics.com/yemen/unemployment-rate. Accessed on July 26, 2022.

8 *Bonyan Organization*, "How Long Has Yemen been in Poverty?," March 28, 2022, available at https://bonyan.ngo/en/poverty-in-yemen/. Accessed on July 26, 2022.

9 Boslaugh, "Yemen," p. 1504.

10 Bruce Riedel, "Who are the Houthis and Why are We at War with Them?," *Brookings*, December 18, 2017, available at https://www.brookings.edu/blog/markaz/2017/12/18/who-are-the-houthis-and-why-are-we-at-war-with-them/. Accessed on July 25, 2022.

11 Micheline R. Ishay, *The Levant Express: The Arab Uprisings, Human Rights, and the Future of the Middle East*, New Haven, CT: Yale University Press, 2019, p. 46.

12 Laurent Bonnefoy, *Yemen and the World: Beyond Security*, translated by Cynthia Schoch, New York: Oxford University Press, 2018, p. 40.

13 Kali Robinson, "Yemen's Tragedy: War, Stalemate, and Suffering," *Council on Foreign Relations*, April 8, 2022, available at https://www.cfr.org/backgrounder/yemen-crisis. Accessed on July 25, 2022.

14 Giulio Coppi, "The Humanitarian Crisis in Yemen: Beyond the Man-Made Disaster," *International Peace Institute (IPI)*, January 2018, pp. 1–34; see p. 1.

15 Stephen J. King, *The Arab Winter: Democratic Consolidation, Civil War, and Radical Islamists*, New York: Cambridge University Press, 2020, p. 217.

16 William L. Cleveland and Martin Bunton, *A History of the Modern Middle East*, 6th ed., New York: Routledge, 2016, p. 471.

17 Ibid.

18 Cammett, Diwan, Richards, and Waterbury, *A Political Economy of the Middle East*, p. 221.

19 Asher Orkaby, *Yemen: What Everyone Needs to Know*, New York: Oxford University Press, 2021, p. 73.

20 United Nations Foundation, "Yemen: A Brief Background," available at https://unfoundation.org/what-we-do/issues/peace-human-rights-and-humanitarian-response/yemen-a-brief-background/. Accessed on July 25, 2022.

21 Adam C. Seitz, "Ties That Bind and Divide: The 'Arab Spring' and Yemeni Civil-Military Relations," in Helen Lackner, ed., *Why Yemen Matters; A Society in Transition*, London: SOAS, 2014, pp. 50–67; see pp. 60–1.

22 Ibid., p. 62.

23 Mahmood Monshipouri, *Middle East Politics: Changing Dynamics*, New York: Routledge, 2019, p. 89.

24 Robert Malley and Stephen Pomper, "Accomplice to Carnage: How America Enables War in Yemen," *Foreign Affairs*, March/April 2021, available at file:///C:/Users/909186933/Documents/Rob%20Malley%20How%20America%20Enables%20War%20in%20Yemen%20Foreign%20Affairs.htm. Accessed on July 28, 2022.

25 Monshipouri, *Middle East Politics*, p. 89.

26 Orkaby, *Yemen*, p. 86.

27 Ibid., p. 2.

28 Ibid., p. 2.

29 Marieke Brandt, *Tribes and Politics in Yemen: A History of the Houthi Conflict*, New York: Oxford University Press, 2017, pp. 346–7.

30 King, *The Arab Winter*, p. 228.

31 Mehran Kamrava, *The Modern Middle East: A Political History Since the First World War*, 3rd ed., Berkeley, CA: University of California Press, 2013, p. 296.

32 Orkaby, *Yemen*, p. 7.

33 United Nations Foundation, "Yemen: A Brief Background."

34 Ibid.

35 Center for Preventive Action, "War in Yemen," *Council on Foreign Relations*, Global Conflict Tracker, May 4, 2025, available at https://www.cfr.org/global-conflict-tracker/conflict/war-yemen. Accessed on July 25, 2022.

36 King, *The Arab Winter*, p. 237.

37 Robert E. Worth, "Can Yemen Survive Saudi Arabia's War," *The New Times Magazine*, November 4, 2018, pp. 48–57, and 69; see pp. 50–1.

38 Ibid., p. 52.

39 Robinson, "Yemen's Tragedy."

40 Marwa Daoudy, *Climate Change and Regional Instability in the Middle East*, Discussion Paper Series on Managing Global Disorder, New York: Council on Foreign Relations, No. 14, October 2023, pp. 1–31; see p. 13.

41 Ibid.

42 Hadi Mohamed, Moosa Elayah, and Lau Schuplen, "Yemen Between the Impact of the Climate Change and the Ongoing Saudi-Yemen War: A Real Tragedy," *Governance and Peace-Building Center*, Sana'a, November 2017, pp. 1–10; see p. 6.

43 Center for Preventive Action, "War in Yemen."

44 Sunayana Samantaray, "Why Yemen's Civil War is also an Environmental One," *The Vegan Review*, September 1, 2021, available at https://theveganreview.com/yemen-civil-war-climate-change-environmental-crisis-human-rights/. Accessed on July 27, 2022.

45 Ibid.

46 Phillips, "Yemen," p. 914.

47 Ibid., p. 914.

48 Massasb Al-Aloosy, "Saudi-Iran Negotiations: Managing the Conflict and Gauging Broader Regional Impact," *The Arab Gulf States Institute in Washington*, May 18, 2022,

available at https://agsiw.org/saudi-iranian-negotiations-managing-the-conflict-and-gauging-broader-regional-impact/. Accessed on July 27, 2022.

49 Orkaby, *Yemen*, p. 106.

50 Ibid., 106.

51 Ibid., 106.

52 Ibid., 107.

53 Ibid., 107.

54 Ibid., 107–8.

55 Ibid., p. 108.

56 Ibid., p. 108.

57 Helen Lackner, "Water Scarcity: Why Doesn't It Get the Attention It Deserves?," in Helen Lackner, ed., *Why Yemen Matters; A Society in Transition*, London: SOAS, 2014, pp. 161–82; see pp. 180–1.

58 Ibid., pp. 181–2.

59 Georgia Grzywacz, "Conflict, Climate, and Catastrophe: Climate Change in Yemen," available at http://www.aksik.org/node/3591. Accessed on July 26, 2022.

60 Ibid.

61 Ibid.

62 Ibid.

63 Michael Hart, "Is Climate Change Driving Conflict in the Middle East?," *International Policy Digest*, October 29, 2016, available at https://intpolicydigest.org/2016/10/29/climate-change-driving-conflict-middle-east/. Accessed on October 11, 2017.

64 C. Breisinger, O. Ecker, P. Al-Riffai, R. Robertson, R. Thiele, and M. Wiebelt, "Climate Change, Agricultural Production and Food Security: Evidence from Yemen," *Kiel Working Papers*, No. 1747, November 2011, pp. 1–33; see p. 12.

65 Ibid., p. 27.

66 Collin Douglas, "A Storm Without Rain: Yemen, Water, Climate Change, and Conflict," *The Center for Climate and Security*, Briefer No. 40, August 3, 2016, pp. 1–8; see p. 1.

67 Ibid., p. 3.

68 Hadil Al-Mowafak, "Yemen's Forgotten Environmental Crisis Can Further Complicate Peacebuilding Efforts," *Yemen Policy Center*, December 2021, available at https://www.yemenpolicy.org/yemens-forgotten-environmental-crisis-can-further-complicate-peacebuilding-efforts/. Accessed on July 27, 2022.

69 Ibid.

70 Ibid.

71 Ashok Swain and Andres Jägerskog, *Emerging Security Threats in the Middle East*, Lanham, MD: Rowman & Littlefield, 2016, p. 125.

72 Hélène Thiollet, "From Migration Hub to Asylum Crisis: The Changing Dynamics of Contemporary Migration in Yemen," in Helen Lackner, ed., *Why Yemen Matters; A Society in Transition*, London: SOAS, 2014, pp. 267–86; see p. 283.

73 Ibid., p. 283.

74 Ibid., p. 284.

75 Swain and Jägerskog, *Emerging Security Threats in the Middle East*, p. 125.

76 Ibid.

77 Ibid.

78 Daoudy, *Climate Change and Regional Instability in the Middle East*, p. 14.

79 Coppi, "The Humanitarian Crisis in Yemen: Beyond the Man-Made Disaster," p. 12.

80 Ibid., p. 279.

81 Ibid., p. 280.

82 Ibid., p. 281.

83 Alexander Betts and Paul Collier, *Refuge: Rethinking Refugee Policy in a Changing World*, New York: Oxford University Press, 2017, p. 27.

84 UNHCR: Global Focus, "Needs Mount as Conflict in Yemen Rages on," April 1, 2022, available at https://reporting.unhcr.org/yemen. Accessed on July 25, 2022.

85 UNHCR, "Yemen Crisis Explained," July 14, 2022, available at https://www.unrefugees.org/news/yemen-crisis-explained/. Accessed on July 28, 2022.

86 Ibid.

87 Ibid.

88 Coppi, "The Humanitarian Crisis in Yemen: Beyond the Man-Made Disaster," p. 12.

89 Bonnefoy, *Yemen and the World: Beyond Security*, pp. 120–1.

90 Ibid., p. 122.

91 Ibid., p. 126.

92 *UNHCR*, "Yemen Situation Regional Refugee and Migrant Response Plan, January-December 2016 | December 2015 [EN/AR]," December 15, 2015, available at https://reliefweb.int/report/yemen/yemen-situation-regional-refugee-and-migrant-response-plan-january-december-2016. Accessed on February 14, 2024.

93 Schâdi Sémnani, "The Road From Yemen: Part 3," *Internal Displacement Monitoring Center (IDMC)*, July 2019, available at https://www.internal-displacement.org/expert-opinion/the-road-from-yemen-part-3. Accessed on July 28, 2022.

94 Coppi, "The Humanitarian Crisis in Yemen: Beyond the Man-Made Disaster," p. 13.

95 Ibid., p. 13.

96 Samaher al-Hadheri and Morgann Barbara Pernot, "The Evolution of Yemeni Migration to Djibouti," *Sana'a Center for Strategic Studies*, April 14, 2022, available at https://sanaacenter.org/publications/main-publications/17445. Accessed on July 28, 2022.

97 Ibid.

98 Ibid.

99 *UN Sustainable Development Program*, "Learning When Displaced: Yemeni Women and Girls Find Empowerment in Education," December 18, 2023, available at https://unsdg.un.org/latest/stories/learning-when-displaced-yemeni-women-and-girls-find-empowerment-education. Accessed on February 14, 2024.

100 Ibid.

101 Ibid.

102 Coppi, "The Humanitarian Crisis in Yemen: Beyond the Man-Made Disaster," p. 34.

103 Thiollet, "From Migration Hub to Asylum Crisis: The Changing Dynamics of Contemporary Migration in Yemen," pp. 267–86, p. 285.

104 Ibid., pp. 285–6.

105 *Intersos Humanitarian Aid*, "The Vulnerabilities of the Migrant Population in Yemen," available at https://www.intersos.org/en/the-vulnerabilities-of-the-migrant-population-in-yemen/. Accessed on May 25, 2024.

106 Razieh Namdar, Ezatollah Karami, and Marzieh Keshavarz, "Climate Change and Vulnerability: The Case of MENA Countries," *International Journal of Geo-information*, November 22, 2021, pp. 1–19; available at file:///C:/Users/909186933/Downloads/ijgi-10-00794.pdf. Accessed on May 24, 2024.

107 Ibid., pp. 12–13.

108 Ibid., p. 14.

109 C. Breinsinger, O. Ecker, P. Al-Riffai, et al., "Climate Change, Agricultural Production and Food Security: Evidence from Yemen, Kiel Working Papers," No. 1747, November 2011, pp. 1-33; see p. 30.

110 al-Hadheri and Pernot, "The Evolution of Yemeni Migration to Djibouti."

111 Ibid.

Chapter 9

1 *Britannica*, "Iran," available at https://www.britannica.com/place/Iran. Accessed on May 25, 2024.

2 Javad Bazrafshan, Somayeh Hejabi, and Saeid Eslamian, "Drought Modeling Examples," in Saeid Eslamian and Faezeh Eslamian, eds., *Handbook of Drought and Water Scarcity: Principles of Drought and Water Scarcity*, New York: CRC Press, 2017, pp. 167–88; see pp. 167–8.

3 Ibid., p. 168.

4 *Reliefweb*, "Iran, MENA | Drought – Emergency Appeal n° MDRIR009," March 3, 2022, available at https://reliefweb.int/report/iran-islamic-republic/iran-mena-drought-emergency-appeal-n-mdrir009. Accessed on May 29, 2022.

5 World Health Organization, "Health and Climate Change: Country Profile 2022," https://iris.who.int/bitstream/handle/10665/352630/WHO-HEP-ECH-CCH-22.01.01-eng.pdf?sequence=1. Accessed on May 25, 2024.

6 Ibid., pp. 167–9.

7 Ibid., p. 181.

8 Shirin Hakim and Karen Makuch, "Springtime in Iran Signals the Renewal of an Environmental Movement," *Atlantic Council*, April 21, 2023, available at https://www.atlanticcouncil.org/blogs/iransource/springtime-in-iran-signals-the-renewal-of-an-environmental-movement/. Accessed on February 14, 2024.

9 Justin Corfield, "Iran," in S. George Philander, ed., *Encyclopedia of Global Warming & Climate Change,* 2nd ed., Los Angeles, CA: Safe, 2012, pp. 799–801; see especially p. 800.

10 Ibid., p. 800.

11 Ibid.

12 Mehdi Shiva and Hassan Molana, "Climate Change Induced Inter-Province Migration in Iran," pp. 1–22; see p. 4, available at file:///C:/Users/909186~1/AppData/Local/Temp/10536_paper_Si3reN4D.pdf. Accessed on June 1, 2022.

13 Ibid.

14 Lauren Elmore, "The Implications of Climate Change in Iran," *American-Iranian Council* (AIC), July 27, 2021, available at http://www.us-iran.org/news/2021/7/2/covid-19-in-iran-r6by5. Accessed on June 1, 2022.

15 Saeid AshrafVaghefi, Malihe Keykhai, Farshid Jahanbakshi, Jaleh Sheikholeeslami, Azadeh Ahmadi, Hong Yong, and Karim C. Abbaspour, "The Future of Extreme Climate in Iran," *Scientific Reports*, Vol. 9, February 6, 2019, pp. 1–11; available at https://www.nature.com/articles/s41598-018-38071-8.pdf. Accessed on June 3, 2022.

16 Elmore, "The Implications of Climate Change in Iran."

17 Ibid.

18 Ibid.

19 Shiva and Molana, "Climate Change Induced Inter-Province Migration in Iran," p. 4.

20 Ibid., p. 5.

21 An interview with the Iranian botanist Hossein Akhani, Ramin Skibba, "Saving Iran's Precious Lake Urmia," *Scientific American*, April 26, 2016, available at https://blogs.scientificamerican.com/guest-blog/saving-iran-s-precious-lake-urmia/. Accessed on June 10, 2022.

22 Mohammad Reza Mansouri Daneshvar, Majid Ebrahim, and Hamid Nejadsoleymani, "An Overview of Climate Change in Iran: Facts and Statistics," *Environmental Systems Research*, Vol. 8, No. 7, 2019, pp. 1–10; see p. 1; available at https://environmentalsystemsresearch.springeropen.com/track/pdf/10.1186/s40068-019-0135-3.pdf. Accessed on June 2, 2022.

23 Ibid., p. 8.

24 Ibid.

25 *US Energy Information Administration*, "Iran," available at https://www.eia.gov/international/analysis/country/IRN. Accessed on June 8, 2022.

26 Ibid.

27 Ibid.

28 Scott Waldman, "Climate Change May Have Helped Spark Iran's Protests," *Scientific American*, January 8, 2018, available at https://www.scientificamerican.com/article/climate-change-may-have-helped-spark-iran-rsquo-s-protests/. Accessed on June 2, 2022.

29 Elmore, "The Implications of Climate Change in Iran."

30 Ibid.

31 Ibid.

32 *Financial Tribune*, "Increase in Climate Migrants Across Iran," December 4, 2016, available at https://financialtribune.com/articles/people/54651/increase-in-climate-migrants-across-iran. Accessed on June 7, 2022.

33 Ibid.

34 Ibid.

35 Ibid.

36 Ibid.

37 Mostafa Mohammadi Dehcheshmeh and Sohrab Ghaedi, "Climate Change and Ecological Migration: A Study of Villages in the Province of Khuzestan, Iran," *Journal of Environmental Research, Engineering and Management*, Vol. 76, No. 1, 2020, pp. 6–19; see especially p. 8.

38 Ibid., p. 12.

39 Garrett Nada, "Iran's Troubled Provinces: Khuzestan," *United States Institute of Peace: The Iran Primer*, July 29, 2020, available at https://iranprimer.usip.org/blog/2020/jul/29/iran%E2%80%99s-challenges-converge-khuzestan. Accessed on June 2, 2022.

40 Dehcheshmeh and Ghaedi, "Climate Change and Ecological Migration," p. 12.

41 Ibid., p. 17.

42 Nada, "Iran's Troubled Provinces: Khuzestan." United States Institute of Peace: The Iran Primer, July 29, 2020, available at << https://iranprimer.usip.org/blog/2020/jul/29/iran%E2%80%99s-challenges-converge-khuzestan>>. Accessed on June 2, 2022.

43 Somayeh Rafiei, "Khuzestan Water Problem Has Much Deeper Roots than a Multi-year Management Period," *ISNA*, July 20, 2021, available at https://www.isna.ir/news/1400042921463. Accessed on September 3, 2021.

44 Amir Reza Khavarian-Garmsir, Ahmad Pourahmad, Hossein Hataminejad, and Rahmatollah Farhoodi, "Climate Change and Environmental Degradation and the Drivers of Migration in the Context of Shrinking Cities: A Case Study of Khuzestan Province, Iran," *Sustainable Cities and Society*, Vol. 47, February 22, 2019, p. 101-480.

45 Ibid.

46 Ibid.

47 *Brussels International Center for Research and Human Rights*, "Iran and Climate Refugees: An Alarming Situation," May 22, 2019, available at https://www.bic-rhr.com/sites/default/files/inline-files/Climate%20Refugee%20and%20Iran%20Water%20Mismanagement%20-%20Final.pdf. Accessed on June 8, 2022.

48 Quoted in Nada, "Iran's Troubled Provinces: Khuzestan."

49 Ibid.

50 *Human Rights Watch*, "Building Collapse in Abadan," May 27, 2022, available at https://www.hrw.org/blog-feed/iranian-society-under-crackdown. Accessed on June 10, 2022.

51 Banafsheh Keynoush, "Iran's Growing Climate Migration Crisis," *Middle East Institute*, January 30, 2023, available at https://www.mei.edu/publications/irans-growing-climate-migration-crisis. Accessed on February 14, 2024.

52 Ibid.

53 Ibid.

54 Ibid.

55 Golnaz Esfandiari and Mohammad Zarghami, "Iran's Climate Migration Crisis Could Turn into Natural Disaster," *Informed Comment*, December 30, 2023, available at https://

www.juancole.com/2023/12/migration-national-disaster.html. Accessed on February 14, 2024.

56 Ibid.

57 Vivian Yee and Leily Nikounazar, "In Iran, Some Are Chasing the Last Drops of Water," *The New York Times*, June 21, 2023, available at https://www.nytimes.com/2023/06/21/world/middleeast/iran-drought-water-climate.html. Accessed on February 14, 2024.

58 Ibid.

59 Nada, "Iran's Troubled Provinces: Khuzestan."

60 Ibid.

61 Ibid.

62 Shima Silavi, "How Iran's Catastrophic Water Policy Can Lead to Waves of Climate Refugees," *Alarabiya News*, June 26, 2018, available at https://english.alarabiya.net/features/2018/06/26/How-Iran-s-catastrophic-water-policy-can-lead-to-waves-of-climate-refugees. Accessed on June 7, 2022.

63 Nada, "Iran's Troubled Provinces: Khuzestan."

64 Daniel Bellamy, "Thousands of Iranian Farmers Protest against Severe Water Shortages," *euronews*, November 20, 2021, available at https://www.euronews.com/2021/11/20/thousands-of-iranian-farmers-protest-against-severe-water-shortages. Accessed on June 2, 2022.

65 This section is based on Javad Heiran-Nia and Mahmood Monshipouri, "Iran's Environmental Woes Could Be Raisi's Achilles Heel," *Atlantic Council*, September 16, 2021, available at https://www.atlanticcouncil.org/blogs/iransource/irans-environmental-woes-could-be-raisis-achilles-heel/. Accessed on February 14, 2024.

66 Jubin Katiraise, "Raisi Will Not Be Traveling to Scotland for the UN Climate Change Conference," *Iran Focus*, October 13, 2021, available at https://www.iranfocus.com/en/human-rights/47806-raisi-will-not-be-travelling-to-scotland-for-the-un-climate-change-conference/. Accessed on June 11, 2022.

67 *Tehran Times*, "President-Elect Raisi Invited to UN Climate Change Meeting," June 23, 2021, available at https://www.tehrantimes.com/news/462361/President-elect-Raisi-invited-to-UN-climate-change-meeting. Accessed on September 4, 2021.

68 Rens van Munster, "Securitization," June 26, 2012, available at https://www.oxfordbibliographies.com/view/document/obo-9780199743292/obo-9780199743292-0091.xml. Accessed on September 7, 2021.

69 *Fars News*, "Second Debate | Raisi: We Will Conduct a Patrol; But for Managers / Hemmati and Mehr Alizadeh Competing for the Appointment of Minister of Women / Zakani: 26 Spies Arrested in the Most Secure Government," June 8, 2021, available at https://www.farsnews.ir/news/14000318000685. Accessed on September 5, 2021.

70 An interview with the Iranian botanist Hossein Akhani, Skibba, "Saving Iran's Precious Lake Urmia."

71 Bruce Pengra, Vahid Garousi, Asef Seyyed Najafi, and Azar Samadi, "The Dying of Iran's Lake Urmia and Its Environmental Consequences," *United Nations Environment Program*, February 2012, available at https://na.unep.net/geas/getuneppagewitharticleidscript.php?article_id=79. Accessed on June 10, 2022.

72 Ibid.

73 *Tasnim News*, "The Environmental Charter of the People's Government was Unveiled," June 13, 2021, available at https://www.tasnimnews.com/fa/news/1400/03/23/2520149. Accessed on September 5, 2021.

74 *Kayhan London,* "Solution of Ali Akbar Mehrabian, Minister of Energy for Waterlessness: Praying for the Rainfall and Sahifa Sajjadieh!," September 6, 2021, available at https://kayhan.london/fa/?p=255178. Accessed on September 6, 2021.

75 *Yale Center for Environmental Law & Policy,* "Environmental Performance Index 2020," 2020, available at https://epi.yale.edu/epi-results/2020/country/irn. Accessed on September 5, 2021.

76 *Donya-e-Eqtesad*, "Mahmood Sadeghi's Quick Response to Raisi's Request," July 6, 2021, available at https://donya-e-eqtesad.com. Accessed on September 5, 2021.

77 *IRNA News*, "The Paris Climate Accords Implemented in the Dark and Light," April 1, 2021, available at https://www.irna.ir/news/84280154. Accessed on September 5, 2021.

78 This is based on the authors' exclusive interview with Javad Heydarian on September 6, 2021.

79 The authors conducted an exclusive interview with Sadra Mohaqeq on September 10, 2021.

80 Joe Lo, "Drought and Water Mismanagement Spark Deadly Protests in Iran," Climate Home News, July 28, 2021.

81 Mohammad Fashaei, "Environmental NGOs, on the Path to Maturity," *Center for Strategic Studies*, October 10, 2018, available at http://css.ir/cckaut. Accessed on June 15, 2022.

82 The eight detainees were Sam Rajabi, Houman Jowkar, Niloufar Bayani, Morad Tahbaz, Morteza Arianejad, Taher Ghadirian, Amir Hossein Khaleghi, and Sepideh Kashani. See *Radio Farda*, "Iran Court Upholds 58 Years Prison Term for Eight Environmentalists," February 18, 2020, available at https://en.radiofarda.com/a/iran-court-upholds-58-years-prison-term-for-eight-environmentalists/30441558.html. Accessed on June 15, 2022.

83 "Final Verdicts Were Issued by Iranian Environmental Activists," *BBC News*, February 18, 2020, https://www.bbc.com/persian/iran-51543401

84 *Radio Farda*, "Iran Court Upholds 58 Years Prison Term for Eight Environmentalists."

85 "Khamenei Enters the File of Detained Environmental Activists," *Deutsche Welle*, October 15, 2019, https://p.dw.com/p/3RINx.

86 "Kalantari: We Do Not Have the Right to Enter the Environmental Judicial File," *Deutsche Welle*, September 2, 2019, https://p.dw.com/p/3OtcT.

87 "Kaveh Madani: I Was Interrogated from the First Day I Arrived in Iran," *BBC News*, April 2, 2018, https://www.bbc.com/persian/iran-43895458.

88 "Opposition of the Supreme Environmental Council to Miankaleh Petrochemical," *Donya-e-Eghtesad Newspaper*, June 12, 2022, https://www.donya-e-eqtesad.com/fa/tiny/news-3872666.

89 "Security Climate of the Islamic Republic: Environmental Activists in a State of Repression," *Radio Zamaneh*, November 6, 2018, https://web.archive.org/web/20181109153437/https://www.radiozamaneh.com/418767.

90 Ibid.

91 Laurence Teillet and Ahmad Ali Shariati, "The Helmand River Dispute: International Legal Perspectives on the Afghan-Iranian Border Conflict," *OpinioJuris*, July 19, 2023, available at http://opiniojuris.org/2023/07/19/the-helmand-river-dispute-international-legal-perspectives-on-the-afghan-iranian-border-conflict/. Accessed on October 21, 2023.

92 Fatemeh Aman, "Water Dispute Escalating Between Iran and Afghanistan," *Atlantic Council: South Asia Center*, August 2016, available at https://www.atlanticcouncil.org/wp-content/uploads/2016/09/Water_Dispute_Escalating_between_Iran_and_Afghanistan_web_0830.pdf. Accessed on May 25, 2024.

93 Ibid.

94 Ibid.

95 Ibid.

96 *Reuters*, "Prolonged Drought Deepens Afghanistan's Humanitarian Crisis," August 15, 2023, available at https://www.reuters.com/world/asia-pacific/prolonged-drought-deepens-afghanistans-humanitarian-crisis-2023-08-11/. Accessed on October 21, 2023.

97 Holly Dagres, "Iran and Afghanistan are Feuding over the Helmand River. The Water Wars Have No End in Sight." *Atlantic Council*, July 7, 2023, available at https://www.atlanticcouncil.org/blogs/iransource/iran-afghanistan-taliban-water-helmand/. Accessed on October 21, 2023.

98 For more on that, see Mahmood Monshipouri and Reza Mohajer, "Climate Change and the Potential for Conflict in the Middle East: Emerging Threats to Security and Human Rights," *The Maghreb Review*, Vol. 49, No. 1, 2024, pp. 14–41; see especially pp. 29–30.

99 *The Iran Primer*, "Iran and Afghanistan Clash over Water Rights," *United States Institute of Peace*, May 30, 2023, available at https://iranprimer.usip.org/blog/2023/may/30/iran-and-afghanistan-clash-over-water-rights. Accessed on June 30, 2023.

100 Ibid.

101 Dagres, "Iran and Afghanistan are Feuding over the Helmand River."

102 Massaab Al-Aloosy, "Water Insecurity: Iran's Formidable Threat," *Fikra Forum*, May 20, 2022, available at https://www.washingtoninstitute.org/policy-analysis/water-insecurity-irans-formidable-threat. Accessed on June 10, 2022.

103 *Brussels International Center for Research and Human Rights*, "Iran and Climate Refugees."

Chapter 10

1 Mohamed Riad, "Geopolitics and Politics in the Arab Gulf States (GCC)," *GeoJournal*, Vol. 13, No. 3, October 1986, pp. 201–10; see especially p. 203.

2 Mohamed Raouf, "Climate Change Threats, Opportunities, and the GCC Countries," *Middle East Institute*, April 1, 2008, available at https://www.mei.edu/publications/climate-change-threats-opportunities-and-gcc-countries. Accessed on June 25, 2022.

3 Jack Sargent, "Climate Change and the Rentier States of the Gulf Cooperation Council," *LSE Blog*, September 8, 2020, available at https://blogs.lse.ac.uk/mec/2020/09/08/climate

-change-and-the-rentier-states-of-the-gulf-cooperation-council/. Accessed on June 25, 2022.

4 Mari Luomi, *The Gulf Monarchies and Climate Change: Abu Dhabi and Qatar in an Era of Natural Unsustainability*, London: Hurst & Company, 2012, p. 1.

5 Ibid., p. 29.

6 Ibid., p. 45.

7 Aisha Al-Sarihi, "The GCC and the Road to Net Zero," *Middle East Institute*, March 2, 2023, available at https://mei.edu/publications/gcc-and-road-net-zero. Accessed on August 5, 2024.

8 John Whiteaker, "Is the GCC Serious About Averting Climate Catastrophe?," *Offshore Technology*, April 22, 2022, available at https://www.offshore-technology.com/analysis/gcc-serious-averting-climate-change-catastrophe/. Accessed on June 22, 2022.

9 Scott Livermore, "The GCC Region Confronts the Climate Change Quandary," *Arabian Business*, May 2, 2021, available at https://www.arabianbusiness.com/opinion/comment/462868-the-gcc-region-confronts-the-climate-change-quandary. Accessed on June 22, 2022.

10 Ibid.

11 Ibid.

12 Instinctif Partners, "COP26: GCC Members Walk the Talk," March 11, 2021, available at https://instinctif.com/thinking/cop26-gcc-members-walk-talk/. Accessed on June 22, 2022.

13 Hilal M. S. Al-Maamary, Hussein A. Kazem, and Miqdam T. Chaichan, "Climate Change: The Game Changer in the Gulf Cooperation Council Region," *Renewable and Sustainable Energy Reviews*, Vol. 76, March 27, 2017, pp. 555–76; see p. 556.

14 Fahad Radhi Alharbi and Denes Csala, "Gulf Cooperation Council Countries' Climate Change Mitigation Challenges and Exploration of Solar and Wind Energy Resource Potential," *Applied Sciences*, Vol. 11, March 16, 2021, pp. 1–23; see p. 3.

15 Ibid., p. 6.

16 Ibid., p. 12.

17 Ibid., p. 14.

18 Mari Luomi, "Gulf States' Climate Change Policies Amid a Global Pandemic," *The Arab Gulf States Institute in Washington*, September 25, 2020, pp. 1–26; see p. 5; available at https://agsiw.org/gulf-states-climate-change-policies-amid-a-global-pandemic/. Accessed on June 22, 2022.

19 Ibid., p. 8.

20 Ibid., p. 13.

21 Ibid., p. 15.

22 Aarti Nagraj, "The Climate Crisis: Is the GCC Doing Enough in the Sustainability Space?," *Gulf Business*, April 11, 2020, available at https://gulfbusiness.com/the-climate-crisis-is-the-gcc-doing-enough-in-the-sustainability-space/. Accessed on June 27, 2022.

23 Luomi, "Gulf States' Climate Change Policies Amid a Global Pandemic," p. 24.

24 Ibid., p. 24.

25 John Whiteaker, "Is the GCC Serious About Averting Climate Catastrophe?," *Offshore Technology*, April 22, 2022, available at https://www.offshore-technology.com/analysis/gcc-serious-averting-climate-change-catastrophe/. Accessed on June 22, 2022.

26 *Reuters*, "UAE Launches Plan to Achieve Net Zero Emissions by 2050," October 7, 2021, available at https://www.reuters.com/world/middle-east/uae-launches-plan-achieve-net-zero-emissions-by-2050-2021-10-07/. Accessed on June 23, 2022.

27 Ibid.

28 Nader Kabbani and Nejla Ben Mimoune, "Economic Diversification in the Gulf: Time to Redouble Efforts," *Brookings*, January 31, 2021, available at https://www.brookings.edu/research/economic-diversification-in-the-gulf-time-to-redouble-efforts/. Accessed on June 25, 2022.

29 UN Convention to Combat Desertification, "UAE Ministry of Climate Change and Environment," 2016, available at https://www.unccd.int/resources/knowledge-sharing-system/uae-ministry-climate-change-and-environment. Accessed on June 27, 2022.

30 Jon B. Alterman, "A New Revolution in the Middle East," *Center for Strategic & International Studies* (CSIS), August 9, 2021, available at https://www.csis.org/analysis/new-revolution-middle-east-0. Accessed on July 5, 2022.

31 United Arab Emirates Ministry of Climate Change and Environment, "Climate Change," available at https://www.moccae.gov.ae/en/knowledge-and-statistics/climate-change.aspx. Accessed on August 9, 2024.

32 Alterman, "A New Revolution in the Middle East."

33 Climate Action Tracker, "UAE," November 9, 2021, available at https://climateactiontracker.org/countries/uae/. Accessed on June 25, 2022.

34 M. Jamil, Farzana Ahmad, and Y. J. Jeon, "Renewable Energy Technologies Adopted by the UAE: Prospects and Challenges—A Comprehensive Overview," an unpublished paper, April 15, 2015.

35 Nikolaus J. Kurmayer, "UAE, Germany Sign 'Blue Hydrogen' Contract in Abu Dhabi," *Euractive.com*, March 21, 2022, available at https://www.euractiv.com/section/energy-environment/news/uae-germany-sign-blue-hydrogen-contract-in-abu-dhabi/. Accessed on June 27, 2022.

36 Nikolaus J. Kurmayer, "UAE's First Hydrogen Derived Shipment Delivered to Germany," *Euractive.com*, October 23, 2022, available at https://www.euractiv.com/section/energy-environment/news/uaes-first-hydrogen-derived-shipment-delivered-to-germany/. Accessed on July 29, 2024.

37 Luomi, "Gulf States' Climate Change Policies Amid a Global Pandemic," p. 121.

38 Ibid., pp. 132–3.

39 Ibid., p. 138.

40 "Qatar: Climatology," Climate Change Knowledge Portal, available at https://climateknowledgeportal.worldbank.org/country/qatar/climate-data-historical. Accessed on June 24, 2022.

41 *AA Energy*, "Qatar Aims to Raise LNG Production by 64% by 2027," November 25, 2019, available at https://www.aa.com.tr/en/energy/general/qatar-aims-to-raise-lng-production-by-64-by-2027/27501. Accessed on June 23, 2022.

42 Ibid.

43 Mona Al-Asmakh and Nadya Al-Awainati, "Counting the Carbons: Assessing Qatar's Carbon Dioxide Emissions," *Qscience*, Vol. 2018, No. 1, March 2018, available at https://www.qscience.com/content/papers/10.5339/qfarc.2018.EEPD592. Accessed on June 23, 2022.

44 Luomi, "Gulf States' Climate Change Policies Amid a Global Pandemic," p. 162.

45 Ibid., p. 167.

46 *IQAir*, "Air Quality in Doha," projected estimate for August 1, 2024. July 30, 2024, available at https://www.iqair.com/us/qatar/baladiyat-ad-dawhah/doha. Accessed on July 29, 2024.

47 Andrew Mills, "Qatar Targets 25% Cut in Greenhouse Gas Emissions by 2030 Under Climate Plan," *Reuters*, October 28, 2021, available at https://www.reuters.com/business/cop/qatar-targets-25-cut-greenhouse-gas-emissions-by-2030-climate-change-plan-2021-10-28/. Accessed on June 24, 2022.

48 *The Peninsula*, "Qatar Enhances Efforts to Address Challenges of Climate Change," December 30, 2021, available at https://thepeninsulaqatar.com/article/30/12/2021/qatar-enhances-efforts-to-address-challenges-of-climate-change. Accessed on June 24, 2022.

49 Ibid.

50 Deepak John, "Qatar Has Placed Air Quality Among Highest Priorities," *The Peninsula*, March 31, 2024, available at https://thepeninsulaqatar.com/article/31/03/2024/qatar-has-placed-air-quality-among-highest-priorities. Accessed on July 29, 2024.

51 Planning and Strategic Authority, *Qatar Voluntary National Review 2021: Report on the Implementation of the 2030 Agenda for Sustainable Development*, available at https://sustainabledevelopment.un.org/content/documents/280362021_VNR_Report_Qatar_English.pdf. Accessed on June 24, 2022.

52 Ibid.

53 Ibid.

54 Ibid.

55 Ibid.

56 Ashok Swain and Anders Jägerskog, *Emerging Security Threats in the Middle East: The Impact of Climate Change and Globalization*, Lanham, MD: Rowman & Littlefield, 2016, pp. 110–11.

57 *US-Saudi Business Council*, "Water in Saudi Arabia: Desalination, Wastewater, and Privatization," January 7, 2021, available at https://ussaudi.org/water-in-saudi-arabia-desalination-wastewater-and-privatization/. Accessed on June 25, 2022.

58 Adam Ramsay, "Will Climate Change Stop Arabia from Being Saudi?," *OpenDemocracy*, October 26, 2021, available at https://www.opendemocracy.net/en/north-africa-west-asia/will-climate-change-stop-arabia-from-being-saudi/. Accessed on June 25, 2022.

59 Ibid.

60 Alharbi and Csala, "Gulf Cooperation Council Countries' Climate Change," pp. 8–9.

61 Abdullah. Saad. Modaihsh and Mohamed Osman Mahjoub, "Falling Dust Characteristics in Riyadh City, Saudi Arabia During Winter Months," *SciVerse ScienceDirect*, January 19–20, 2013, available at www.sciencedirect.com. Accessed on June 25, 2022.

62 Giorgio Cafiero, "Vicious Cycle: Storms Intensify in the Gulf as Climate Changes," *Al Jazeera*, June 2, 2022, available at https://www.aljazeera.com/news/2022/6/2/a-vicious-cycle-intensifying-storms-in-gulf-as-climate-changes. Accessed on June 25, 2022.

63 Banafsheh Keynoush, "Severe Sand and Dust Storms Are an Underrated Risk in the Gulf Region, Despite Mitigation Measures," *Middle East Institute*, March 12, 2022, available at https://www.mei.edu/publications/severe-sand-and-dust-storms-are-underrated-risk-gulf-region-despite-mitigation. Accessed on June 25, 2022.

64 Abeer Abdulkareem and Amgad Ellaboudy, "Saudi Arabia's CO2 Emissions Have Increased Over 3x Since 1990," *Climate Scorecard*, July 11, 2021, available at https://www.climatescorecard.org/2021/07/saudi-arabias-co2-emissions-have-increased-over-3x-since-1990/. Accessed on June 25, 2022.

65 Abdullah Al-Badi and Imtenan AlMubarak, "Growing Energy Demand in the GCC Countries," *Arab Journal of Basic and Applied Sciences*, Vol. 26, No. 1, 2019, pp. 488–96; see especially p. 495.

66 Ibid., p. 495.

67 *Arab News*, "GCC Countries Call for United Effort to Tackle Climate Change," April 1, 2022, available at https://www.arabnews.com/node/2055211/saudi-arabia. Accessed on June 25, 2022.

68 John Whiteaker, "Is the GCC Serious About Averting Climate Catastrophe?" *Offshore Technology*, April 22, 2022, available at https://www.offshore-technology.com/analysis/gcc-serious-averting-climate-change-catastrophe/. Accessed on June 22, 2022.

69 *Climate Change Tracker*, "Saudi Arabia," November 4, 2021, available at https://climateactiontracker.org/countries/saudi-arabia/. Accessed on June 25, 2022.

70 This section is based largely on Mahmood Monshipouri, "The Middle East Post-Petroleum: Averting the Storm," *Middle East Policy*, Vol. XXVI, No. 3, Fall 2019, pp. 77–91.

71 Sean L. Yom, "Introduction: The Middle East and North Africa in Comparative Perspective," in Mark Gasiorowski and Sean L. Yom, eds., *The Government and Politics of the Middle East and North Africa*, Boulder, CO: Westview Press, 2017, pp. 1–43; see p. 31.

72 Mehran Kamrava, *Troubled Waters: Insecurity in the Persian Gulf*, Ithaca, New York: Cornell University Press, 2018, p. 148.

73 Kristian Coates Ulrichsen, *Insecure Gulf: The End of Certainty and the Transition to the Post-Oil Era*, New York: Columbia University Press, 2011, p. 1.

74 Gregory Aftandilian, "A Saudi-Iranian Rapprochement Still Has a Long Way to Go," *Arab Center Washington, DC*, September 21, 2021, available at https://arabcenterdc.org/resource/a-saudi-iranian-rapprochement-still-has-a-long-way-to-go/. Accessed on May 27, 2024.

75 Mahmood Sarioghalam, "Transparency and Good Governance: A Road Map to Job Creation and Inclusive Growth," World Economic Forum, Regional Agenda, in *Accelerating Economic Reforms in the Middle East and North Africa: A Private Sector Perspective*, January 2017, available at http://www3.weforum.org/docs/Media/AM17/Accelerating_Reforms_MENA_.pdf. Accessed on November 14, 2018.

76 For a highly educational and illustrative view on this topic, see Mehtap Isik, "Getting Ready for the Post-Oil Age: The Entrepreneurship as the Rising Star of Development Policies in the Gulf States," *Modern Economy*, Vol. 9, No. 1, 2018, pp. 125–48.

77 Melani Cammett, Ishac Diwan, Alan Richards, and John Waterbury, *A Political Economy of the Middle East*, 4th Edition, Boulder, CO: Westview Press, 2015, p. 337.

78 Ibid., p. 6.

79 Ibid., p. 6.

80 International Monetary Fund, "Economic Diversification in Oil-Exporting Arab Countries," Annual Meeting of Arab Ministers of Finance, Manama, Bahrain, April 2016, available at https://www.imf.org/external/np/pp/eng/2016/042916.pdf. Accessed on November 18, 2018.

81 Ibid.

82 Dana El Baltaji and Glen Carey, "Saudi Arabia Looks to Post-Oil Age in Great Takeover: QuickTake," *The Washington Post*, November 9, 2017, available at https://www.washingtonpost.com/business/saudi-arabia-looks-to-post-oil-age-in-great-makeover-quicktake/2017/11/10/743e25dc-c56a-11e7-9922-4151f5ca6168_story.html?utm_term=.1495cfd01794. Accessed on November 4, 2018.

83 David Hobbs, "How Saudi Arabia Can Become a Post-oil Economy," *World Economic Forum*, May 19, 2017, available at https://www.weforum.org/agenda/2017/05/saudi-arabia-post-oil-economy-david-hobbs/. Accessed on July 30, 2024.

84 Ali Asghar Pourezzat, Mostafa Nejati, Ghazaleh Taheri Attar, and Seyed Mahdi Sharifmousavi, "Scenario Planning and Futurology of the Persian Gulf Post-Oil Economy," *Foresight*, Vol. 13, No. 6, 2011, pp. 18–33; see p. 21.

85 Ibid., pp. 30–1.

86 Ibid., p. 25.

87 Banafsheh Keynoush, "Climate-Induced Migration in the GCC States: A Looming Challenge," *Middle East Institute*, January 13, 2023, available at https://www.mei.edu/publications/climate-induced-migration-gcc-states-looming-challenge. Accessed on March 25, 2024.

88 Ibid.

89 Ibid.

90 *Human Rights Watch*, "Qatar: Gulf's First Refugee Asylum Law," October 30, 2018, available at https://www.hrw.org/news/2018/10/30/qatar-gulfs-first-refugee-asylum-law. Accessed on March 22, 2024.

91 Ibid.

92 *Integral Human Development*, "Country Profiles: Saudi Arabia," available at https://migrants-refugees.va/country-profile/saudi-arabia/. Accessed on March 22, 2024.

93 Ibid.

94 Ibid.

95 Ibid.

96 Ibid.

97 Ibid.

98 *Integral Human Development*, "Country Profiles: United Arab Emirates," available at https://migrants-refugees.va/country-profile/united-arab-emirates/. Accessed on March 22, 2024.

99 Ibid.

100 Ibid.

101 Ibid.

102 Georgina Rannard, "Deadly Dubai Floods Made Worse by Climate Change," *BBC*, April 25, 2024, available at https://www.bbc.com/news/science-environment-68897443. Accessed on May 27, 2024.

103 Ibid.

104 UNHCR, "The Sultanate of Oman," available at file:///C:/Users/909186933/Downloads/UNHCR_UPR23_OMN_E_Main.pdf. Accessed on March 25, 2024.

105 Ibid.

106 *Integral Human Development*, Country Profiles: Oman, available at https://migrants-refugees.va/country-profile/oman/. Accessed on March 25, 2024.

107 Ibid.

108 Ibid.

109 Timothy Mitchell, *Carbon Democracy: Political Power in the Age of Oil*, New York: Verso Books, 2013, p. 253.

110 Ibid., p. 254.

111 Ibid., p. 233.

112 Maria Sarraf, "Assessing the Costs of Environmental Degradation in the Middle East and North Africa Region," The World Bank: Environment & Strategy, No. 9, April 2004; available at http://documents.worldbank.org/curated/en/780231468774275451/pdf/298630PAPER0As1Env0Strat0Note0no-09.pdf. Accessed on November 8, 2018.

113 *MENA Herald*, "The Middle East: Rise of the Post-Oil Era," June 21, 2016, available at https://www.menaherald.com/en/business/events-services/middle-east-rise-post-oil-era. Access on November 13, 2018.

114 Mehran Kamrava, *Inside the Arab State*, New York: Oxford University Press, 2018, p. 32.

115 Ibid., p. 148.

116 Banafsheh Keynoush, "Climate-Induced Migration in the GCC States."

117 *Petro Service Company (PSC)*, "History and Evolution of Medicine," February 2, 2018, available at https://petroleumservicecompany.com/blog/history-evolution-medicine-2/. Accessed on May 28, 2024.

118 Alharbi and Csala, "Gulf Cooperation Council Countries' Climate Change," p. 19.

Chapter 11

1 Climate Expert, "Morocco," available at https://www.climate-expert.org/en/home/business-adaptation/morocco. Accessed April 19, 2024.

2 For further details about this, see Raphael Chijioke Njoku, *Culture and Customs of Morocco*, Westport, CT: Greenwood Press, 2006.

3 C. R. Pennell, *Morocco: From Empire to Independence*, Oxford: Oneworld Publications, 2003, p. xiii.

4 Ibid., p. xiv.

5 Ibid.

6 Harold D. Nelson, ed., *Morocco: A Country Study*, Washington, DC: American University, 1985, p. xiv.

7 Ibid., p. 101.

8 International Energy Agency (IEA), "Climate Resiliency for Energy Transition in Morocco," available at https://www.iea.org/reports/climate-resilience-for-energy-transition-in-morocco. Accessed on April 19, 2024.

9 Ibid.

10 Ibid.

11 Mohamed Behnassi, "Environmental Change and Human Conflict," in Sean Yom, ed., *The Societies of the Middle East and North Africa: Structures, Vulnerabilities, and Forces*, New York: Routledge, 2019, pp. 207–34; see p. 222.

12 Lore Van Praag, Loubna Ou-Salah, Elodie Hut, and Caroline Zickgraf, *Migration and Environmental Change in Morocco*, New York: Springer Publishing, 2021; see Chapter 3: "Environmental Change and Migration in Morocco: What Has Been Done So Far?" pp. 35–59; see especially pp. 36–7.

13 Ibid., p. 39.

14 Hein de Haas, "Morocco: Setting the Stage for Becoming a Migration Transition Country," *Migration Policy Institute (MPI)*, March 19, 2014, available at https://www.migrationpolicy.org/article/morocco-setting-stage-becoming-migration-transition-country. Accessed on April 23, 2024.

15 Ibid.

16 Ibid.

17 Integral Human Development, "Country Profiles: Migrants/Refugees: Morocco," available at https://migrants-refugees.va/it/wp-content/uploads/sites/3/2021/05/2020-CP-Morocco.pdf. Accessed on April 23, 2024.

18 Gilles Olakunlé Yabi, "Climate Change and Migration Prospects Between West and North Africa," *Carnegie Endowment for International Peace*, available at https://carnegieendowment.org/2023/07/06/climate-change-and-vulnerability-in-middle-east-pub-90089. Accessed on April 19, 2024.

19 Ibid.

20 UNHCR, "Morocco," September 2023, available at file:///C:/Users/909186933/Downloads/Morocco%20-%20Factsheet_ENG_September%2020231.pdf. Accessed on April 23, 2024.

21 Ibid.

22 Ibid.

23 Ibid.

24 Integral Human Development, "Country Profiles: Migrants/Refugees: Morocco."

25 Ursula Lindsey, "Culture is the Solution: As Political Reform Stalls, Moroccan Activists Engage Citizens Through Art," in Thanassis Cambanis and Michael Wahid Hanna, eds., *Arab Politics Beyond the Uprisings: Experiments in an Era of Resurgent Authoritarianism*, New York: The Century Foundation Press, 2017, pp. 193–212; see pp. 202–3.

26 Integral Human Development, "Country Profiles: Migrants/Refugees: Morocco."

27 Ibid.

28 Ibid.

29 See CDP Worldwide, "2023 Morocco Earthquake," October 25, 2023, available at https://disasterphilanthropy.org/disasters/2023-morocco-earthquake/.

30 Global Edge, "Morocco: Government," available at https://globaledge.msu.edu/countries/morocco/government. Accessed on April 22, 2024.

31 Freedom House, "Morocco," available at https://freedomhouse.org/country/morocco/freedom-world/2022. Accessed on April 22, 2024.

32 Ibid.

33 Ibid.

34 Bahey eldin Hassan, "Reflections on Human Rights Before and After the Arab Spring," in Anthony Tirado Chase, ed., *Routledge Handbook on Human Rights and the Middle East and North Africa*, New York: Routledge, 2017, pp. 492–501; see especially p. 495.

35 Amy Hamblin, "Morocco: The Struggle for Political Legitimacy," in I. William Zartman, ed., *Arab Spring: Negotiating in the Shadow of the Intifadat*, Athens, GA: The University of Georgia Press, 2015, pp. 182–208; see p. 204.

36 Driss Maghraoui and Saloua Zerhouni, "Morocco," in Ellen Lust, ed., *The Middle East*, 14th ed., Thousand Oaks, CA: CQ Press, 2017, pp. 674–706; see p. 678.

37 Driss Maghraoui, "The Dynamics of Civil Society in Morocco," in Ellen Lust-Okar and Saloua Zerhouni, eds., *Political Participation in the Middle East*, Boulder, CO: Lynne Rienner Publishers, 2008, pp. 193–216.

38 Ibid.

39 Ibid.

40 Saloua Zerhouni, "The Moroccan Parliament," in Ellen Lust-Okar and Saloua Zerhouni, eds., *Political Participation in the Middle East*, Boulder, CO: Lynne Rienner Publishers, 2008, pp. 217–38.

41 Ibid.

42 Ibid.

43 NTS-Asia, "About Non-Traditional Security," available at https://rsis-ntsasia.org/about-nts-asia/. Accessed on April 23, 2024.

44 Michael Mason, "Climate Change, Environmental Degradation and Renewable Energy," in Richard Gillespie and Frédéric Volpi, eds., *Routledge Handbook of Mediterranean Politics* (New York: Routledge, 2018), 268–78; see 273.

45 Ibid., 273.

46 Melani Cammett, Ishac Diwan, Alan Richards, and John Waterbury, *A Political Economy of the Middle East*, Fourth Edition, Boulder, CO: Westview Press, 2015, p. 510.

47 Ibid., 512.

48 Van Praag, Ou-Salah, Hut, and Zickgraf, *Migration and Environmental Change in Morocco*, see especially p. 42.

49 Justin Corfield, "Morocco," in S. George Philander, ed., *Encyclopedia of Global Warming and Climate Change*, 2nd ed., Los Angeles, CA: Sage, 2012, pp. 958–9.

50 Van Praag, Ou-Salah, Hut, and Zickgraf, *Migration and Environmental Change in Morocco*, p. 45.

51 Ibid., p. 46.

52 Ibid.

53 Nelson, *Morocco*, p. 102.

54 Marvine Howe, *Morocco: The Islamist Awakening and Other Challenges*, New York: Oxford University Press, 2005, p. 115.

55 Ibid., pp. 207–8.

56 Ibid., p. 208.

57 Ibid., p. 274.

58 Ibid., p. 293.

59 Ibid., p. 338.

60 Corfield, "Morocco," pp. 958–9; see especially p. 958.

61 Ibid., p. 958.

62 World Data Atlas Morocco Environment, "Morocco: CO2 Emissions Per Capita," available at https://knoema.com/atlas/Morocco/CO2-emissions-per-capita. Accessed on April 18, 2024.

63 Corfield, "Morocco," pp. 958–9.

64 Ibid., p. 959.

65 International Energy Agency (IEA), "Climate Resiliency for Energy Transition in Morocco."

66 Ibid.

67 Kahime Kholoud, Behnassi Mohamed, Messouli Mohammed, Boussaa Samia, and Ali Boumezzough, "Infectious Diseases and Climate Case of Morocco," *Journal of Coastal Life Medicine*, Vol. 4, No. 8, 2016, pp. 659–62; see p. 659.

68 Kahime Kholoud, Sereno Denis, Bounoua Lahouari, Moulay Abdelmonaim, El Hidan, and Bouhout Souad, "Management of Leishmaniases in the Era of Change in Morocco," *International Journal of Environmental Research and Public Health*, July 20, 2018, available at https://pdfs.semanticscholar.org/020c/4dfcc2bce830a4851cea97a2613b788fd75b.pdf?_gl=1*1vuhlim*_ga*MjA0NjY1NTcxOS4xNzEzOTk5Mjcx*_ga_H7P4ZT52H5*MTcxMzk5OTI3MC4xLjAuMTcxMzk5OTI5Ny4zMy4wLjA. Accessed on April 24, 2024.

69 International Energy Agency (IEA), "Climate Resiliency for Energy Transition in Morocco."

70 Ibid.

71 Climate Action Tracker, "Morocco," April 17, 2023, available at https://climateactiontracker.org/countries/morocco/. Accessed on April 21, 2023.

72 Ibid.

73 Lahcen Haddad, "Global Food Security and the Role of Morocco," *Atalayar*, June 22, 2023, available at https://www.atalayar.com/en/opinion/lahcen-haddad/global-food-security-and-role-morocco/20221015152923136824.html. Accessed on April 20, 2024.

74 Hafed Al-Ghwell, "Morocco's Struggle with Water Scarcity and Food Security," *Arab News*, February 24, 2024, available at https://www.arabnews.com/node/2465921. Accessed on April 19, 2024.

75 Ibid.

76 Ibid.

77 Richard Javad Heydarian, *How Capitalism Failed the Arab World: The Economic Roots and Precarious Future of the Middle East Uprisings*, London: Zed Books, 2014, p. 86.

78 Ibid., 89.

79 Ibid., 89.

80 Daniel Workman, "Wheat Imports by Country," World's Top Exports," July 20, 2023, available at https://www.worldstopexports.com/wheat-imports-by-country/#google_vignette. Accessed on April 19, 2024.

81 Qussama Aamari, "UN Report: 2.1 Million Moroccans are Undernourished," *Morocco World News*, April 1, 2023, available at https://www.moroccoworldnews.com/2023/04/354796/un-report-2-1-million-moroccans-are-undernourished. Accessed on April 19, 2024.

82 Ibid.

83 Marie-Therese Nanlong, "Increasing Efforts to Ensure Food Security in Morocco," *Vanguard*, April 4, 2023, available at https://www.vanguardngr.com/2023/04/increasing-efforts-to-ensure-food-security-in-morocco/. Accessed on April 20, 2024.

84 Ibid.

85 Anne Barre, Irene Bankelman, Anke Stock, Eleanor Blomstrom, and Bridget Burns, "From Marrakesh to Marrakesh: The Rise of Gender Equality in the Global Climate Governance and Climate Action," in Sébastien Duyck, Sébastien Jodoin, and Alyssa Johl, eds., *Routledge Handbook of Human Rights and Climate Governance*, New York: Routledge, 2018, pp. 239–50; see especially pp. 239–40.

86 Ibid., p. 241.

87 Ibid., p. 243.

88 Amy Elizabeth Young, "Gender," in David S. Sorenson, ed., *Interpreting the Middle East*, Boulder, CO: Westview Press, 2010, pp. 219–48; see p. 227.

89 Barre, Bankelman, Stock, Blomstrom, and Burns, "From Marrakesh to Marrakesh," p. 246.

90 Ibid., p. 247.

91 Ibid., p. 247.

92 Aya Kamil, "The Climate-Gender Nexus: A Gap in Moroccan Policy," *Carnegie Endowment for International Peace*, August 3, 2023, available at https://carnegieendowment.org/sada/90321. Accessed on April 20, 2024.

93 Ibid.

94 Ibid.

95 Ibid.

96 Thomas O. Falk, "Morocco: MENA's Rare Climate Change Success Story," *Al Jazeera*, November 10, 2021, available at https://www.aljazeera.com/news/2021/11/10/morocco-leads-the-fight-against-climate-change-in-the-middle-east. Accessed on May 10, 2023.

97 Ibid.

98 Ibid.

99 Ibid.

100 Climate Action Tracker, "Morocco."

101 Van Praag, Ou-Salah, Hut, and Zickgraf, *Migration and Environmental Change in Morocco*, p. 47.

102 Behnassi, "Environmental Change and Human Conflict," p. 230.

103 Climate Action Tracker, "Morocco."

104 Van Praag, Ou-Salah, Hut, and Zickgraf, *Migration and Environmental Change in Morocco*, p. 51.

105 Ibid., p. 48.

106 Ibid., pp. 53–4.

107 Siham Kasmi, Maria Snoussi, Otamane Khalfaoui, Rajaa Aitali, and Latifa Flayou, "Increasing Pressures, Eroding Beaches and Climate Change in Morocco," *Journal of African Earth Sciences*, Vol. 164, April 2020, available at https://www-sciencedirect-com.jpllnet.sfsu.edu/science/article/pii/S1464343X20300479?via%3Dihub. Accessed on April 22, 2024.

108 Ibid.

109 Ibid.

110 Ibid.

111 International Energy Agency (IEA), "Climate Resiliency for Energy Transition in Morocco."

112 Ibid.

113 Ibid.

114 Ibid.

115 Youness Hrour et al., "A Framework to Assess Future Water-Resource Under Climate Change in Northern Morocco Using Hydro-Climatic Modelling and Water-Withdrawal Scenarios," *Journal of Hydrology: Regional Studies*, Vol. 48, August 2023, available at https://www.sciencedirect.com/science/article/pii/S2214581823001520?via%3Dihub. Accessed on April 22, 2024.

116 Climate Action Tracker, "Morocco."

117 Van Praag, Ou-Salah, Hut, and Zickgraf, *Migration and Environmental Change in Morocco*, p. 54.

Chapter 12

1 Sandeep Sengupta, "Climate Change, International Justice and Global Order," *International Affairs*, Vol. 99, No. 1, January 2023, pp. 121–40; see pp. 138–9.

2 Marwa Daoudy, "Climate Change and Regional Instability in the Middle East," *Council on Foreign Relations*, Discussion Paper Series on Managing Global Disorder, No. 14, October 2023, pp. 1–31; see especially p. 5.

3 Ibid., pp. 5–6.

4 Ibid.

5 Mariëlle Snel, Nikolas Sorensen, and Reed Power, *Climate Change and Water Scarcity in the Middle East: A Transitional Approach*, London: Taylor & Francis Human Geography, 2025, p. 23.

6 Lahcen Haddad, "Global Food Security and the Role of Morocco," *Atalayar*, June 22, 2023, available at https://www.atalayar.com/en/opinion/lahcen-haddad/global-food-security-and-role-morocco/20221015152923136824.html. Accessed on April 20, 2024.

7 Mia Prange, "Climate Change Is Fueling Migration. Do Climate Migrants Have Legal Protections? *Council on Foreign Relations*, December 19, 2022, available at https://www.cfr.org/in-brief/climate-change-fueling-migration-do-climate-migrants-have-legal-protections. Accessed on October 21, 2023.

8 Ibid.

9 Joseph Romm, *Climate Change: What Everybody Needs to Know*, 2nd ed., New York: Oxford University Press, 2018, p. 133,

10 *UNHCR, 2021 Global Compact on Refugees: Indicator Report*, available at https://www.unhcr.org/global-compact-refugees-indicator-report/wp-content/uploads/sites/143/2021/11/2021_GCR-Indicator-Report_spread_web.pdf. Accessed on July 10, 2023.

11 Alexander Betts, *The Wealth of Refugees: How Displaced People Can Build Economies*, New York: Oxford University Press, 2021, p. 117.

12 Ibid.

13 Ibid., p. 140.

14 Alexander Betts and Paul Collier, *Refuge: Rethinking Refugee Policy in a Changing World*, New York: Oxford University Press, 2017, pp. 3–5.

15 Rawan Arar and David Scott FitzGerald, *The Refugee System: A Sociological Approach*, Cambridge: Polity Press, 2023, p. 241.

16 Ibid.

17 Ibid., p. 244.

18 UNHCR, *Convention and Protocol Relating to the Status of Refugees*, 2011, available at file:///C:/Users/909186933/Downloads/3b66c2aa10.pdf. Accessed on March 28, 2024.

19 USA for UNHCR, "Syria Refugee Crisis Explained," UNHCR, March 13, 2024, available at https://www.unrefugees.org/news/syria-refugee-crisis-explained/. Accessed on March 28, 2024.

20 Ibid.

21 M. Murat Erdogan, "Syrian Refugees in Turkey: A Burden or Benefit," DCAF: A Center for Security, Development, and the Rule of Law, 2016, available at https://www.dcaf.ch/sites/default/files/publications/documents/DCAF-STRATIM-Paper-2.pdf.

22 Betts and Collier, *Refuge*, pp. 6–7.

23 Ibid., pp. 156–7.

24 Volker Türk and Rebecca Dowd, "Protection Gaps," in Elena Fiddian-Qasmiyeh, Gil Loescher, Katy Long, and Nando Sigona, eds., *The Oxford Handbook of Refugee and Forced Migration Studies*, New York: Oxford University Press, 2014, pp. 278–89; see especially p. 283.

25 Roger Zetter and James Morrissey, "The Environment-Mobility Nexus: Reconceptualizing the Link Between Environmental Stress, (Im)Mobility, and Power," in Elena Fiddian-Qasmiyeh, Gil Loescher, Katy Long, and Nando Sigona, eds., *The Oxford Handbook of Refugee and Forced Migration Studies*, New York: Oxford University Press, 2014, pp. 342–54; see especially pp. 343–4.

26 Robert A. McLeman, *Climate and Human Migration: Past Experiences, Future Challenges*, New York: Cambridge University Press, 2014, p. 47.

27 Ibid., p. 75.

28 Ibid., p. 109.

29 Mia Prange, "Climate Change Is Fueling Migration. Do Climate Migrants Have Legal Protections?," *Council on Foreign Relations*, December 19, 2022, available at https://www.cfr.org/in-brief/climate-change-fueling-migration-do-climate-migrants-have-legal-protections. Accessed on July 10, 2023.

30 Ibid.

31 National Geographic, "Environmental Refugee," available at https://education.nationalgeographic.org/resource/environmental-refugee/. Accessed on July 10, 2023.

32 Benjamin Freedman, "Getting Ahead of the Middle East's Climate Refugee Conundrum," *Middle East Institute*, April 26, 2023, available at https://www.mei.edu/publications/getting-ahead-middle-easts-climate-refugee-conundrum. Accessed on July 10, 2023.

33 Ibid.

34 Jeannie Sowers, "Environmental Activism in the Middle East and North Africa," in Harry Verhoeven, ed., *Environmental Politics in the Middle East*, New York: Oxford University Press, 2009, pp. 27–52; see especially p. 28.

35 Nilüfer Göle, "Gezi—Anatomy of a Public Square Movement," *Insight Turkey*, Vol. 15, No. 3, Summer 2013, pp. 7–14; see p. 9.

36 Ibid.

37 Coşkun Taştan, "The Gezi Park Protests in Turkey: A Qualitative Field Research," *Insight Turkey*, Vol. 15, No. 3, Summer 2013, pp. 27–38; see pp. 33–4.

38 Naoual Belakhdar, "Algeria is not for Sale!" Mobilizing Against Fracking in the Sahara," Institute for Middle East Studies, Fall 2020, available at https://imes.elliott.gwu.edu/public-outreach-and-educator-resources/impop/environmental-exploitation-triggers-protests-in-algeria/. Accessed on April 11, 2024.

39 Sowers, "Environmental Activism in the Middle East and North Africa," p. 34.

40 Ibid.

41 Ibid., p. 35.

42 See https://arab.org/directory/environmental-groups/. Accessed on April 11, 2024.

43 Will Todman, “Environmental Civil Society in the Middle East,” *Center for Strategic and International Studies* (*CSIS*), June 9, 2023, available at https://www.csis.org/analysis/environmental-civil-society-middle-east. Accessed on April 11, 2024.

44 Ibid.

45 Ibid.

46 This section is based on Mahmood Monshipouri and Reza Mohajer, “Climate Change and the Potential for Conflict in the Middle East: Emerging Threats to Security and Human Rights,” *The Maghreb Review*, Vol. 49, No. 1, 2024, pp. 14–41; see especially pp. 33–6.

47 Thomas O. Falk, “Morocco: MENA's Rare Climate Change Success Story,” *Al Jazeera*, November 10, 2021, available at https://www.aljazeera.com/news/2021/11/10/morocco-leads-the-fight-against-climate-change-in-the-middle-east. Accessed on May 10, 2023.

48 Ibid.

49 *Climate Action Tracker*, available at https://climateactiontracker.org/countries/. Accessed on May 14, 2023.

50 Falk, “Morocco.”

51 Aisha Al-Sarihi, “The GCC and the Road to Net Zero,” *Middle East Institute*, March 2, 2023, available at https://www.mei.edu/publications/gcc-and-road-net-zero. Accessed on May 10, 203.

52 Mari Luomi, *The Gulf Monarchies and Climate Change: Abu Dhabi and Qatar in an Era of Natural Unsustainability*, London: Hurst & Company, 2012, pp. 233–4.

53 Joshua Meltzer, Nathan Hultman, and Claire Langley, “Low-Carbon Energy Transitions in Qatar and the Gulf Cooperation Council Region,” *Global Economy and Development at Brookings*, February 2014, pp. 1–84; see pp. 12–13.

54 Ibid., p. 13.

55 Shakeel Ahmad Ramay, “Vision 2030: How Far Has Saudi Arabia Come?,” *Tribune*, December 26, 2022, available at https://tribune.com.pk/story/2392846/vision-2030-how-far-has-saudi-arabia-come. Accessed on May 2, 2023.

56 Tariq A. Al Maeena, “The Road to Saudi Vision 2030,” *Gulf News*, March 4, 2023, available at https://gulfnews.com/opinion/op-eds/the-road-to-saudi-vision-2030-1.94216169. Accessed on April 30, 2023.

57 International Monetary Fund, “Saudi Arabia: Staff Concluding Statement of the 2022 Article IV Mission,” June 22, 2022, available at https://www.imf.org/en/News/Articles/2022/06/17/saudi-arabia-staff-concluding-statement-of-the-2022-article-iv-mission. Accessed on April 23, 2023.

58 Caroline Wright, “How the UAE is Committing to Climate Change Action,” *Harvard Business*, February 2022, available at https://www.harvardbusiness.org/how-the-uae-is-committing-to-climate-change-action/. Accessed on May 11, 2023.

59 Mohammed Alardhi, “The UAE Can Lead the Way in Climate Adaptation, not Just Mitigation,” *The National*, January 19, 2023, available at https://www.thenationalnews.com/opinion/comment/2023/01/19/the-uae-can-lead-the-way-in-climate-adaption-not-just-mitigation/. Accessed on May 10, 2023.

60 Wright, “How the UAE is Committing to Climate Change Action.”

61 Ibid.

62 Government of Dubai, "Renewable Energy Shaping the Future of Sustainability," January 2, 2020, available at https://www.dewa.gov.ae/en/about-us/media-publications/latest-news/2020/01/renewable-energy-shaping-the-future-of-sustainability. Accessed on April 30, 2023.

63 John H. Knox, "Climate Change and Human Rights Law," *Virginia Journal of International Law*, Vol. 50, No. 1, September 30, 2009, pp. 1–59; Brooke A. Ackerly, Mujibul Anam, Jonathan Çilligan, and Steven Çoodbred, "Climate and Community: The Human Rights, Livelihood and Migration Impacts of Climate Change," in Dimitra Manou, Andrew Baldwin, Dug Cubie, Anja Mihr, and Teresa Thorp, eds., *Climate Change, Migration and Human Rights: Law and Policy Perspectives*, New York: Routledge, 2017, pp. 189–202; Simon Behrman and Avidan Kent, eds., *Climate Refugees: Beyond the Legal Impasse*, New York: Routledge, 2018.

64 Cher Weixia Chen and Alison Dundes Renteln, *International Human Rights: A Survey*, New York: Cambridge University Press, 2023, pp. 293–7.

65 Bonny Ibhawoh, "Inclusion Versus Exclusion," in Anja Mihr and Mark Gibney, eds., *The Sage Handbook of Human Rights*, Vol. 1, Los Angeles, CA: Sage Reference, 2014, pp. 324–39; see especially p. 337.

66 Gita Parihar and Kate Dooley, "Human Rights, Differentiated Responsibilities? Advancing Equity and Human Rights in the Climate Change Regime," in Sébastien Duyck, Sébastien Jodoin, and Alyssa Johl, eds., *Routledge Handbook of Human Rights and Climate Governance*, New York: Routledge, 2018, pp. 266–79; see especially p. 276.

67 Justin Dargin, "Beyond 'Green Pledges': Saudi Arabia and Society-Centered Climate Reform," in Frederic Wehrey et al., eds., *Climate Change and Vulnerability in the Middle East*, Carnegie Endowment for International Peace, available at https://carnegieendowment.org/2023/07/06/climate-change-and-vulnerability-in-middle-east-pub-90089. Accessed on April 12, 2024.

68 Ibid.

69 Ibid.

70 Marwan Muasher, "Costs of Delaying Improvements in Climate Change Governance in Jordan," in Frederic Wehrey et al., eds., *Climate Change and Vulnerability in the Middle East*, Carnegie Endowment for International Peace, available at https://carnegieendowment.org/2023/07/06/climate-change-and-vulnerability-in-middle-east-pub-90089. Accessed on April 12, 2024.

71 Maha Yahya and Issam Kayssi, "Lebanon: Can a Green Economy Pave the Way out of Economic Collapse?" in Frederic Wehrey et al., eds., *Climate Change and Vulnerability in the Middle East*, Carnegie Endowment for International Peace, available at https://carnegieendowment.org/2023/07/06/climate-change-and-vulnerability-in-middle-east-pub-90089. Accessed on April 12, 2024.

72 Ibid.

73 Zaha Hassan, Madison Andrews, and Mathew Madain, "Palestine and U.S.-Supported Regional Approaches to Address Climate Change and Water Scarcity," in Frederic Wehrey et al., eds., *Climate Change and Vulnerability in the Middle East*, Carnegie Endowment for International Peace, available at https://carnegieendowment.org/2023/07/06/climate-change-and-vulnerability-in-middle-east-pub-90089. Accessed on April 12, 2024.

74 Ibid.

75 Assma Maad, "Why the Gaza Health Ministry's Death Count is Considered Reliable," *Le Monde*, October 13, 2024, available at https://www.lemonde.fr/en/les-decodeurs/article/2024/10/13/why-the-gaza-health-ministry-s-death-count-is-considered-reliable_6729264_8.html. Accessed on October 27, 2024.

76 Mohammad Al-Mailam and Amr Hamzawy, "Rural Vulnerability and Resource Governance in Water-Scarce Egypt," in Frederic Wehrey et al., eds., *Climate Change and Vulnerability in the Middle East*, Carnegie Endowment for International Peace, available at https://carnegieendowment.org/2023/07/06/climate-change-and-vulnerability-in-middle-east-pub-90089. Accessed on April 12, 2024.

77 Ibid.

78 Sarah Yerkes and Haley Clasen, "Federal Barriers to Tunisia's Climate Momentum," in Frederic Wehrey et al., eds., *Climate Change and Vulnerability in the Middle East*, Carnegie Endowment for International Peace, available at https://carnegieendowment.org/2023/07/06/climate-change-and-vulnerability-in-middle-east-pub-90089. Accessed on April 12, 2024.

79 Ibid.

80 Ibid.

81 Sengupta, "Climate Change, International Justice and Global Order," p. 139.

82 Doreen Stabinsky, "Climate Justice and Human Rights," in Sébastien Duyck, Sébastien Jodoin, and Alyssa Johl, eds., *Routledge Handbook of Human Rights and Climate Governance*, New York: Routledge, 2018, pp. 280–90; see especially p. 287.

83 Ibid., p. 289.

84 Center for the Study of Human Rights, *Twenty-Four Human Rights Documents*, New York: Columbia University, 1992, pp. 6–9; see especially 8.

85 Robert Guest, *Borderless Economics: Chinese Sea Turtles, Indian Fridges and The New Fruits of Global Capitalism*, New York: St. Martin's Griffin, 201; see p. 10.

86 Sari Hanafi, "Forced Migration in the Middle East and North Africa," in Elena Fiddian-Qasmiyeh, Gil Loescher, Katy Long, and Nando Sigona, eds., *The Oxford Handbook of Refugee and Forced Migration Studies*, New York: Oxford University Press, 2014, pp. 585–98; see especially pp. 592–3.

87 Jon Barnett, "Human Security," in John S. Dryzek, Richard B. Norgaard, and David Schlosberg, eds., *Oxford Handbook of Climate Change and Society*, New York: Oxford University Press, 2011, pp. 267–77.

88 Ashok Swain and Anders Jägerskog, *Emerging Security Threats in the Middle East: The Impact of Climate Change and Globalization*, Lanham, MD: Rowman & Littlefield, 2016, pp. 5–6.

89 *The World Bank*, "Investing in Human Capital for a Greener Future," May 25, 2023, available at https://www.worldbank.org/en/news/feature/2023/05/25/investing-in-human-capital-for-a-greener-future. Accessed on April 15, 2024.

90 Ibid.

91 Tim Dobermann, "How Investing in Human Capital Can Help Adapt to Climate Change," The *International Growth Center (IGC)*, October 9, 2023, available at https://

www.theigc.org/blogs/climate-priorities-developing-countries/how-investing-human-capital-can-help-adapt-climate. Accessed on April 15, 2024.

92 Mike Berners-Lee, *There Is No Planet B: A Handbook for the Make or Break Years*, Updated Edition, New York: Cambridge University Press, 2021, pp. 222–4.

93 S. Duygu Sever, "Climate Change and the Energy Transition in the MENA Region," in Damilola S. Olawuyi, ed., *Climate Change Law and Policy in the Middle East and North Africa Region*, New York: Routledge, 2023, pp. 82–105; see especially pp. 96–7.

94 James L. Gelvin, *The New Middle East: What Everyone Needs to Know*, New York: Oxford University Press, 2018, pp. 144–5.

95 Heba Gowayed, "Climate Change and Migration in the Middle East and North Africa," *Arab Center*, Washington DC, September 29, 2022, available at https://arabcenterdc.org/resource/climate-change-and-migration-in-the-middle-east-and-north-africa/. Accessed on July 10, 2023.

96 Betts, *The Wealth of Refugees*.

97 Robert A. McLeman, *Climate and Human Migration: Past Experiences, Future Challenges*, New York: Cambridge University Press, 2014, p. 230.

98 Masood Ahmad, "2022 Global Refugee Work Rights Report, *Refugees International*, July 28, 2022, available at https://www.refugeesinternational.org/reports-briefs/2022-global-refugee-work-rights-report/. Accessed on July 10, 2023.

99 Carmen Gómez Martin, "Rethinking the Concept of 'Durable Solution': Sahrawi Refugee Camps Four Decades On," *Ethics and International Affairs*, Vol. 31, No. 1, Spring 2017, pp. 31–45; see especially p. 43.

100 Ibid.

101 Ibid.

102 Snel, Sorensen, and Power, *Climate Change and Water Scarcity in the Middle East*, p. 3.

103 Ibid., pp. 36–7.

104 Nils Gilman, Doug Randall, and Peter Schwartz, "Climate Change and Security," in John D. Dryzek, Richard B. Norgaard, and David Schlosberg, eds., *Oxford Handbook of Climate Change and Society*, New York: Oxford University Press, 2011, pp. 251–66; see especially p. 263.

105 *World Economic Forum*, "Digital and Climate," available at https://initiatives.weforum.org/digital-transformation/digital-and-climate. Accessed on April 17, 2024.

106 Climate Change Knowledge Portal, "Egypt," available at https://climateknowledgeportal.worldbank.org/country/egypt/vulnerability. Accessed on May 10, 2024.

107 Zainab Mehdi, "Preserving Iraq's Mesopotamian Marshes in the Face of Climate Change," in Carnegie Endowment, *Climate Change and Vulnerability in the Middle East*, July 6, 2023, available at https://carnegieendowment.org/posts/2023/07/climate-change-and-vulnerability-in-the-middle-east?lang=en#iraq. Accessed on July 26, 2024.

108 Mari Luomi, *The Gulf Monarchies and Climate Change: Abu Dhabi and Qatar in an Era of Natural Unsustainability*, London: Hurst & Company, 2012, p. 236.

109 Eleftherios Giovanis and Oznur Ozdamur, "The Impact of Climate Change on Budget Balances and Debt in the Middle East and North Africa (MENA) Region," *Climate Change*, Vol. 172, article # 34, pp. 1–127, available at https://link.springer.com/article/10.1007/s10584-022-03388-x. Accessed on May 11, 2024.

110 Office of the Director of National Intelligence, "Global Trends: 2040," available at https://www.dni.gov/index.php/gt2040-home/gt2040-5-year-regional-outlooks/mena. Accessed on May 10, 2024.

111 Saber Osman, "Assessing Climate Adaptation Plans in the Middle East and North Africa," *Carnegie Endowment for International Peace*, April 15, 2024, available at https://carnegieendowment.org/research/2024/04/assessing-climate-adaptation-plans-in-the-middle-east-and-north-africa?lang=en. Accessed on May 12, 2024.

Selected Bibliography

Abou Hadid, Ayman F., "Climate Change and Food Sovereignty in the Nile Valley and Red Sear," in Mannava V. K. Sivakumar, Rattan Lal Ramasamy Selvaraju, and Ibrahim Hamdan, eds., *Climate Change and Food Security in West Asia and North Africa*," Cham: Springer, 2013, pp. 193–206.

Ackerly, Brooke A., Mujibul Anam, Jonathan Gilligan, and Steven Goodbred, "Climate and Community: The Human Rights, Livelihood, and Migration Impacts of Climate Change," in Dimitra Manou, Andrew Baldwin, Dug Cubie, Anja Mihr, and Teresa Thorp, eds., *Climate Change, Migration, and Human Rights: Law and Policy Perspectives*, New York: Routledge, 2017, pp. 189–202.

Alharbi, Fahad Radhi, and Denes Csala, "Gulf Cooperation Council Countries' Climate Change Mitigation Challenges and Exploration of Solar and Wind Energy Resource Potential," *Applied Sciences*, Vol. 11, March 16, 2021, pp. 1–23.

Allan, John Anthony, *Virtual Water: Tackling the Threat to Our Planet's Most Precious Resource*, New York: I. B. Tauris, 2011.

Al-Maamary, Hilal M. S., Hussein A. Kazem, and Miqdam T. Chaichan, "Climate Change: The Game Changer in the Gulf Cooperation Council Region," *Renewable and Sustainable Energy Reviews*, Vol. 76, March 27, 2017, pp. 555–76.

Amery, Hussein A., *Arab Water Security: Threats and Opportunities in the Gulf States*, New York: Cambridge University Press, 2015.

Ari, Izzet, "Türkiye's Climate Policy: Aligned with International Climate Politics," in Muhittin Ataman, and Gloria Shkurti Özdemir, eds., *Different Dimensions of Environmental Security in Türkiye and Beyond*, Ankara: SETA, 2023, pp. 129–51.

Arar, Rawan, and David Scott FitzGerald, *The Refugee System: A Sociological Approach*, London: Polity, 2023.

Arraf, Jane, "War and Climate Change Ravage Syria's Farms," *The New York Times*, February 20, 2022, p. 4.

Ataman, Muhittin, and Gloria Shkurti Özdemire, eds., *Different Dimensions of Environmental Security in Türkiye and Beyond*, Ankara: SETA, 2023.

Ataman, Muhittin, and Ibrahim Efe, eds., *On Climate Change: Exploring Cases from Türkiye and Beyond*, Ankara: SETA, 2023.

Atapattu, Sumudu, *Human Rights Approaches to Climate Change: Challenges and Opportunities*, New York: Routledge, 2016.

Atapattu, Sumudu, "The Paris Agreement and Human Rights: Is Sustainable Development the "New Human Rights?," *Journal of Human Rights and Development*, Vol. 9, No. 1, March 2018, pp. 68–88.

Babar, Zahra, and Suzi Mirgani, eds., *Food Security in the Middle East*, New York: Oxford University Press, 2014.

Baldwin, Andrew, "Conclusion: On the Politics of Climate Change, Migration, and Human Rights," in Dimitra Manou, Andrew Baldwin, Dug Cubie, Anja Mihr, and Teresa Thorp, eds., *Climate Change, Migration, and Human Rights: Law and Policy Perspectives*, New York: Routledge, 2017, pp. 219–25.

Barnett, Jon, "Human Security," in John S. Dryzek, Richard B. Norgaard, and David Schlosberg, eds., *Oxford Handbook of Climate Change and Society*, New York: Oxford University Press, 2011, pp. 267–77.

Barre, Anne, Irene Dankelman, Anke Stock, Eleanor Blomstrom, and Bridge Burns, "From Marrakesh to Marrakesh: The Rise of Gender Equality in the Global Climate Governance and Climate Action," in Sébastien Duyck, Sébastien Jodoin, and Alyssa Johl, eds, *Routledge Handbook of Human Rights and Climate Governance*, New York: Routledge, 2018, pp. 239–250.

Bazrafshan Javad, Somayeh Hejabi, and Saeid Eslamian, "Drought Modeling Examples," in Saeid Eslamian and Faezeh Eslamian, eds., *Handbook of Drought and Water Scarcity: Principles of Drought and Water Scarcity*, New York: CRC Press, 2017, pp. 167–88.

Blau, Judith, *The Paris Agreement: Climate Change, Solidarity, and Human Rights*, New York: Palgrave/Macmillan, 2017.

Behrman, Simon, and Avidan Kent, *Climate Refugees: Beyond the Legal Impasse?* New York: Routledge, 2018.

Betts, Alexander, *The Wealth of Refugees: How Displaced People Can Build Economies*, New York: Oxford University Press, 2021.

Betts, Alexander, and Paul Collier, *Refuge: Rethinking Refugee Policy in a Changing World*, New York: Oxford University Press, 2017.

Bidinger, Sarah, "Syrian Refugees and the Right to Work: Developing Temporary Protection in Turkey," *Boston University International Law Journal*, Vol. 33, No. 223, 2015, pp. 224–49.

Biermann, Frank, and Ingrid Boas, "Preparing for a Warm World: Towards a Global Governance System to Protect Climate Refugees," *Global Environmental Politics*, Vol. 10, No. 1, February 2010, pp. 60–88.

Brandt, Marieke, *Tribes and Politics in Yemen: A History of the Houthi Conflict*, New York: Oxford University Press, 2017.

Bronen, Robin, "Climate-Induced Community Relocations: Institutional Challenges and Human Rights Protections," in Robert McLeman and François Gemenne, eds., *Routledge Handbook of Environmental Displacement and Migration*, New York: Routledge, 2018, pp. 338–402.

Brönnimann, Stefan, "Global Warming (1970-Present)," in Sam While, Christian Pfister, and Franz Mauelshagen, eds., *The Palgrave Handbook of Climate History*, New York: Palgrave/Macmillan, 2018, pp. 321–8.

Cleveland, William L., and Martin Bunton, *A History of the Modern Middle East*, 6th ed., New York: Routledge, 2016.

Corfield, Justin, "Iran," in S. George Philander, *Encyclopedia of Global Warming & Climate Change*, 2nd ed., Los Angeles, CA: Safe, 2012, pp. 799–801.

Cubie, Dug, "Human Rights, Environmental Displacement and Migration," in Robert McLeman and François Gemenne, eds., *Routledge Handbook of Environmental Displacement and Migration*, New York: Routledge, 2018, pp. 329–41.

Degroot, Dagomar, 'Climate Change and Conflict," in Sam White, Christian Pfister, and Franz Mauelshagen, eds., *The Palgrave Handbook of Climate History*, New York: Palgrave/Macmillan, 2018, pp. 367–85.

Dryzek, John D., Richard B. Norgaard, and David Schlosberg, eds., *Oxford Handbook of Climate Change and Society*, New York: Oxford University Press, 2011.

Evans, Jason P., "21st Century Climate Change in the Middle East," *Climatic Change*, Vol. 92, 2009, pp. 418–32.

Ferris, Elizabeth, "Governance and Climate-Induced Mobility: International and Regional Frameworks," in Dimitra Manou, Andrew Baldwin, Dug Cubie, Anja Mihr, and Teresa

Thorp, eds., *Climate Change, Migration, and Human Rights: Law and Policy Perspectives*, New York: Routledge, 2017, pp. 11–27.

Fiddian-Qasmyeth, Elena, Gil Loescher, Katy Long, and Nando Sigona, eds., *The Oxford Handbook of Refugee and Forced Migration Studies*, New York: Oxford University Press, 2014.

Forsythe, David P., *Advanced Introduction to the Politics of International Human Rights*, Cheltenham: Edward Elgar, 2021.

Foster, Michelle, *International Refugee Law and Socio-Economic Rights: Refuge From Deprivation*, Cambridge: Cambridge University Press, 2007.

Francioni, Francesco, and Ottavio Quirico, "Untying the Gordian Knot: Toward the Human Rights to a Climatically Sustainable Environment," in Ottavio Quirico and Mouloud Boumghar, eds., *Climate Change and Human Rights: An International and Comparative Law Perspective*, New York: Routledge, 2016, pp. 133–56.

Gelvin, James L., *The New Middle East: What Everyone Needs to Know*, New York: Oxford University Press, 2018.

Gilman, Nils, Doug Randall, and Peter Schwartz, "Climate Change and Security," in John D. Dryzek, Richard B. Norgaard, and David Schlosberg, eds., *Oxford Handbook of Climate Change and Society*, New York: Oxford University Press, 2011, pp. 251–66.

Gleick, Peter H., "Water, Drought, Climate Change, and Conflict in Syria," *American Meteorological Society*, Vol. 6, July 2014, pp. 331–40.

Göle, Nilüfer, "Gezi—Anatomy of a Public Square Movement," *Insight Turkey*, Vol. 15, No. 3, Summer 2013, pp. 7–14.

Hanafi, Sari, "Forced Migration in the Middle East and North Africa," in Elena Fiddian-Qasmyeth, Gil Loescher, Katy Long, and Nando Sigona, eds., *The Oxford Handbook of Refugee and Forced Migration Studies*, New York: Oxford University Press, 2014, pp. 585–98.

Heydarian, Richard Javad, *How Capitalism Failed the Arab World: The Economic Roots and Precarious Future of the Middle East Uprisings*, London: Zed Books, 2014.

Hiskes, Richard P., *The Human Right to a Green Future: Environmental Rights and Intergenerational Justice*, New York: Cambridge University Press, 2009.

Hiskes, Richard P., *Suffer the Children: A Theoretical Foundation for the Human Rights of the Child*, New York: Oxford University Press, 2021.

Hoffman, Peter J., and Thomas G. Weiss, *Humanitarianism, War, and Politics: Solferino to Syria and Beyond*, Lanham, MD: Rowman & Littlefield, 2018.

Howe, Marvine, *Morocco: The Islamist Awakening and Other Challenges*, New York: Oxford University Press, 2005.

Humphreys, Stephen, ed., *Human Rights and Climate Change*, New York: Cambridge University Press, 2009.

Kamrava, Mehran, *Troubled Waters: Insecurity in the Persian Gulf*, Ithaca, NY: Cornell University Press, 2018.

Khavarian-Garmsir, Amir Reza, Ahmad Pourahmad, Hossein Hataminejad, and Rahmatollah Farhoodi, "Climate Change and Environmental Degradation and the Drivers of Migration in the Context of Shrinking Cities: A Case Study of Khuzestan Province, Iran," *Sustainable Cities and Society*, Vol. 47, February 22, 2019. p. 101-480.

Kešeljevic, Aleksandar, and Rok Spruk, "Estimating the Effects of the Syrian Civil War," *Empirical Economics*, Vol. 66, 2024, pp. 671–703.

Khedr, Mosaad, "Challenges and Issues in Water, Climate Change, and Food Security in Egypt," in Abdelazim M. Negm, ed., *Conventional Water Resources and Agriculture in Egypt*, Cham: Springer International Publishing, 2019, pp. 229–44.

Kibaroğlu, Ayşegül, "Türkiye's Water Security Policy: Energy, Agriculture and Transboundary Issues," in Muhittin Ataman and Globria Shkurti Özdemir, eds., *Different Dimensions of Environmental Security in Türkiye and Beyond*, Ankara: SETA, 2023, pp. 179–97.

Luomi, Mari, *The Gulf Monarchies and Climate Change: Abu Dhabi and Qatar in an Era of Natural Unsustainability*, London: Hurst & Company, 2012.

Maghraoui, Driss, and Saloua Zerhouni, "Morocco," in Ellen Lust, ed., *The Middle East*, 14th ed., Thousand Oaks, CA: CQ Press, 2017, pp. 674–706.

Manou, Dimitra B., "Climate Change and Human Rights," in Anja Mihr and Mark Gibney, ed., *The Sage Handbook of Human Rights*, Vol. 1, Los Angeles, CA: Sage Reference, 2014, pp. 238–52.

Mayer, Benoît. "Definitions and Concepts," in Robert McLeman and François Gemenne, eds., *Routledge Handbook of Environmental Displacement and Migration*, New York: Routledge, 2018, pp. 323–8.

McAdam, Jane, *Climate Change, Forced Migration, and International Law*, New York: Oxford University Press, 2012.

McAdam, Jane, "Human Rights and Forced Migration," in Elena Fiddaian-Qasmiyeh, Gil Loescher, Katy Long, and Nando Sigona, eds., *The Oxford Handbook of Refugee and Forced Migration Studies*, New York: Oxford University Press, 2014, pp. 203–14.

McAdam, Jane, "Swimming Against the Tide: Why a Climate Change Displacement Treaty is Not *the* Answer," *International Journal of Refugee Law*, Vol. 23, No. 1, March 2011, pp. 2–27.

McInerney-Lankford, Siobhan, Mac Darrow, and Lavanya Rajamani, *Human Rights and Climate Change: A Review of the International Legal Dimensions*, Washington, DC: The World Bank, 2011.

McLeman, Robert A., *Climate and Human Migration: Past Experiences, Future Challenges*, New York: Cambridge University Press, 2014.

Mihr, Anja, "Climate Justice, Migration, and Human Rights," in Dimitra Manou, Andrew Baldwin, Dug Cubie, Anja Mihr, and Teresa Thorp, eds., *Climate Change, Migration, and Human Rights: Law and Policy Perspectives*, New York: Routledge, 2017, pp. 45–67.

Mohamed, Hadi, Moosa Elayah, and Lau Schuplen, "Yemen Between the Impact of the Climate Change and the Ongoing Saudi-Yemen War: A Real Tragedy," Governance and Peace-Building Center, Sana'a, November 2017, pp. 1–10.

Monshipouri, Mahmood, *Middle East Politics: Changing Dynamics*, New York: Routledge, 2019.

Monshipouri, Mahmood, Burcu Ellis, and Cassidy Renee Yip, "Managing the Refugee Crisis in the Era of the COVID-19 Pandemic," *Insight Turkey*, Vol. 22, No. 4, Fall 2020, pp. 179–200.

Monshipouri, Mahmood, and Reza Mohajer, "Climate Change and the Potential for Conflict in the Middle East: Emerging Threats to Security and Human Rights," *The Maghreb Review*, Vol. 49, No. 1, 2024, pp. 14–41.

Mostafa, Soha M., Osama Wahed, Walaa Y. El-Nashar, Samia M. El-Marsafawy, Martina Zeleňáková, and Hany E. Abd-Elhamid, "Potential Climate Change Impact on Water Resources in Egypt," *Water*, Vol. 13, No. 1715, June 21, 2021, pp. 1–14.

Mougou, Raoudha, Ayman Abou-Hadid, Ana Iglesias, Mahmoud Medany, Amel Nafti, Riadh Chetali, Mohsen Mansour, and Helmy Eid, "Adapting Dryland and Irrigated Cereal Farming to Climate Change in Tunisia and Egypt," in Neil Leary, James Adejuwon, Vicente Barros, Ian Burton, Jyoti Kulkarni, and Rodel Lasco, eds., *Climate Change and Adaptation*, London: Earthscan, 2009, pp. 181–95.

Orkaby, Asher, *Yemen: What Everyone Needs to Know*, New York: Oxford University Press, 2021.
Ouda, Samiha A. H., Tahany Noreldin, and Alia Amer, "Rain Fed Areas in Egypt: Obstacles and Opportunities," in Samiha A. H. Ouda and Abd El-Hafeez Zohry, eds., *Management of Climate Induced Drought and Water Scarcity in Egypt: Unconventional Solutions*, Cham: Springer, 2016, pp. 27–46.
Panter-Brick, Catherine, "Inclusive Partnerships: Building Resilience Humanitarianism with Syrian Refugee Youth in Jordan," in Marcia C. Inhorn and Lucia Volk, eds., *Un-Settling Middle Eastern Refugees: Regimes of Exclusion and Inclusion in the Middle East*, Europe, and North America, New York: Berghahn, 2021, pp. 261–80.
Parihar, Gita, and Kate Dooley, Human Rights, Differentiated Responsibilities? Advancing Equity and Human Rights in the Climate Change regime, in Sébastien Duyck, Sébastien Jodoin, and Alyssa Johl, eds., *Routledge Handbook of Human Rights and Climate Governance*, New York: Routledge, 2018, pp. 266–79.
Parks, Bradley C., and J. Timmons Roberts, "Environmental and Ecological Justice," in Mechele M. Bestill, Kathryn Hoxhstetle, and Dimitris Stevis, eds., *Palgrave Advances in International Environmental Politics*, New York: Palgrave/Macmillan, 2006, pp. 329–60.
Pérez, Beatriz Felipe, "Beyond the Shortcomings of International Law: A Proposal for the Legal Protection of Climate Migrants," in Simon Behrman and Avidan Kent, eds., *Climate Refugees: Beyond the Legal Impasse*, New York: Routledge, 2018, pp. 214–29.
Poon, Jenny, "Drawing Upon International Refugee Law: The Precautionary Approach to Protecting Climate Change-Displaced Persons," in Simon Behrman and Avidan Kent, eds., *Climate Refugees: Beyond the Legal Impasse*, New York: Routledge, 2018, pp. 157–71.
Praag, Lore Van, Loubna Ou-Salah, Elodie Hut, and Caroline Zickgraf, *Migration and Environmental Change in Morocco*, New York: Springer Publishing, 2021; see Chapter 3: "Environmental Change and Migration in Morocco: What Has Been Done So Far?" pp. 35–59.
Rabinowitz, Dan, *The Power of Deserts: Climate Change, The Middle East, and the Promise of a Post-Oil Era*, Stanford, CA: Stanford University Press, 2020.
Romm, Joseph, *Climate Change: What Everyone Needs to Know*, New York: Oxford University Press, 2018.
Şahin Mencütek, Zeynap, "Forces Behind Lebanese Governance Patterns and Refugee Politics," in Zeynep Şahin Mencütek, ed., *Refugee Governance, State and Politics in the Middle East*, New York: Routledge, 2018, pp. 165–80.
Sengupta, Sandeep, "Climate Change, International Justice and Global Order," *International Affairs*, Vol. 99, No. 1, January 2023, pp. 121–40.
Skillington, Tracey, *Climate Justice and Human Rights*, New York: Palgrave/Macmillan, 2017.
Snel, Mariëlle, Nikolas Sorensen, and Reed Power, *Climate Change and Water Scarcity in the Middle East: A Transitional Approach*, London: Taylor & Francis Human Geography, 2025.
Sowers, Jeannie, "Environmental Activism in the Middle East and North Africa," in Harry Verhoeven, ed., *Environmental Politics in the Middle East*, New York: Oxford University Press, 2009, pp. 27–52.
Springborg, Robert, "The Challenge of Squaring the Energy-Environment-Growth Triangle," in Robert E. Looney, ed., *Handbook of Transitions in Energy and Climate Security*, New York: Routledge, 2017, pp. 272–83.
Stabinsky, Doreen, "Climate Justice and Human Rights," in Sébastien Duyck, Sébastien Jodoin, and Alyssa Johl, eds., *Routledge Handbook of Human Rights and Climate Governance*, New York: Routledge, 2018, pp. 280–90.

Swain, Ashok, and Anders Jägerskog, *Emerging Security Threats in the Middle East: The Impact of Climate Change and Globalization*, Lanham, MD: Rowman & Littlefield, 2016.
Thiollet, Hélène, "From Migration Hub to Asylum Crisis: The Changing Dynamics of Contemporary Migration in Yemen," in Helen Lackner, ed., *Why Yemen Matters*; A Society in Transition, London: SOAS, 2014, pp. 267–86.
Thomas, Alice, "Human Rights and Climate Displacement and Migration," in Sébastein Duyck, Sébastien Jodoin, and Alyssa Johl, eds., *Routledge Handbook of Human Rights and Climate Governance*, New York: Routledge, 2018, pp. 110–27.
Türk, Volker, and Rebecca Dowd, "Protection Gaps," in Elena Fiddian-Qasmiyeh, Gil Loescher, Katy Long, and Nando Sigona, eds., *The Oxford Handbook of Refugee and Forced Migration Studies*, New York: Oxford University Press, 2014, pp. 278–89.
Wallace-Wells, David, "Energy 'Too Cheap to Meter' Has Long Been a Sci-fi Dream. But as the Cost of Solar Power Plummets Toward Zero, It Just Might Become a Reality," *The New York Times Magazine*, September 1, 2024, pp. 18–19.
Wallace-Wells, David, "World Burns," *The New York Times Magazine*, October 29, 2023, pp. 26–48.
Wendle, John, "Middle East: The Boiling Point of Climate Change and National Security," in John R. Wennersten and Denise Robbins, eds., *Rising Tides: Climate Refugees in the Twenty-First Century*, Bloomington, IN: Indiana University Press, 2017, pp. 157–88.
Woerz, Eckart, "Historic Food Regimes and the Middle East," in Zahra Babar and Suzi Mirgani, eds., *Food Security in the Middle East*, London: Hurst & Company, 2014, pp. 19–38.
World Bank Group, *Beyond Scarcity: Water Security in the Middle East and North Africa*, Washington, DC: World Bank Group, 2018.
Worth, Robert F., "They Break Us or We Break Them," *The New York Times Magazine*, November 4, 2018, pp. 48–57, and 69.
Yilmaz, Ömer, "Turkey's Buffer Zone in Northern Syria and Its Impact on Irregular Migration," in Muhittin Ataman and Ibrahim Efe, eds., *On Climate Migration: Exploring Cases from Turkey and Beyond*, Ankara: SETA Publications, 2023, pp. 101–22.
Zerhouni, Saloua, "The Moroccan Parliament," in Ellen Lust-Okar and Saloua Zerhouni, eds., *Political Participation in the Middle East*. Boulder, CO: Lynne Rienner Publishers, 2008, pp. 217–38.
Zetter, Roger, and James Morrissey, "The Environment-Mobility Nexus: Reconceptualizing the Link Between Environmental Stress, (Im)Mobility, and Power," in Elena Fiddian-Qasmiyeh, Gil Loescher, Katy Long, and Nando Sigona, eds., *The Oxford Handbook of Refugee and Forced Migration Studies*, New York: Oxford University Press, 2014, pp. 342–54.

INDEX

www.ingramcontent.com/pod-product-compliance
Ingram Content Group UK Ltd.
Pitfield, Milton Keynes, MK11 3LW, UK
UKHW021905150926
475560UK00003B/200